2019
서울도시건축비엔날레
SEOUL BIENNALE
OF ARCHITECTURE
AND URBANISM

집합도시
COLLECTIVE
CITY

KB151933

2019 서울도시건축비엔날레 총감독
임재용, 프란시스코 사닌

2019 서울도시건축비엔날레 큐레이팅 팀
— 주제전
큐레이터.∴.베스 휴즈
협력큐레이터.∴.김효은
보조큐레이터.∴.리비아 왕, 이자벨 옥덴,
제프리 킴, 이유진
코디네이터.∴.황희정
영상상영코디네이터.∴.안나 리비아 바슬
— 도시전
큐레이터.∴.임동우, 라파엘 루나
협력큐레이터.∴.김유빈
코디네이터.∴.조웅희
관람경험 디자인.∴.NOLGONG
— 글로벌 스튜디오
큐레이터.∴.최상기
협력큐레이터.∴.이희원
보조큐레이터.∴.최영민
전시 디자인.∴.(주)건축사사무소오드투에이
— 현장 프로젝트
큐레이터.∴.장영철
협력큐레이터.∴.유아람, 최주연, 홍주석
— 서울마당
총괄.∴.임재용
협력 큐레이터.∴.강민선
보조 큐레이터.∴.정진우

국외 총감독 지원.∴.강민선

그래픽디자인.∴.홍은주, 김형재 (전수민 도움)
사진.∴.김태윤(Step by snap), 진효숙,
김용순(나르실리온 포토그래피)

SBAU 2019 Co-directors
Jaeyong Lim, Francisco Sanin

SBAU 2019 Curatorial Team
— Thematic Exhibition
Curator.∴.Beth Hughes
Associate Curator.∴.Hyoeun Kim
Assistant Curator.∴.Livia Wang,
Isabel Ogden, Jeffrey Kim, Yoojin Lee
Coordinator.∴.Heejung Hwang
Film researcher.∴.Anna Livia Vørsel
— Cities Exhibition
Curator.∴.Dongwoo Yim, Rafael Luna
Associate Curator.∴.YouBeen Kim
Coordinator.∴.Tony Woonghee Cho
Visitor Experience Design.∴.NOLGONG
— Global Studio
Curator.∴.Sanki Choe
Associate Curator.∴.Heewon Lee
Assistant Curator.∴.Youngmin Choi
Exhibition design.∴.ODETO.A
— Live Projects
Curator.∴.Young Chul Jang
Associate Curator.∴.Aram you, Jooyeon
Choi, Jooseok Hong
— Seoul Madang
Director.∴.Jaeyong Lim
Associate Curator.∴.Minsun Kang
Assistant Curator.∴.Jinwoo Jung

Foreign Directorial Assistant.∴.Minsun
Kang

Graphic Design.∴.Eunjoo Hong and
Hyungjae Kim (Assistant Sumin Jeon)
Photography.∴.Taeyoon Kim, Hyosook Jin,
Yongsun Kim

2019
서울도시건축비엔날레
SEOUL BIENNALE
OF ARCHITECTURE
AND URBANISM

집합도시
COLLECTIVE
CITY

가이드북　　　Guide Book

반갑습니다.

서울도시건축비엔날레가 올해 2회를 맞았습니다. 이번 서울비엔날레의 주제는 '집합도시'로, 오늘날 도시의 공간적·사회적 집합 유형을 토대로 새로운 차원의 협업과 모델을 연구하여 그 결과를 담고 담론을 이어나갈 수 있는 장을 마련했습니다. 이러한 과정과 결과는 결국 도시의 새로운 전략과 비전을 제시할 것이며, '함께 만들고 함께 누리는 도시'로 자라나는 토양이 될 것입니다.

이번 서울비엔날레는 도시의 다양한 담론이 오가는 장입니다. 동시에 세계 여러 도시에서 겪고 있는 도시의 불균형 또는 첨단 기술과 함께 진화하는 도시의 시스템 등 다양한 도시의 연구들을 들여다볼 수 있는 전시 프로그램이 서울 도심 곳곳에서 펼쳐집니다.

더불어 시민들이 자신의 공간인 도시에 관한 이슈, 정책, 연구 등에 다가가는 과정을 보다 더 쉽게 제안하고 즐길 수 있도록 돕는 다양한 시민참여 프로그램을 준비했습니다. 건축가, 만화가, 미디어 아티스트, 방송작가 등 다양한 분야의 사람들에게 도시의 이야기를 듣거나, 평소 쉽게 가 볼 수 없는 대사관 및 관저를 탐색하고, 과거 혹은 현재의 서울이라는 도시를 감상할 수 있는 프로그램 등이 많은 시민들의 참여를 기다리고 있습니다.

시민은 도시를 만들고, 도시가 다시 시민을 만들듯이 서울도시건축비엔날레를 통해 모든 사람들이 서울을 재발견하고 새로운 경험을 할 수 있기를 바랍니다.

감사합니다.

박원순
서울특별시장

Greetings,

This year marks the second edition of the Seoul Biennale of Architecture and Urbanism. Under the theme of the "collective city," this year's event provides a platform for the discussion and exploration of new levels of collaboration and models in architecture and urbanism, based on contemporary spatial and social forms of urban collectivity. These explorations and their outcomes will present new urban visions and strategies, while serving as the foundation for "the city that is collectively made and collectively enjoyed."

The 2019 Seoul Biennale serves as an important arena for diverse discussions on architecture and urbanism. Exhibition programs organized across the city will grant a closer look at diverse topics in urban research, such as the imbalances affecting many cities across the globe today as well as new urban systems evolving alongside technological advances. Moreover, a wide range of public programs invite citizens to familiarize and engage themselves with the issues, policies and research regarding their cities. For example, they offer a venue for architects, cartoonists, media artists, screenwriters and other experts to share their urban stories. A tour program takes participants on a journey into embassies and official residences that are generally not open to the public. Another connects participants with the past and present of Seoul. Exciting programs such as these await the participation and interest of many people.

Just as citizens and cities shape one another, it is my sincere hope that the 2019 Seoul Biennale will inspire all participants to rediscover Seoul and partake in brand new experiences.

Thank you.

Park Won-soon
Mayor of Seoul Metropolitan City

GREETINGS

비엔날레 가이드

2019 서울건축도시비엔날레(SBAU)는 기본적인 명제를 제시합니다. 곧, 집단성이 도시의 근원이자 본질 그 자체에 있다는 것입니다. 이는 도시가 곧 권리이며 도시 프로젝트는 집단적인 틀을 구성하는 행위이자, 시민을 위한 공간과 자원을 운용하는 활동임을 제언합니다. 그러나 전세계 도시들이 직면한 난제들, 즉, 사회적 정의 및 환경적 정의에 관한 문제, 주택난, 공간적, 그리고 경제적 내몰림, 취약한 교육 및 의료 접근성 등을 보고 있노라면, 우리는 '도시는 우리 공동의 프로젝트다'라는 주장을 다시 펼치기 위해 정치적 대리인이 필요하다는 명제로 오늘날의 '집단성'을 이해해야 할 필요가 있다는 것을 깨닫게 됩니다. 우리는 사회적, 정치적, 공간적 프로젝트를 통해 도시를 되찾는 행위가 오늘날 우리 사회에 지닌 함의가 무엇인지 알아보고자 하며 어떠한 방안과 전략을 취해야 하는지, 주체는 누구이며 어떠한 절차를 내포하는지 다 함께 자문하고 고민하는 시간을 갖고자 합니다. 이러한 맥락에서 우리는 실용학문이나 직업적인 범주를 넘어서 건축이 지닌 잠재적 역할을 알아보고, 지식의 보고이자 행위의 선도자로서의 가능성을 살펴보고자 합니다.

2019 서울도시건축비엔날레는 전시물의 집결지이며, 담론과 토론의 향연장입니다. 우리는 연구와 실험을 통해 지식을 구축하고자 하며 전세계에서 가장 흥미진진하고 연관성있는 경험을 한데 모아 집단성을 갖춘 도시를 되찾고 기본 권리를 수호하고자 합니다. 건축의 역할을 재구성하고 확장한 실험들은 오늘날 집합성을 구축하는데 있어 하나의 규율이 될 뿐만 아니라 이에 수반되는 도구로서의 역할을 수행하게 될 것입니다.

임재용, 프란시스코 사닌
2019 서울도시건축비엔날레 총감독

A GUIDE TO THE BIENNALE

The Seoul Biennale of Architecture and Urbanism (SBAU) 2019 puts forward a basic proposition: the collective is at the very heart of the origin and nature of the city. It claims that the city is a right and that the project of the city is that of the construction of a collective framework, space, and resource for its citizens. However, when we look at the challenges facing cities across the globe: social and environmental justice, the housing crisis, spatial and economic marginalization, access to education, health etc., we realize there is a need to understand the collective today as having political agency in reclaiming the city as a collective project. We ask what it means today to reclaim the city as a social, political, and spatial project; what are the tactics, strategies as well as actors and processes engaged in this project; and in this context, what is the potential role of architecture beyond the boundaries of the profession—as a practice-based discipline, a form of knowledge, and actions upon the city.

SBAU 2019 is conceived not only as an exhibition but also as a space of discourse and debate. It is interested in the construction of knowledge through research and experimentation. It brings together the most interesting and relevant experiences from around the world that aims to reclaim the city as a collective condition and a basic right. It calls for experiments that reframe and expand the role that architecture can play as a discipline in the construction of the collective and the tools that are necessary to do so.

Jaeyong Lim, Francisco Sanin
Co-directors, Seoul Biennale of Architecture & Urbanism 2019

PREFACE

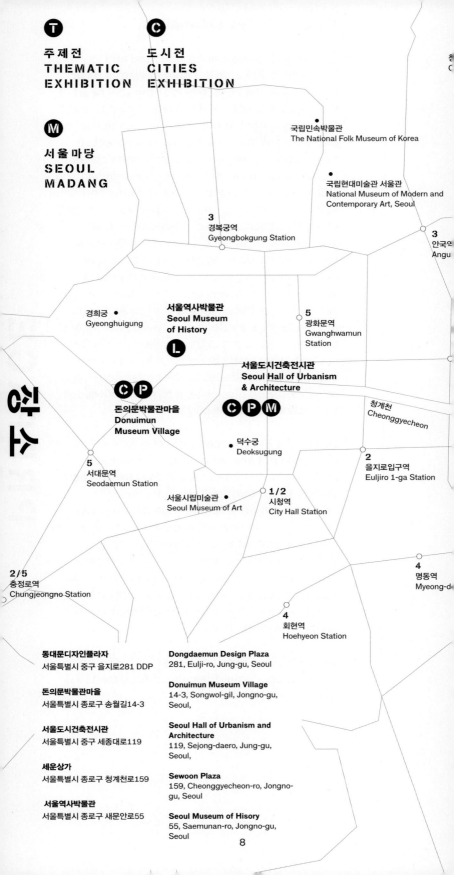

T
주제전
THEMATIC
EXHIBITION

C
도시전
CITIES
EXHIBITION

M
서울마당
SEOUL
MADANG

국립민속박물관
The National Folk Museum of Korea

국립현대미술관 서울관
National Museum of Modern and
Contemporary Art, Seoul

3
경복궁역
Gyeongbokgung Station

3
안국역
Angu

경희궁 ·
Gyeonghuigung

서울역사박물관
Seoul Museum
of History

L

5
광화문역
Gwanghwamun
Station

서울도시건축전시관
Seoul Hall of Urbanism
& Architecture

C P
돈의문박물관마을
Donuimun
Museum Village

C P M

청계천
Cheonggyecheon

덕수궁
Deoksugung

5
서대문역
Seodaemun Station

2
을지로입구역
Euljiro 1-ga Station

서울시립미술관 ·
Seoul Museum of Art

1/2
시청역
City Hall Station

2/5
충정로역
Chungjeongno Station

4
명동역
Myeong-d

4
회현역
Hoehyeon Station

동대문디자인플라자
서울특별시 중구 을지로281 DDP

돈의문박물관마을
서울특별시 종로구 송월길14-3

서울도시건축전시관
서울특별시 중구 세종대로119

세운상가
서울특별시 종로구 청계천로159

서울역사박물관
서울특별시 종로구 새문안로55

Dongdaemun Design Plaza
281, Eulji-ro, Jung-gu, Seoul

Donuimun Museum Village
14-3, Songwol-gil, Jongno-gu,
Seoul,

**Seoul Hall of Urbanism and
Architecture**
119, Sejong-daero, Jung-gu,
Seoul,

Sewoon Plaza
159, Cheonggyecheon-ro, Jongno-
gu, Seoul

Seoul Museum of Hisory
55, Saemunan-ro, Jongno-gu,
Seoul

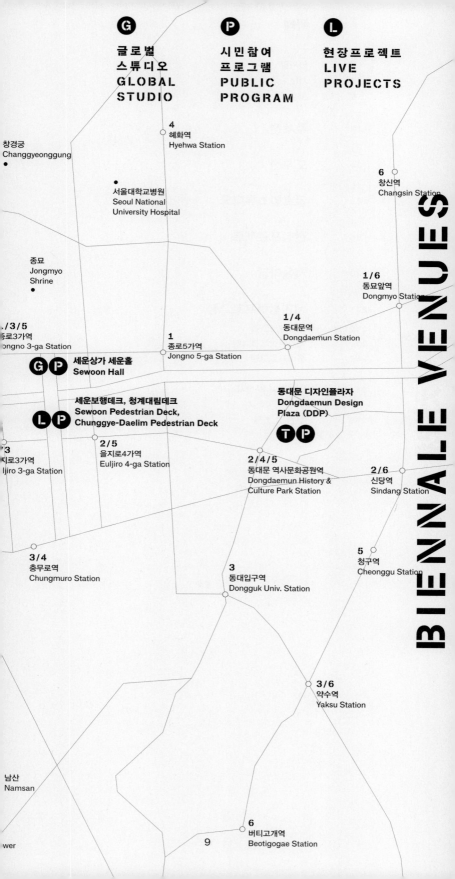

차례

차례

CONTENTS

CONTENTS

일부 프로젝트가 미처 완료되기 이전에
가이드북 인쇄가 진행되었습니다.
이에 따라, 몇몇 작품에 대한 설명이 일부
정확하지 않을 수 있습니다. 독자 분들의 깊은
양해 부탁드립니다.

We regret that due to the publication
timeline we had to print the guidebook
before a certain number of the projects
had taken their definitive form.
We can only request the reader's
understanding as some of the
descriptions may no longer be entirely
accurate.

주제전
THEMATIC
EXHIBITION

2019 서울도시건축비엔날레의 주제전은 "집합도시"를 주제로 집합적 실천과 행위가 어떻게 도시의 개발 패러다임을 변화시키고 공간 생산의 지배적 시스템에 저항할 수 있는지 질문을 던진다. 이번 서울비엔날레는 건축과 도시, 환경의 대안적 개념을 제시하고 건축의 정치적 동력을 탐색하기 위해서 공존, 사회적 실천, 거버넌스, 연구 및 추측의 새로운 모델을 반추하고자 한다. 주제전은 현재의 도시 구성을 재해석하라는 권유이자, 우선순위를 재배열해보자는 자극제다. 이때 부동산 투기와 토지 상품화를 통한 개인 및 자본의 성공으로부터 집합적 권리와 도시가 공유 투자라는 논점으로 초점이 변화된다.

전세계에서 참여하는 건축가들이 이번 서울비엔날레에서 보여주는 작업들은 현대적인 도시화 과정, 새로운 영토를 확장하고 정의하는 생태·사회기반적 시스템 및 생산, 물질과 생산에 관한 문제, 개발의 대안적 모델과 유형적 혁신, 새로운 유형의 거주권과 토지 소유권 그리고 행동주의나, 시위, 중재, 참조로서의 건축을 재고하는 등 매우 다양하다.

주제전의 전시배치는 연속적인 공간에 연구 결과물과 명제, 추측을 한데 모아 연결과 중첩을 가능하게 한다. 관객들에게 제시하는 특별한 관람동선이나 관람방법은 없고 오늘날 전세계에서 이뤄지고 있는 도시 건축 행위들의 규모와 형태에 몰입하도록 돕는다. 관람객은 오늘날 세계가 마주하고 있는 과제와 우리의 거주 방식 변화의 시급성, 그리고 이러한 상황에서 건축과 형태의 역할에 대해 자신만의 방식으로 이해하고 탐험해 볼 수 있다.

작가들의 전시와 함께 주제전 큐레이팅 팀이 준비한 영상 시리즈가 상영된다. 이 영상물은 도시를 형성하고 도시에 의해 만들어지는 다양한 집합체의 삶과 행동양식을 보여준다. 기록영상과 다큐멘터리, 예술 영화와 연구프로젝트를 통해 세계 현대 도시들의 상황과 집합성을 선보이고, 시민과 그들의

INTRO

'Collective City' seeks to question how modes of collective practice and action can challenge the current paradigms of city development and offer resistance to the dominant systems of spatial production. The Seoul Biennale reflects on new models of co-existence, social practice, governance, research and speculation, to suggest alternative concepts of architecture, the city and the environment. The Thematic Exhibition is an invitation to radically reimagine the structure of our cities, a provocation to fundamentally reprioritize, shifting focus from the success of the individual and capital through real-estate speculation and the commodification of land, to foreground collective rights and to claim the city as a shared investment.

Participating architects have contributed from all around the world. Together they present varied critiques on the contemporary processes of urbanization, exploration of ecological and infrastructural systems expanding and defining new territories, questions of material and production, alternative models of development and typological innovation, new forms of tenure and landownership to architecture as a form of activism, protest, mediation and consultation.

The layout of the exhibition positions research, proposition and speculation in close proximity to one another in a continuous space, allowing for connections and overlap. There is no defined sequence of experience or orientation, rather the exhibition is intended as an immersion within the many scales and forms of action currently active in the global practice of architecture and urbanism. In this saturated space, the viewer can navigate their own encounter to best understand the challenges facing our world today, the urgent need for transformation of our occupation of the planet, and the potential of architecture and form to engage meaningfully within that context.

Accompanying the installations from invited architects, is a body of film-work exhibited by the curatorial team of the Thematic Exhibition,

삶을 담으며, 도시의 복잡성과 인간과 사회 구조를
소개한다.

큐레이터˙.˙.˙베스 휴즈
협력큐레이터˙.˙.˙김효은
보조큐레이터˙.˙.˙리비아 왕, 이자벨 옥덴, 제프리 킴, 이유진
코디네이터˙.˙.˙황희정
영상상영코디네이터˙.˙.˙안나 리비아 바슬

INTRO

that seeks to use moving image to capture the lives and rituals of the communities that shape and are shaped-by our cities. A selection of archive footage, documentaries, artist films and research projects reflect upon the contemporary urban condition and collectives around the world, exposing the complexity of cities and human and societal structures, recording the lives lived within them.

Curator∴Beth Hughes
Associate Curator∴Hyoeun Kim
Assistant Curator∴Livia Wang, Isabel Ogden, Jeffrey Kim, Yoojin Lee
Coordinator∴Heejung Hwang
Film Researcher∴Anna Livia Vørsel

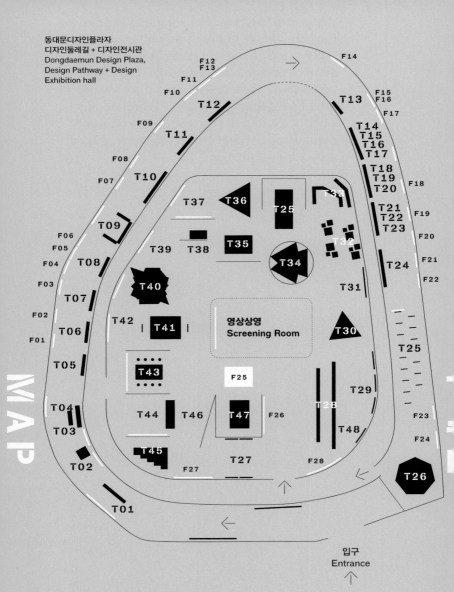

동대문디자인플라자
디자인둘레길 + 디자인전시관
Dongdaemun Design Plaza,
Design Pathway + Design
Exhibition hall

영상상영
Screening Room

MAP

FILMS

F20 플로리안 제이팡, 리사 슈미츠코리네트, 알렉산더 슈 무거, 지상의 조직, 2015, 22분

F21 가스통 퀴리베, 탈주 미래주의자, 1924, 12분

F22 J.B. 홈즈, 중세시대 마을 (렉스톤), 1935, 18분

F23 공모 후보작: 안드레아 루카 짐머만, 집, 몽상, 2015, 83분

F23 공모 후보작: 나딘 베크다쉬, 베이험 거리: 장소의 서사를 매핑하기, 2014, 35분

F23 공모 후보작: 조지나 힐, 작동방식 복사하기, 2019, 6분 26초

F23 공모 후보작: 리사 장 리, 서구의 꿈, 2018, 15분 37초

F24 공모 당선작: 페드로 페트리쉐, 민중의 집, 2019, 11분

F25 로라 우에르타스 밀란, 에콰도르, 2012, 19초

F26 판 류, 보 왕, 보이지 않는 도시의 흔적: 홍콩에 관한 세 개의 노트, 2016, 70분

F27 브리티쉬 파쉐, "강철 배짱!", 1931, 1분

F28 카라빙 영화 콜렉티브, 나이트 타임 고, 2017, 31분

F16 Anson Dyer, Day in Liverpool, 1929, 33mins, BFI National Archive, British Film Institute

F17 Reuters, Sri Lanka: Capital City, Colombo, Undergoes Facelift, 1979, 2min, British Pathé

F18 Sune Lund-Sørensen, The Junk Playground, 1967, 11min, Danish Film Institute, Danmarkpaafilm.dk

F19 Florian Zeyfang, Lisa Schmidt-Colinet, Alexander Schmoeger, Microbrigades, 2013, 30min

F20 Florian Zeyfang, Lisa Schmidt-Colinet, Alexander Schmoeger, Institute Above-Ground, 2015, 22min

F21 Gaston Quiribet, Fugitive Futurist, 1924, 12mins, BFI National Archive, British Film Institute

F22 J.B. Holmes, Medieaval Village (Laxton), 1935, 18min, BFI National Archive, British Film Institute

F23 OPEN CALL RUNNER UP: Andrea Luka Zimmerman, Estate, a Reverie, 2015, 83min

F23 OPEN CALL RUNNER UP: Nadine Bekdache, Beyhum Street: Maping Place Narratives, 2014, 35min

F23 OPEN CALL RUNNER UP: Georgina Hill, Coping (with) Mechanisms, 2019, 6min 26sec

F23 OPEN CALL RUNNER UP: Lisa Chang Lee, Western Dream, 2018, 15min 37sec

F24 OPEN CALL WINNER: Pedro Petriche, Casa do Povo, 2019, 11min

F25 Laura Huertas Millán, Aequador, 2012, 19min

F26 Pan Lu and Bo Wang, Traces of an Invisible City: Three Notes on Hong Kong, 2016, 70mins

F27 British Pathé, "Nerves of Steel!", 1931, 1min, British Pathé

F28 Karrabing Film Collective , Night Time Go, 2017, 31min

FILMS

더불어 사는 일상
이엠에이건축사무소(주)

T01

인구 감소 시대와 고령사회에 접어든 지금, 소도시와 농어촌은 공동화로 소멸 위기에 처해 있다. 낡고 오래된 지역은 외면 받고 주변의 값싼 전답은 전원주택 단지로 바뀌고 있다. 정착 지역의 배타적 분위기 속에서 공동체를 형성하기에는 이주민들 서로간의 결속력이 약하다. 지역으로의 이주는 공동체와 함께 집합적 거주 계획으로 지역에 정착하고 활력을 만들어가는 방식이 대안이 될 수 있다. 개인에게는 기존 삶이 있던 도시를 떠나 새로운 사회의 구성원이 되는 과정적 모델이기도 하다. 오시리가름 협동조합주택, 눈뫼가름 협동조합주택, 의성고운마을 프로젝트들은 주거기반공동체와 지역공동체 형성을 위한 대안적 모델로서 지역활성화의 가능성을 발견하고자 한다.

COLLECTIVE FORM OF THE EVERYDAY
EMA architects & associates

T01

Faced with a declining and aging population, small cities are losing their inhabitants and are under threat of disappearing. The old established regions are going disregarded, while the suburban farmlands are turning into residential complexes. Furthermore, the new urban inhabitants remain disconnected from their newly inhabited regions and also possess a marginal bonding power, failing to form relationships or communal understanding with one another. In order to bring vitality to shrinking rural towns, a population move can be considered an alternative solution, organized under a collective residential plan that can encourage communities to settle and build a life together. By three projects, Osiri gareum Cooperative Housing, Nunmoe gareum Cooperative Housing and Uiseong Gowoonmaeul, a regional revitalization plan could discover its potential through a certain process: an alternative model comes to form a residential-based community and takes root in the regional community.

Nunmoe Gareum

Uiseong Gowoonmaeul

SNS 집합도시

(주)건축사사무소SAAI

T02

현대도시의 집합성을 물리적인 상황을
넘어서 SNS에서 교환되는 밀도에 주목하여
살펴보고자 한다. 프로젝트가 이루어진
홍대 근처의 상업시설은 간선도로에서
살짝 벗어나 있다. 그럼에도 불구하고 이
상업시설에 관한 이야기가 SNS에서 활발하게
전파되고, 사람들이 장소를 방문하고, 그
이미지가 다시 SNS를 통해 교환, 소비된다.
이제는 상업가로의 주요한 지점을 선점하여
그 이미지가 가로를 통행하는 유동인구를
중심으로 소비되기 보다는, 사람들이 SNS를
통해 유통되는 이미지와 밀도를 소비하는 것이
더 익숙한 시대다. 우리는 이번 비엔날레를
통해 SNS에서 드러나는 건축물(어쩌다가게,
어쩌다집, 이안북스사옥)의 이미지와 밀도를
전시하고 디지털시대의 건축의 의미를 다시
생각해 보고자 한다.

SNS COLLECTIVE CITY

Architects Office SAAI

T02

This project looks beyond physical notions
of the collective, focusing instead on the
vast volumes of exchange that occur
on Social Network Services (SNS). The
commercial facilities around Hongdae
are sited off the main road, yet they are
actively shared on social networks. People
are encouraged to visit and share images,
which are exchanged and consumed
through social networks. This is a time
when people consume the city through
images and social media 'density', rather
than through visiting major commercial
streets. Through this exhibition,
SAAI presents the way four buildings
(Uhjjuhdah Shops@Mangwon, Uhjjuhdah
Shops@Seogyo, Uhjjuhdah House, IANN
Books) are represented on SNS, to
reconsider the meaning of architecture in
the digital age.

T02

Uhjjuhdah Shops @ Mangwon

두레주택
(주)조진만 건축사사무소

산새마을은 일제 강점기에 공동묘지로 사용
되었다가 1960년대 후반 철거민과 수재민을
위한 이주 택지로 거듭났다. '산새'라는 명칭은
이곳이 주민참여형 주거재생사업 시범마을로
선정되면서 붙여진 것으로, 사업을 계기로
마을은 주민참여가 활발한 공동체로 세간의
주목을 받기도 하였다.
두레주택은 서울시에서 운영하는 임대용
공유주택으로 청년층과 독거노인 등의 주거
문제를 해결할 목적으로 세워졌다. 우리는
주택 내 세부 공간의 역할을 분석하고
관계맺음의 중요성을 감안해 마을의 개념을
주택에 접목하고, 주택을 다시 마을 단위로
확장시키고자 하였다. 즉, 5명의 거주자가
생활하는 주택이지만 울타리가 없어서
거실과 정원이 골목을 마주하고 외부
도로와도 연결된다. 뿐만 아니라 이 도로는
윗 층에 위치한 다섯 개 방과 다시 연결되어
전체적으로 마을공동체가 활짝 열린 형태를
띤다. 이로써 건축은 거주자와 이웃주민이 다
함께 만들어가는 마을풍경으로 자리매김하게
된다.

DU-RE HOUSE
Jo Jinman Architects (JJA)

Sansae Village, previously a public burial
ground during the Japanese colonial era
was developed as residential land for
relocating the evicted and flood victims
in the late 1960s. The village was named
when it was selected as pilot project of
Participatory Residential Regeneration
and has come into spotlight as a village
with an active local community.
Dure House, rental housing operated
by Seoul Metropolitan Government, is a
share house which can accommodate
young employees and elderly people
living alone. We intend to propose new
architectural type which introduces village
into house and extend house to village
by analyzing the space occupation and
sociality of an existing local community.
Though only five people live there, the
garden and living room face the street
without fences whilst an extended path
weaves together five rooms in the upper
part, allowing the house to be open
towards the village community. Through
this, architecture can exist as part of the
village landscape, to be completed by
residents and neighbors together.

Dure House, Concept Model

건축의 공적 역할
정기용
T04

건축가이자 도시 설계자, 교수,
칼럼니스트이며 운동가였던 정기용(1945 –
2011). 그는 우리 사회의 공공 건축가들을
옹호하는 데에 자신의 삶과 커리어를 바쳤다.
그의 대표작 중 하나는 '기적의 도서관'이라고
불리는 아동을 위한 도서관으로, 시민운동과
공공단체, 지방행정과 건축가 공동의 노력과
거버넌스의 성공적인 사례라고 할 수 있다.
정기용이 택한 건축의 접근방식은 '성취된
것이 아닌 되는 것'이다. "도시는 디자인하는
것이 아니라 생성되는 것이며, 사람과 자연,
역사가 어우러져서 생성되는 것"이라는 그의
말에서 도시를 하나의 생명체로 인식하는
건축가의 비전을 엿볼 수 있다. 도시는
끊임없이 변화한다. 그럼에도 불구하고 도시는
시민들의 집단적 기억을 담아 공동의 문화가
된다. 즉, 모든 도시가 우리 모두의 것이 되는
것이다.
"문제도 이 땅에 있고, 해법도 이 땅에 있다."
만약 그가 오늘날까지 살아있었다면 그의
모토 역시 생생하게 울려 퍼졌을 것이다.
또한 "건축의 핵심은 한마디로 이야기하면
현장성이라고 할 수 있다. 도시든 건축이든,
현장의 역사, 지형, 사람, 현장성이다.
가상적으로 혼자 꿈꾸는 것이 아닌 집단적
현실"이라고 덧붙이지 않았을까. (정구노)

THE PUBLIC ROLE OF ARCHITECTURE
Guyon Chung
T04

As an architect, urban planner, professor,
columnist and militant, Guyon Chung
(1945 – 2011) has dedicated his life and
career to advocating for the public role of
architects in our society.
Among his major public works, the
series of children's libraries, known as
the 'miracle libraries', are an example
of a successful joint endeavor and good
governance between citizen movement,
public authorities, local administration
and the architect.
Guyon Chung's approach of architecture
as a process of becoming rather than an
achieved result, mirrors his vision of the
city as an organism that "isn't designed
but grown from the mixture of people,
nature and history". And despite their
constant transformation, it is because
cities embody the collective memory of its
inhabitants and becomes their common
culture that he has always assessed that
"all cities belong to everyone".
If he was still among us today he would
probably remind us of his motto: "both
the problem and the solution are inherent
in the land." And he might add that
"the essence of architecture may be
found in that realm, whether urban or
architectural, the history, geography, and
people of a site form this visceral reality. It
is not a virtual or an individual dream…",
but a collective reality. (Gounaud Chung)

서울 이야기(2008, 현실문화) 중에서 (사진: 김재경)

난민 헤리티지
DAAR (탈식민건축미술레지던시) / 알레산드로 페티

T05

난민촌(refugee camp)은 철거를 염두에 두고 세워진다. 정치적 실패를 전형적으로 보여주는 난민촌에는 역사도 미래도 없으며, 잊혀질 운명만이 존재할 뿐이다. 국가와 인도주의 단체, 국제기구들은 난민촌의 역사를 계속 지우려고 하며, 난민촌 공동체조차 향후 자신들의 귀향할 권리가 침해될 수 있다는 우려로 난민촌의 존재를 부정한다. 이런 난민촌 공동체들 내에서 유일하게 인정되는 역사는 폭력과 치욕일 뿐이다. 그러나 난민촌 자체가 가지고 있는 도시구조 속, 풍부한 이야기들이 깃들어 있는 장소이기도 하다.

난민 헤리티지는 고난과 실향(失鄕)의 아픔 외에도 다양한 난민의 역사를 추적, 기록, 표현, 공개함으로써 인도주의적 관점 이상의 '난민다움(refugeeness)'을 제시하고 실천하고 있다. 또한, 본 프로젝트는 2년에 걸쳐 기관, 개인, 정치인, 보존 전문가, 활동가, 정부 및 비정부 대표 및 인근 주민들과 함께 데이셰 난민촌(Dheisheh Refugee Camp)의 유네스코 세계문화유산(UNESCO World Heritage Site) 등재에 대해 논의하였다. 이번 비엔날레의 전시는 데이셰 난민촌의 유네스코 세계문화유산 등재 신청서 내용을 담은 시청각물로 구성된다.

REFUGEE HERITAGE
DAAR (Decolonizing Architecture Arts Residency) / Alessandro Petti

T05

Refugee camps are established with the intention of being demolished. As a paradigmatic representation of political failure, they are meant to have no history and no future; they are meant to be forgotten. The history of refugee camps is constantly erased, dismissed by states, humanitarian organizations, international organizations and even self-imposed by refugee communities in fear that any acknowledgement of the present undermines a future right of return. The only history that is recognized within refugee communities is one of violence and humiliation. Yet the camp is also a place rich with stories narrated through its urban fabric.

In tracing, documenting, revealing and representing refugee history beyond the narrative of suffering and displacement, Refugee Heritage is an attempt to imagine and practice refugeeness beyond humanitarianism. Over the course of two years, organizations and individuals, politicians and conservation experts, activists, governmental and non-governmental representatives and proximate residents gathered to discuss the implications of nominating Dheisheh Refugee Camp as a UNESCO World Heritage Site. The project constitutes of an audio/visual installation that narrates the UNESCO nomination dossier of Dheisheh Refugee Camp as a World Heritage Site.

Capuano, Luca & Favero, Carlo. 2016. Refugee Heritage. Photography. Dheisheh Refugee Camp in Palestine.

두 개의 변신 전략

알레한드로 에체베리,
호르헤 페레스-하라미요

T06

본 프로젝트는 메데인 시의 집합적
변천(collective process)에 기여한 다양한
규모의 실험들을 탐구하고자 한다. 메데인
시의 역사상 가장 결정적인 시기였던
1980년대 후반부터 현재까지, 특히 지난
20년간 진행되어 온 도시의 변화 과정에
집중하였다.
이 프로젝트에서는 규모가 서로 다르지만,
상호보완적인 두 가지 다른 유형의
개입을 두 가지 관점으로 과정과 결과를
종합하고 분석하고자 한다. 첫 번째 방식은
도시·광역시계획과 전략적 프로젝트에
기반을 둔 접근법으로, 사업규모가 큰 것이
특징이다. 이 방법은 영향력이 큰 제도적
개입(intervention)과 도시계획에 대한
의사결정을 내리는데 사용된다. 한편, 사회적
도시주의(Social Urbanism) 전략은 주로
중소규모 사업에 적용된다. 메데인 시에서
주거환경이 가장 열악한 지역을 대상으로
진행된 이 사업은 공동체 구성원들과 협업을
진행하고 주민들의 삶의 질을 향상시켰다.
본 프로젝트는 도시사업의 역할을 연구하고,
도시사업이 크고 작은 도시 관리의 수단으로
활용 될 때 나타나는 상호의존적 특성을
연구한다. 두 가지 접근법 모두 변화를
유도하는 도구라는 전략을 추구한다. 이에
따라, 시민 참여와 민주적으로 도출된 합의와
공동의 비전이 강화되는 변화가 생겨난다.
우리는 본 프로젝트를 통해 비판적 시각으로
메데인 시의 변화 과정을 보다 이해하게 될
것이다. 이해를 바탕으로 오늘날까지의 교훈,
향후 주요 과제, 현재 진행 중인 모범사례 등을
살펴봄으로써 오늘날 세계 도시들이 직면한
과제 해결에 메데인 시의 사례가 도움이
되기를 희망한다.

TWO TRANSFORMATION STRATEGIES
Alejandro Echeverri & Jorge Pérez-Jaramillo

T06

We propose to explore diverse scales
of experimentation that contributed to
the collective process of Medellín, from
the critical years of the late 1980s up
to now, with a special emphasis on the
transformation process led during the
past twenty years.
We intend to analyze, synthesize and
assess the results and progress achieved
through two points of view that defined
two different, but complementary scales
of intervention; the large scale of urban
and metropolitan planning and strategic
projects, with decision making processes
in terms of city planning and institutional
interventions with strong impact, versus
the intermediate and small scale of the
Social Urbanism strategy in the most
critical neighborhoods of the city, that
use collaborative processes with the
community to improve their everyday
itinerary. Studying these roles of the
urban project and their interdependence
as a tool for different scales of urban
management, the two voices present
strategies as agents of change that
strengthen agreements and shared
visions, from citizen participation and
democratic development.
We hope to consolidate a critical
reflection on the evolution of the city,
which allows us to present lessons
learned, prevalent challenges and good
practices implemented in Medellin that
can contribute to solve global issues of
the contemporary city.

T06

Social Urbanism strategy and Integral Urban
Project -PUI-: Co-Led by Alejandro Echeverri
and the team of Empresa de Desarrollo Urbano
-EDU-during his Strategic Urban Projects
Direction in Medellín, between 2004 – 2008.

건물들과 그 영역
토니 프레튼 아키텍트
T07

도시는 정치적, 경제적, 문화적 개념들로
형성되지만, 사람들은 이 외의 각기
다양한 생각을 기준으로 도시를 점유하고
경험하고 다시 상상한다. 이러한 생각들은
때로는 개인적이거나 난해할 수 있지만,
이 생각들이 사회·문화적 인식을 토대로
사회 전반에서 통용되는 개념이라면
소통구조(communicative architecture)의
기반이 될 수 있다. 소통구조는 도시의
건물들은 물론이고 신문, 영화, 소설 등의
대중매체가 건물을 활용, 변형, 해석하는
과정에서 건물들에 각인된 흔적 속에서 찾아볼
수 있다.
이 프로젝트에서는 도시를 구성하는 주 요소인
이 개념을, 인간관계와 이해의 토대가 되는
'공감'이라는 감정과 함께 실제로 활용하고자
한다. 프로젝트를 통해, 도시를 되찾고 건축의
개념을 재정립한다기보다, 시민들이 도시를
영유할 수 있는 새로운 방식을 공유하고자
한다.

BUILDINGS AND THEIR TERRITORIES
Tony Fretton Architects
T07

While cities are shaped by political,
financial and cultural concepts, people
occupy, experience and reimagine them
according to very different ideas. Some of
these ideas are personal and inaccessible,
but those which lie in the broad social and
cultural agreements of a society can form
the basis of communicative architecture.
We find them evident in buildings that
exist and in the imprint on them of use,
alteration and interpretation by mass
media such as newspapers, film and
other types of fiction.
It is with the inventive use of these things
that we work as a practice, using the
rough forms of empathy that are the basis
of human relations and understanding.
So we are not so much 'Reclaiming
the city and redefining architecture',
as working with the ways that cities
are constantly and persistently being
reclaimed by their citizens.

The politicians see their City and the
City sees its politicians

기후변화대응조치 2.0:
건축학적 연대

카담바리 백시

T08

기후변화대응조치 2.0 프로젝트 프로젝트는
기후정상회의(Climate Summit)에서
발생한 시위와 환경파괴에 대한 전례 없는
법적 소송 사례 등을 소개함으로써, 다양한
집단들의 기후변화 대응조치를 강화하기
위해 뉴욕 국제연합(UN) 본부와 워싱턴DC의
미국대법원에 작품을 설치할 것을 제안하였다.
이에 따라, UN 본부에서는 지난 10년간
세계 9개 도시에서 개최한 UN기후변화협약
당사국총회(COP)에서의 시위현장 영상을
캡처한 이미지를 파노라마 형태로 제작하여
시위 당시 모습을 재현한다. 이 설치작품을
통해 제3세계가 요구하는 기후 공정(climate
justice)의 중요성이 강조되고, 법적 효력이
없는 조약과 자발적 협약의 비효율성이
여실히 드러날 전망이다. 대법원의 공간적
전시 배치는 복잡한 기후변화 소송 사례들을
서술한다. 환경오염을 유발하는 석탄 발전소에
대한 세계은행(World Bank)의 투자를
둘러싼 법적 분쟁, 안정적인 미래 기후에 대한
청년층의 권리 보장을 명목으로 미국 정부에
제기된 소송 사례 등을 다룬다. 각 사건의
타임라인과 주요 변론 내용들이 갖는 공공성이
전시공간적으로 표현이 된다.
이에 연계되어 영상 몽타주들은 공공건물을
매개로 기관들의 내부사정을 드러내면서
하나의 단면을 보여준다. 이를 통해
관람객들은 규제와 합법적인 방법들을
기반으로 하는 독창적인 형태의 저항운동과
기후운동들을 눈여겨 볼 수 있을 것이다.
아울러 이 프로젝트는 도시의 랜드마크에
전시를 제안함으로서 기후변화에 대한
대응조치를 가속화할 수 있는 '건축적 연대'를
제공한다.

CLIMATE ACTIONS 2.0:
ARCHITECTURAL
SOLIDARITIES

Kadambari Baxi

T08

Climate Actions 2.0 proposes
installations at two sites: The United
Nations Complex in New York and the
Supreme Court in Washington D.C.
They amplify diverse forms of climate
actions, including targeted protests at
climate summits and unprecedented
litigation on environmental degradation.
At the UN, excerpted video-clips of
protests at the UN-COP summits held
in nine global cities over the last decade
unfold as a panorama of marches and
disruptive actions. It highlights climate
justice demands from the Global South
and emphasizes the ineffectiveness
of non-binding treaties and voluntary
agreements. At the Supreme Court,
spatial displays narrate complex
climate lawsuits: including one that
challenges The World Bank over its
financing of a coal powerplant that
creates environmental pollution, and
another that sues the US government
over youth's rights to a stable future
climate. The lawsuit timelines and major
arguments are made public as spatial
interfaces. Exhibited as paired video-
montages showing architectural mediatic
interventions, the project focuses on
public buildings as interfaces to reveal
aspects of institutional inner-workings.
It draws attention to inventive forms
of resistance and climate mobilization
based on regulatory frameworks and
legal mechanisms. By proposing tactical
installations at civic landmarks, the
project offers architectural solidarities
towards accelerated collective climate
actions.

T08

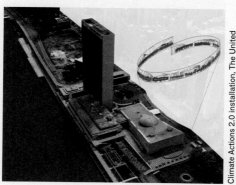

Climate Actions 2.0 installation, The United
Nations Complex, New York

멕시코 주거 도시화
엘 시엘로
T09

최근 도시의 팽창과 사회 공간의 분리가
문제시됨에 따라 수많은 개발도상국의
도시외곽에서 저소득계층 대상의 주거
프로젝트가 크게 확장되고 있다. 이들
프로젝트는 문화적 차이에 따라 각기 고유성을
갖는 한편, 각 프로젝트의 가장 핵심적인
의도가 주거 문제 해소라는 점에서 유사성을
지니기도 한다.
자원이 동일하다고 해서 결과물이 같다는
법은 없다. 다수 이해관계자들의 의사결정과
자원분배방식에 따라 저마다 다른 결과물이
도출될 수도 있다. 따라서 수단과 도구를
응집하고 관계성을 통합하는 방식으로 건축에
접근할 필요가 있다.
본 프로젝트는 향후 멕시코 도시환경의 형태를
결정지을 공공주거정책과 금융 상황, 정치적
안건, 사회참여와 건축 간의 상관관계를
알아보고자 한다.

HOUSING URBANISM MEXICO
El Cielo
T09

In recent times, low income housing
projects have emerged as gigantic
footsteps at the outskirts of many cities
in the global south, premonitory of urban
expansion and sociospatial segregation.
The narratives out of which these projects
come about are very similar at their core
intentions across nations, in culturally
unequal contexts.
However, the resources available, the
processes and the distribution of the
decision making capacity of the several
stake holders take different arrangements
that have impact on built form.
With different degrees of success,
architectural practice has to navigate
through this weaving of relationships to
claim its' relevance as the coalescing
instrument, the tool for embodiment,
specially when the collective has to be
represented.
Categorizing projects by collective
by government, labor, individual and
empowerment, El Cielo Architecture´s
participation at the Seoul Biennale of
Architecture and Urbanism 2019 looks
into the relationship of public housing
policy, finance, political agenda, social
engagement and architecture, as it
determines the shape of Mexican urban
environments for years to come.

Salinas Victoria, Photo by Jorge Taboada

디지털 광장
페드로 앙리크 데 크리스토 / +D

T10

디지털 광장은 참여 민주주의를 실천하는 물리적인 가상의 공간으로, 다수의 참여를 도모하고 의견을 수렴하여 구체적인 공공정책을 통합함으로써 함께 만드는 도시를 실현하는 데 목적을 둔다. 본 프로젝트는 직접민주주의와 새로운 체계, 통합, 선진기술 등을 활용하여 글로벌 추세에 부합하는 미래형 도시 운영법을 도출해보고자 한다. 인류는 14세기부터 오늘날까지 지식의 습득을 통해 공간과 에너지를 변형시켜 이를 기술 자원으로 활용해왔다. 이는 곧 오늘날 디지털 공간의 탄생으로 이어지기도 한다. 하지만 미래는 공공정책, 공간, 기술을 통합하는 새로운 제도적 설계뿐 아니라 다양한 이해관계자들의 참여가 필수적이라고 할 수 있다. 본 프로젝트는 디지털 광장을 진보적인 도시, 즉 오늘날의 소통불가형 도시 형태에 대한 대안으로 제시하고자 한다.

DIGITAL AGORA
Pedro Henrique de Cristo / +D

T10

The Digital Agora is a concept program of a physical + digital space for participatory democracy that responds to the global demand for more participation on the public decision making of cities by integrating specific public policies for instruments of direct democracy, spaces for systematization, synthesis and articulation, and effective technologies to generate a new calibration between representative and direct democracy at the city level. From the creation of the scientific method in the 17th century to today, humankind intensified mastering the knowledge to transform space and energy into technological instruments, resulting in the creation of a new form of space: the digital. Democracy now is in acute need of a new institutional design that integrates public policy, space and technology in the pursuit of the ideal calibration of representative and direct democracy in this new world. The way our societies, economies and democracies work has resulted in the spread of the divided city program all over the globe. A direct translation of inequality and politics into space, this reality demands bringing together activism and institutional roles as well as to redefine architecture's disciplinary boundaries to create a specific program capable of adapting democracy to the challenges of our time. As there are schools for education and hospitals for health, the Digital Agora is proposed as the central democratic instrument to overcome the predominant divided city program towards progressive cities.

Digital Agora at Vidigal, Rio de Janeiro, RJ, BR

31

멀티플라이도시
아틀리에 얼터너티브 아키텍처
T11

멀티플라이도시는 중국 선전에 거주하는
노동이민자들의 주거형태를 도시라는
맥락에서 분석하고 조명하였다. 본 프로젝트는
폭스콘 롱화 공장 인근에 위치한 기숙사와
마을을 연구사례로 삼아, '도시 속 어떠한
물리적 건축 척도가 사회에 활력이 되고
공동체 의식을 함양하는가' 라는 질문에
주력하여 도면을 설계하고 모형을 구축하였다.
멀티플라이도시는 마을주민, 개발자,
공무원들이 공중권 소유에 대해 열린 협상을
진행한다면 도시 공동체가 활기를 가질 뿐
아니라 지역의 물리적, 감성적인 역사도
보존할 수 있다는 가정을 하였다.
가상의 미래는 450,000 종에 달하는 부품이
조립 라인을 타고 대량생산되어 반출되고,
공장 내부에서 노동자들의 숙식까지
한꺼번에 이루어지는 폭스콘 공단의 모습과는
대조된다. 멀티플라이도시는 이주 노동자를
위한 적응형 사회 및 물리적 인터페이스가
갖는 초(한자)근대성을 넘어서며, 기생적인
구조를 넘어 공동체가 성장할 수 있는 미래를
상상한다.

MULTIPLICITY
Atelier Alternative Architecture
T11

MultipliCity is an analysis and projection
of migrant worker housing in Shenzhen,
contextualized within the urban village.
Using the dormitories and villages
surrounding Foxconn's Longhua campus
as a case study, the drawings and model
ask a single question: by what parameters
might the physical architecture of the
urban village reflect its social vibrancy
and sense of community? MultipliCity
hypothesizes that through processes of
air right negotiation between villagers,
developers and the government, the
urban village can grow as a vibrant
organism while preserving layers of
physical and emotional history. The
imagined future is held in contrast to the
interior of the Foxconn campus, in which
the condition of the assembly line is
multiplied across 450,000 bodies through
processes of manufacturing, assembling,
eating, sleeping, and exportation.
MultipliCity imagines a future in which
community grows upwards through
parasitic structures, outcompeting the
hyper-modernity of the globalized factory
to create an adaptive social and physical
interface for Chinese migrant workers.

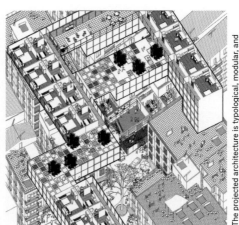

The projected architecture is typological, modular, and incremental, with interventions alternatively acting as public program, residences, restaurants, shops, and gardens.

고시원 엿보기
블랙스퀘어 / 마리아 지우디치
T12

고시원 엿보기에서는 도시와 건축에서 공과
사, 집단과 개인 등이 어떻게 상호작용할 수
있는지 살펴보고자 한다. 이를 통해 공동 혹은
개인 소유라는 이분법적인 논리에서 벗어나 또
다른 형태의 도시공간을 제시하고자 한다.
본 프로젝트는 지난 수십 년 간 한국의 두
가지 주거 양식에서 영감을 받아 기획한 총
6가지 시나리오로 구성되어 있다. 여기서
말하는 주거 양식은 고시원과 다양한 종류의
방을 말한다. 고시원 엿보기는 여러 용도의
방이 밀집된 형태의 고시원을 파헤치며,
목욕탕, PC방, 수다방 등을 갖춘 남다른
한국형 주거형태가 전 세계 여타 도시에
적용가능한지 여부를 살펴보고자 한다. 아울러
본 프로젝트가 기존의 주거공간에 도전장을
내밀고 미래형 주거공간을 기획하는 데 참고가
되길 기대한다.

OUT WITH A BANG
Black Square / Maria Giudici
T12

Out with a Bang seeks to reimagine an
urban and architectural model that works
with levels of ownership, privacy, sharing,
and jurisdiction that are different from
the public/private divide which shapes
most contemporary cities. The project
is composed of six projective scenarios
for new domestic forms inspired by two
architectural archetypes developed in
South Korea in the past few decades:
the goshiwon, or boarding house, and
the bang, or function-specific rentable
rooms. Out with a Bang envisages a
goshiwon model that is punctuated by the
insertion of large scale bang complexes
— spa centers, gaming rooms, meeting
rooms. The scenarios seek to learn
from the contemporary South Korean
city to formulate projects that can fit in
other metropolises as well. Ultimately,
we see the bang as a lever to challenge
conventional domesticity while looking for
new forms of agency in the design of our
spaces for living.

T12

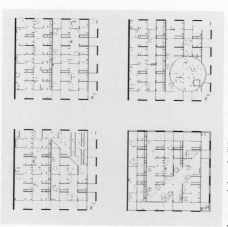

Large-scale bang facilities punctuate the boardinghouse prototype

33

도시의 경기장
NP2F

도시가 끊임없이 진화하면서, 경기장은 거대한 집합공간이 형태화된 도시적 오브제와 같아진다. '경기장(stadium)'이라는 단어는 계단 15단의 거리를 가리키는 고대 단어에서 유래하였다. 이후에는 경주 시합이 개최되는 장소를 의미하게 되었고, 오늘날에는 여러 행사가 가능한 다목적 장소가 되었다. 우리는 이렇게 긴 역사를 지닌 장소를 통해 스포츠 경기의 진화 과정과 스포츠 경기와 도시 간의 관계를 연구하고자 한다.

경기장은 구조상 다양하게 활용될 수 있다. 본 프로젝트는 개방적이고 어느 정도 집합공간의 성격을 갖는 광장을 통해 경기장과 도시의 상호작용을 중점적으로 다룬다. 여러 스포츠 종목이 공존하는 경기장은 스포츠가 진화하는 장(場)의 역할을 하여 공간의 역할이 강화되는 결과를 낳는다.

대형 건축물들은 공간을 자유롭게 한다. 다양한 용도를 지니며 많은 인원을 수용할 수 있는 공간(capable spaces)은 넓은 면적을 포괄한다. 이러한 의미에서 경기장은 도시의 일부 공간을 차지하여 경기장 안쪽에 텅 빈 도시 공간을 새롭게 만들어 낸다고 할 수 있다. 도시의 일부인 동시에, 직관적이고 복합적인 공간의 특성을 강화하는 도구가 되는 것이다. 하지만 오늘날의 경기장들은 도심의 중심에서 멀어져 점점 교외로 향하고 있다. 따라서 우리는 경기장의 위치와 도시적 상황에 대한 의문을 다시 던지게 된다. 대규모의 바닥 면적을 차지하는 경기장은 도시의 중심을 벗어났지만, 경기장 내부적으로는 또 다른 도시를 형성하고 있는 이중성을 띄고 있는 것이다.

URBAN ENCLOSURE
NP2F

Stimulated by the permanent evolution of the city, the stadium — a large enclosure, is an urban object that gives form to a collective space. During the antiquity the first meaning for stadium was a measure of length of 15 steps, then it designated the space where the race took place and today it has become a field for mixed-purpose events. We question this evolution of sport activities and their relationship with the city by using this archetypal object. The architecture of stadiums generates a multiplicity of uses. We are particularly interested in the interaction of the stadium with the city through the square as an open and properly collective space. On the other side of the enclosure, the sport scene is an evolutionary ground where several practices can coexist, allowing a spatial amplification. Large structures free the ground. This analogy illustrates their ability to enclose large areas of capable spaces dedicated to multifunctionality. As 'other space' the stadium creates active and firm urban voids. As an urban machine, the stadium becomes an object capable to support programmatic and spatial hybridity. The city stadiums of today seem to gravitate in a suburban system, in total disconnection with city centers. Hence, we re-question their place and urban condition. With their vast footprints they generate a duality inside/out: the stadium has been ejected from the city and in the same time it continues to produce the city itself.

Aguilera Stadium, Biarritz

집합도시: 인도 현대 건축의 집합적 수행의 형태

사미프 파도라 건축연구소(sP+a/sPare) 기획: 아키텍처 레드 RED, 아비지트 무쿨 키쇼어 & 로한 시브쿠마르, 반드라 콜렉티브, 버스라이드 디자인 스튜디오, St+art 인디아 재단, 아빈 디자인 스튜디오, 피케이 다스 앤 어소시에이츠, 공간적 대안을 위한 모임, 아다르카르 어소시에이츠
T14 – T23

인간 활동의 산물인 도시화는 다양한 폭과 깊이로 진행되고 있으며, 인류는 그 도시화의 공간에서 살아가고 있다. 2019 서울도시건축비엔날레는 도시라는 주제에 상응하여 도시집합(collectives)라는 개념을 간직하고 현실화하려는 전세계 건축 활동가들을 한데 모으는 — 반드시 필요한 — 일을 하려고 한다. 전시 기획자 노트에서 다음 몇 가지 중요한 질문을 추려볼 수 있다. 집합성(the collective)의 추진력은 무엇일까? 집합성이라는 개념에 대한 정의가 현대 도시 생활 인하여 바뀌었을까? 집합성은 어디에서 시작되고 끝나는가? 집합성이 포함하거나 포함하지 않는 대상과 사람은 무엇이고 누구일까? 집합성은 얼마나 지속될까? 도시집합들은 필요에 따라 집합성을 보이지만, 자신만의 특징을 뽐낼 특별한 도시를 만들겠다는 희망을 품고 있다. 이 자리에 모인 인도의 건축 활동가들은 다양한 형태의 집합을 이렇게 제시한다.

1. 기념/특정 시기/단기적 행사와 관련된 집합
2. 공간적 집합
3. 기억/역사와 관련된 집합
4. 행동주의/저항과 관련된 집합

모든 집합을 이 네 가지 범주로 한정할 수는 없으며, 범주들이 명확하게 구분되는 것도 아니다. 오히려 이 집합들은 서로 복잡하게 중첩되고 얽혀 있다. 그러나 각 범주에 따른 건축 활동은 특정 단체 및 다양한 가능성과 연관된다. 디자인, 예술, 법률 문서 연구, 소외된 공동체를 위한 노력, 특정 시기에 진행되는 행사, 도시 보존, 심지어는 도발적 문제 제기까지도 포함하는 다양한 활동이 한데 모여 도시 공간의 다원성을 활발하게 유지시켜 나가는 원동력이 된다.

COLLECTIVE CITIES:
Notes on Forms of Collective Practices in Contemporary Indian Architecture

Ten Practices: India. Curated by Sameep Padora Architecture and Research (sP+a/sPare): Architecture RED, Avijit Mukul Kishore & Rohan Shivkumar, Bandra collective, Anthill design, The Burside design studio, St+art India Foundation, Abin Design Studio, PK Das & Associates, Collective for Spatial Alternatives (CSA), Adarkar Associates
T14 – T23

SBAU 2019 seeks to build a global inventory of necessary practices that engage with and through the idea of collectives in cities in response to the varying intensities and amplitudes of urbanization produced and inhabited by humanity at large.

A few important questions that can be sieved from the curatorial note for SBAU 2019 are: what is the impetus for the collective today? Have our contemporary urbanities redefined the notion of the collective? Where does it begin and where does it end? What or who does it include or exclude? How long does it last?

The city regurgitates forms of collectives in formats that it needs and the collectives exist in the hope of making a particular kind of city of their specificity. The practices presented here from India demonstrate a variety of collective forms classified within four operative formats:

1. Celebratory/Temporal/Ephemeral Collectives
2. Spatial Collectives
3. Memory/Historical Collectives
4. Activist/Resistant Collectives.

The collectives are not limited by these four categories, nor are these categorizations watertight, they have multiple overlaps and entangled trajectories. However, each of the forms of practices bring with them particular agencies and projective capacities - design, art, research legal instruments, mobilization for marginalized communities, temporal events, urban conservation or even provocations- which together keep the plurality of the urban space active and engaging.

타인과 어울리는 방법
제너럴 아키텍처 콜라보레이티브 (GAC)
T24

제너럴 아키텍처 콜라보레이티브(GAC)는
2008년부터 르완다의 마소로(Masoro)
마을의 여러 프로젝트를 진행하고 있다.
프로젝트의 규모와 복잡성은 다양하지만,
지역 공동체의 참여, 연구, 교육을 통한
협업 디자인을 추구하는 점은 공통적이다.
타인과 어울리는 방법 디자인 프로젝트는
타인, 장소, 환경, 사물, 문화, 기술 등과의
관계에서 영감을 받아 진행되었다. 이를 통해
디자이너와 협력자들은 남다른 공동체를
형성할 수 있었다. 이러한 접근법으로 타인과
어울리는 방법 프로젝트는 "관계성", "사물",
"과정"이라는 세 가지 결과물을 도출하였다.
먼저, "관계성"은 "프로젝트 지도"로 나타난다.
지도는 네 가지 프로젝트를 통해 개인, 조직,
기관, 정부, 관습 간의 연결성을 소개한다.
"사물"은 눈금 표기된 도표와 그림을 활용하여
연결망으로 도출해 낸 프로젝트를 통해
선보여진다. "과정"은 네 가지 프로젝트를
수행하며 동고동락한 이들의 소중한 순간과
활약상을 엮은 영상을 통해 전시된다.

PLAYING WELL WITH OTHERS
General Architecture Collaborative (GAC)
T24

Since 2008, GAC has been working
in and around the village of Masoro in
Rwanda. Our projects have varied in scale
and complexity, but each has focused on
collaborative design practices through
community engagement, research and
education. Playing Well With Others
describes an approach to design that
has grown from the relationships we
have made with other people, places,
environments, materials, cultures and
techniques. It is an approach that has
allowed us to develop a unique form of
agency for designers and those we work
with. Playing Well With Others presents
three outputs from this approach:
Relationships, Products, and Processes.
Relationships are illustrated through
a Project Map, where four projects are
used to reveal the connections between
individuals, organizations, institutions,
governments and our practice. Products
shows the projects that have resulted
from this network through scaled
diagrams and drawings. Processes is
a series of films which show the people
we have worked with and the various
ways they have participated in these four
projects.

The Masoro House, completed in 2014 is
a community built project which combines
local materials and techniques with new
scales and forms of production to create a
novel yet grounded housing prototype for
rural Rwanda. Photo by Bruce Engel.

공간적 가치의 창조
CBC (차이나 빌딩 센터)
T25

농촌 개발은 효율적이고 중앙집중적인 하향식 접근 방식의 개발 형태보다는 좀 더 깊고 섬세한 인간적인 요소들을 고려할 필요가 있다. 차이나 빌딩 센터(CBC)는 사회 발전, 경제, 문화, 생태계 간의 세심한 균형을 도출하고자 일련의 연구를 수행 하였으며 이를 통하여 도시와 농촌의 관계 속에서 국가가 어떠한 새로운 역할을 할 수 있는지 살펴보았다. 공간적 가치의 창조는 그간의 연구에 대한 집약적 산물이라고 할 수 있다. 핵심 전략은 디자인과 건축의 특징과 장점을 끌어 모으는 것이었다. 즉, 공공주택의 장점과 디자인, 문화, 공동체적인 요소를 한데 모아 하의상달식 접근을 구현할 뿐 아니라 도시와 농촌 간의 문제를 관철하고 이에 혁신적으로 대응하고자 하였다. 본 전시는 CBC의 감독이자 도시환경디자인(UED) 매거진의 편집장인 펑 리 샤오(PENG LIXIAO)가 총괄했다. 루나 국제건축인의 마을(LOUNA INTERNATIONAL ARCHITECTS' VILLAGE)과 시아무탕 마을(XIAMUTANG VILLAGE)이 혁신형 농촌 개선 마을의 사례로 선정되었다.

CREATION OF SPATIAL VALUE
CBC (China Building Centre)
T25

Taking into account efficient, centralized, top-down development models, CBC believes that urban and rural development needs to prioritize more profound, subtle human factors. The research conducted by CBC reflects specifically on how to achieve a delicate balance between issues of social development, economy, culture and ecology; an approach that hopes to obtain a new position outside of the traditional urban-rural relation and to generate new possibilities in the field. CBC summarized this series of actions as the Creation of Spatial Value. The main strategy is to exert the power of design, and architects as a whole; taking advantage of community building and promotion, whilst drawing upon resources of design, culture and community, in order to explore a more bottom-up and acupunctural approach and therefore address urban and rural issues in an innovative way. The exhibition is curated by Peng Lixiao, Director of CBC (China Building Centre), Chief Editor of Urban Environment Design (UED) Magazine. Two villages, Louna International Architects' Village and Xiamutang Village were selected as examples to showcase innovative rural revitalization models.

T25

Xiamutang Pond Theatre -2 © CBC

일곱가지 서적을 올린 제단

Baukuh

T26

글로벌 자본주의가 양산한 도시들의 한계와 약점이 드러나기 시작하면서 심각한 논란의 대상이 되고 있다. 그 중에서도 기후변화에 대한 인식과 시민을 소비자로만 볼 수 없음은 다각도로 그 핵심적 가치에 대한 반론에 직면하게 된 것이다. 지난 수십년 동안 도시설계는 변화과정의 가장자리만 다듬을 뿐 사실상 영리 창출이 주 목적이었다. 즉 공동의 노력이 필요하다는 합의와 식견이 부족했던 것이다. 더 이상 도시에 건축은 없다. 그렇기에, 문화가 중심을 되찾아야 하며, 모든 도시가 일반화되어 자본주의의 힘에 의해 소비되고 마는 반복적인 게임에서 탈피해야 한다. 각 도시의 고유성과 도시라는 동질성은 단순히 주택의 담보물로서의 가치, 지역적 가치 혹은 노동 자본이라는 의미를 넘어서 대화를 통해 재정비 되어야 한다.

본 프로젝트는 일곱가지 서적을 올린 제단을 제안한다. 이는 도시에서 활용 가능한 공유지를 제시한 시범형으로, 지나친 구조적 기교와 색을 포함하는 디자인의 총력이 담긴 지붕구조를 가진 정자의 전통을 따른 것이다. 제단은 고풍스런 7각형 구조를 띠며 아름다운 색감의 바닥재와 비계판에 고정된다. 또한 49가지 형광 튜브로 덧씌우고 플라스틱 단장이 하단에서 위로 올라오는 형태로 꾸몄다. 방문객은 몸을 굽혀야 조용한 실내로 들어올 수 있다. 바닥에는 실로 짠 둥근 깔개가 놓였고 주변 낮은 제단 위에는 도시에 관한 일곱 권의 건축 서적이 놓여있다. 이는 포스트 자본주의 시대에 걸맞는 도시건축을 위해 '공동의 지식'을 기원하는 마음으로 개인적 헌물을 바치는 행위라고 할 수 있다.

ALTAR OF THE SEVEN BOOKS

Baukuh

T26

The generic city produced by globalized capitalism starts to show its limits and fragilities, increasingly becoming a heavily contested field; among other factors, the acknowledgment of climate change and the impossibility to reduce citizens to mere consumers confront its core values from different directions. Decades of urban design practices centered on the creation of profit while smoothing the edges of the transformation processes (simulacra of communities or bits of participation, green-washing or pink-washing...) left a tangible void in the professional knowledge which is necessary to understand and build the city as a collective endeavor. No more architecture of the city. Culture has to regain centrality, opposing the idea of the city as a homogeneous flat terrain totally at disposal of capitalistic forces to endlessly re-play the same game. A consistent yet multifaceted literature on the city and urbanism needs to be reintroduced into the discourse, beyond the vulgate about cost of mortgages, positional value or working capital cycle.

For the Seoul Biennale of Architecture and Urbanism 2019, we propose the Altar of the Seven Books, a model prototype of a possible public space for the city, in the tradition of the Korean garden pavilions with their impressive roofing structures, where all of the design intensity is typically accumulated (obsessive use of iterative structural parts, colorful decoration extensively applied to the timber members). The Altar is an archaic heptagonal architecture, supported by a scaffolding frame clad by colorful portions of floor carpet. It is crowned by 49 fluorescent tubes and rises from a base made by pedestrian plastic fly-curtains. Visitors need to bend in order to enter the calm inner space, whose floor is constituted by a circular long hair carpet. An off-center low altar sports seven architecture books of the city, as a personal offer of collective knowledge to the construction of post-capitalistic urbanity.

매니
켈러 이스터링
T27

'문제(problems)'에서 출발해야 한다. 문제는 바람직하다. 집합도시는 문제를 더 늘리는 경향이 있다. 집합도시는 문제를 해결하기 위해 계획이나 해결책, 알고리즘을 완전히 익히기보다, 서로 반응하며 변화를 일으키는 문제들을 종합해 "상호작용(interplay)"하는 장치들을 설계한다. 정치가들과 시장이 자리에 남긴 문제들은 자원이 된다. 도시공간이 정치적으로나 경제적으로 순기능을 하지 못하면, "상호작용"이 영향력을 발휘할 차례다. 법적, 경제학적 개념은 차치하고서라도 실패한 도시공간 역시 심오하고 물리적이며 인지적인 가치를 발현한다. 즉, 실패한 도시공간은 도시와 환경이라는 맥락에 대한 근접성, 편향성, 위험성을 내포한다는 것이다. 상호작용에는 다섯 가지 방법이 있으며 필요, 재난, 위험, 난개발(overdevelopment)이 세부 요소로 고려된다. MANY는 필요에 따라 사람들의 이주를 돕는 온라인 플랫폼이다. 도시들은 MANY를 통해 필요한 인재를 유입할 수 있고, 이주를 희망하는 인재들의 필요 역시 고려함으로써 상호 이익을 추구한다. 소셜 캐피털 크레딧(Social Capital Credits)을 통해서는 지역사회들이 사회적 선행을 일종의 통화로 전환해 교육, 보건 등 필요에 따라 사용할 수 있다. 참여형 토지구획정리(Participatory Land Readjustment)는 공간을 재배치하여 부동산 가치를 높인다. 다른 두 가지 방법은 민감한 장소나 기후변화로 위기에 처한 지역에 기존 개발과는 완전히 상반되는 개발 방식을 제공한다.

MANY
Keller Easterling
T27

Start with problems. Problems are good. The collective city likes to multiply problems. It designs, not master plans, solutions and algorithms to eliminate problems, but rather organs of interplay to combine problems. Problems leaven and catalyze each other. Problems that politicians and markets leave in their wake are resources. When Urban spaces fail politically or financially, they become available for interplay. Shedding legal or econometric abstractions, they offer a parallel portfolio of heavy, physical, situated values embedded in spaces — proximities, dispositions and risks in a wetter, hotter world. Five protocols of interplay are considered here. They start with need, disaster, risk and overdevelopment as raw materials. MANY is an online platform to facilitate migration through an exchange of needs. It allows cities to attract a changing influx of talent matching their needs to the needs of mobile people for mutual benefit. Social Capital Credits allow communities to consider needs as currencies rather than deficits. Participatory Land Readjustment produces increased property value through spatial rearrangement. Two subtraction protocols demonstrate how to put the development machine into reverse in sensitive landscape and areas of climate risk.

MANY interface, 2018

약속의 땅, 저가형 주거지와 건축에 관하여
도그마 + 뉴아카데미
T28

본 프로젝트는 현대사회의 주택 위기에 대응하는 새로운 저가형 주거형식을 탐구하고자한다. 이를 위해 우리는 런던, 브뤼셀, 헬싱키 등 세 개의 유럽 도시를 선정하여 토지 소유, 건축, 소유권, 지형을 중심으로 주택과 관련된 사회·경제적 문제를 살펴보았다. 그뿐만 아니라 건축 환경에서 서로 통합되지 못한 부분은 없는지 짚어보며 지형이 진화함에 따른 건축 정책의 개정 역시 필요한지 알아보고자 하였다. 프로젝트는 전통적인 틀에서 벗어나 협력적 대안을 모색하였으며 다양한 이해관계자들의 입장도 고려하였다. 이 연구는 도시별 규제법, 수요, 문제점을 토대로 설계 전략을 수립하고 있으며, 이를 통해 관련 지식을 축적할 뿐 아니라, 대안에 관한 지침을 마련하여 향후 보다 활발한 논의와 연구가 이뤄질 수 있도록 한다.

PROMISED LAND, RETHINKING TYPOLOGY AND CONSTRUCTION OF AFFORDABLE HOUSING
Dogma + New Academy
T28

'Promised Land' focuses on collective habitation not as an idealist projection, but as a structural re-consideration of the entire housing production process: from land procurement to home ownership, from construction to typology. 'Promised Land' is not a solution to the current housing crisis, but an attempt to consider what it is to build affordable housing today in post-welfare Europe. Our research focuses on three cities — London, Brussels and Helsinki — where the lack of affordable housing is becoming a pressing problem. In each of these cities we have team up with a local actor: a housing co-operative in London, a community land trust association in Brussels and homeless housing association in Helsinki. In collaboration with these actors we have developed three pilot projects that can be built in multiple versions and on different sites which are unused public or private land. In different ways, these pilot projects attempt to withdraw housing from the market and compose a body of knowledge that furthers non-commercial housing initiatives, bringing together diverse responses to regulations, necessities, and problems regarding affordable housing. The table displays the designs for three pilot project plus interviews with the representatives of the housing associations with whom we have collaborated. The wall displays twenty-four applications of the 'pilot projects' in London, Brussels and Helsinki.

Evolutive housing: possible variations in dimension and subdivision of different typologies

PARK

신경섭

T29

산업화와 경제발전에 따라 도시 밀도가 증가하면서 도심지역의 환경여건 개선에 대한 요구가 나날이 늘어가고 있다. 자연에 기반한 도심 공간은 인류가 자연을 재생산하는 유익하고 효율적인 방편이다. Park 시리즈를 통해 인공도심공원이 어떠한 방식으로 여가생활은 물론, 활동적이고도 감성적인 생활양식을 도모하고, 공원을 주로 이용하는 시민들이 이 공간을 어떻게 소비하는지 살펴볼 수 있다.

PARK

Kyungsub Shin

T29

Heightening urban density following industrialization and economic growth has seen an incessant and increasing call to improve and introduce nature into the urban environment. Nature-based spaces found in the urban environment have always been considered beneficial, functioning as civilization's method to reproduce nature. The Park Series is a record of artificially created urban public parks promoting leisure-based, active and emotionally affective lifestyles, and how these spaces are consumed by targeted citizens.

Kyungsub Shin, Park No.08, 152cmX224cm, Pigment Print, 2017

세 도시의 현장조사
볼스 + 윌슨
T30

세 개의 벽면에 삼각형 구도로 결과물 전시:

A.   도쿄 패러다임:
　　　30여 년간의 도쿄 모습을 새로운 도시 형식으로 담았다. 기존 유럽 도시들의 고정된 기하학적 구조와는 상반된 유동적이고 역동적인 패러다임이 우리의 상상을 자극한다.

B.   유로랜드샤프트 패러다임:
　　　최근 도시가 자발적으로 진화하면서, 이전에는 특정 지역에 집중되었던 도시의 기능들이 유럽 전역으로 확산되고 있다. 이제 현장 네트워크는 고속도로뿐 아니라 디지털로 연결된다. 볼스 + 윌슨은 이처럼 예전보다 밀도가 낮아진 새로운 형태의 도시를 예측하고, 정기적으로 체험하고, 도시 지도를 만든다.

C.   코르차 마스터플랜:
　　　알바니아 산맥의 고지에 자리 잡은 소도시 코르차는 (2012년 마스터플랜 국제공모전에서 우승한 이후로) 도심을 세부 지역으로 재조정하는 볼스 + 윌슨의 '도시 침구학' 프로젝트의 실험실이 되어, 도쿄의 사례로부터 배우게 된 유의미한 장소들이 확산될 수 있게 하였다.

3 URBAN FIELD RESEARCHES
BOLLES + WILSON
T30

A triangular display with three walls of research:

A.   The Paradigm of Tokyo:
Some thirty years Tokyo, by mutating into a new type of city, caught the imagination of the west — fluid and dynamic — the anti-thesis of the fixed geometries of the historic European City.

B.   The Paradigm of the Eurolandschaft:
Recent spontaneous evolutions have dispersed previously concentrated urban functions across the entire European landscape. A networked field, connected not only by freeways but also by digital communication. BOLLES + WILSON regularly traverse, map and project new typologies into this thin urban soup.

C.   Korça Masterplan:
The small city of Korça, high in the Albanian mountains has (since winning the 2012 Masterplan Competition) become a laboratory for BOLLES + WILSON to re-script the city center with localized interventions — urban acupunctures. The fragmentary character of these interventions produces a dispersed field of signification (learning from Tokyo).

Tokyo Derive (1994) / © Peter Wilson

전환기 카이로의 공공 공간
CLUSTER
(카이로도시교육환경연구소)
T31

카이로도시교육환경연구소(CLUSTER)는
도시가 자아내는 격식과 거리감을 허물고
건축가, 기획가, 정책자들이 보다 자유롭게
소통할 수 있는 대안적인 체제를 추구하는
카이로 중심부와 주변부의 액션프로젝트
다수를 제시한다.
카이로도시교육환경연구소는 2011년부터
카이로에 급습한 정치적 격변과 도심의
변화를 기록해왔으며 카이로의 비공식적인
경제활동과 도시의 권리에 주목하며 이에 대한
담화가 이루어질 수 있는 공간을 모색하는
연구를 진행하고자 한다.

NEGOTIATING PUBLIC SPACE IN CAIRO DURING A TIME OF TRANSITION
CLUSTER (Cairo Lab for Urban Studies, Training and Environmental Research)
T31

CLUSTER (Cairo Lab for Urban Studies, Training and Environmental Research) presents a number of its action projects in both downtown Cairo and the informal periphery that seek to analyze and understand the ordinary urban landscape, providing alternative frameworks for architects, planners and policy makers to engage informality on its own terms. Since 2011, CLUSTER has been documenting a time of rapid political and urban change in Cairo. CLUSTER's research raises questions concerning informal economies in Cairo and the right to the city, while focusing on the spatial manifestation of these practices.

CLUSTER's Street Vendor Initiative

세상이 이토록 거대할 줄이야!

스틸스.언리미티드(아나 조코치,
마크 닐렌), 제르 커즈마니크,
프레드라그 밀리크

T32

2029년을 기점으로 과거를 다루는 일련의
포스터를 살펴보면 이 미래형 공상 소설은
자칭 "부적응자"집단의 현실을 언뜻
보여준다. 2008년 세계경제 위기의 여파로
이들은 집단을 이루어 색다른 미래를 위한
기반시설을 구축하기 시작했다. 이는 곧
새로운 삶의 방식을 탐색하고 생존에만
국한되지 않은 직업, 성장에 목마르지 않은 경제,
획기적인 에너지 자원의 전환, 공동의 이익을
우선시하는 미래인 것이다. 하지만 오늘날에는
"부적응자"집단은 기후변화에 따른 인류의
파멸이 실현되기 이전 그들만의 결속을
다지려는 듯 그저 베오그라드, 스플리트,
로테르담과 변두리 지역을 무리 지어
떠돌아다닐 뿐이다. "오히려 앞으로의 10년은
우리가 잘 적응하는지 지켜보는 일종의
시험 기간인지도 모른다."라는 문구는 향후
숙명적으로 펼쳐질 일종의 주문이자, 다가올
미래를 관찰하는 방식이기도 하다.

I DID NOT KNOW THE WORLD IS SO BIG

STEALTH.unlimited (Ana
Džokić and Marc Neelen), Jere
Kuzmanić, Predrag Milić

T32

Looking back from 2029, through a
series of posters, this future fiction
gives a glimpse into the reality of a self-
proclaimed pool of 'misfits'. Following
the Global Financial Crisis (2008),
they started building infrastructures to
collectively break ground for a different
future — for new forms of living, for work
beyond the reality of jobs, for an economy
beyond growth, for a radical energy
transition, for cities based in commons.
However, today these are merely
enterprises of collective survival floating
between Belgrade, Split, Rotterdam
and beyond, trying to strengthen the
ties before the promise of the climate
collapse comes true. "If anything, the
coming ten years are testing if we're well
accompanied. Well enough to make it
through." These words indicate a spell,
bound to unfold, and indeed, that is how
they observe much of the future lying
ahead.

다카: 백만가지 이야기
마리나 타바시움 + 벵갈
인스티튜트 포 아키텍처,
랜드스케이프 앤 세틀먼츠
T33

만약 도시에서의 집합이, 공적영역의 토대가
되는 열린 공간과 작은 틈새들로 구성된다면
다카라는 도시는 구조적 모순을 보여준다고
할 수 있을 것이다. 다카는 틀에 얽매인
듯하면서도 창의적인 도시다. 곳곳에 세워진
높은 장벽이 공과 사를 구별하고 법규의
적용을 받는 곳과 그렇지 않은 곳을 나누기
때문이다. 우리는 장벽들로 구성된 다카를
사각지대의 집단성이 발현되는 공간으로
묘사하였다. 장벽은 추진력과 전술을
발휘하여 제도적 법규에 배타적이면서도
생활공간으로서는 협조적인 양상을 띤다.

DHAKA: A MILLION STORIES
Marina Tabaussum +
the Bengal Institute for
Architecture, Landscapes and
Settlements
T33

If the collective in a city is constituted by
porosity and openness, the fundaments
of the public realm, Dhaka city presents
a structured contrariness. Dhaka is both
restrictive and creative. Much of this is
constituted by the presence and intensity
of walls that separate the private from the
public, and the seemingly legal from the
undocumented. In our presentation, we
re-describe the walls of Dhaka city as a
meta-site against which the unregistered
collective unfolds. The walls provide
a spatial reference against which to
conduct actual situationist and tactical
operations: subversive to city ordinances,
but essential for negotiating life in the
city.

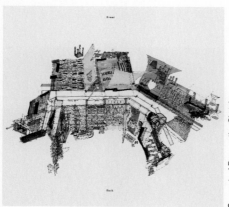

Conceptual Perspective Sketch

집 없는 문명

아미드.세로9

T34

미래의 급진적인 도시형 거주의 전형은 우리가
지금 알고있는 '집'의 형태가 사라지고 공적인
공동체적 가사활동이 지배적일 것이다. 한국의
'방' 문화에 대한 깊은 연구를 기반으로,
찜질방이 전통적인 가정의 모습과는 완전히
다른 모습으로 발전한 점을 볼 때, 미래에는
비(非) 소유와 공동체 생활에 기반을 둔 집단
거주(collective housing) 모델이 등장할
것으로 전망된다. 전통적인 가정의 개념이
완전히 무너지고 공공성(publicness)이
이를 대체하는 거대한 공동체 주택 생활이
예상된다. 수면과 휴식은 24시간 깨어있는
도시의 리듬과 생산성 향상에 저항하려는
집단의식으로 행해질 것이다. 본 프로젝트는
함께 산다는 것에 대한 새로운 의미를
제시하고 개인의 가정과 공공장소 간의 관계를
연구한다. 또한 집에 대한 과거의 개념을
지우고, 도시의 근본적인 동력인 집합에 대한
근본적인 해석을 제시한다.

A CIVILIZATION WITHOUT HOMES

amid.cero9

T34

A radical prototype for inhabiting cities in
the future, celebrating the disappearance
of homes as we know them and
taking command of public interiors for
communal domestic activities. Based on
a deep examination on Korean Bangs,
and how the presence of Jimjilbangs has
shifted drastically the interiors towards
a non-conventional domesticity, the
prototype presents a model for collective
housing based on dispossession and
communal living. A big communal house
where the conventional understanding
of the domestic is subverted merging it
with publicness. The act of sleeping and
resting is performed ritualistically and
collectively as the last realm of resistance
to the acceleration of production and the
24/7 urban rhythms. The project presents
a renewed idea of cohabitation that
examines the relation between the private
domestic realm and public space, the
eradication of former notions of the home
and a radical take on the collective as the
fundamental force to shape our cities.

Fluor Dreams. A Civilization without Homes, 2019

생산의 장소, 알루미늄

누라 알 사예, 안네 홀트롭

T35

걸프 지역 최초의 알루미늄 제련소는 1968년 바레인에 건설되었다. 현재 이 제련소는 단일 제련소 중 세계에서 네 번째 규모로, 기원전 제3천년기(BC 3,000~2,001년)로 거슬러 올라가는 금속 무역의 역사를 오늘날에도 이어가고 있다. 알루미늄 제련소 건설 이후, 바레인 인근 지역에서는 알루미늄과 관련된 공식·비공식 경제부문이 발전하기 시작했다. 알루미늄 부산물을 생산하는 다국적기업의 생산기지 주변에도 생산량이 상대적으로 적은 알루미늄 공장들이 설립되었다. 알루미늄은 오늘날 도시 현대화의 주요한 상징 중 하나가 되었다. 새로 건설되는 건물의 입면(facade)과 오래된 건물의 개보수 작업 모두에 알루미늄이 사용되면서, 도시의 현대성을 대표하는 이미지로 자리매김한 것이다. 알루미늄 생산 공정에 대한 조사를 바탕으로 한 본 설치작품은 필름, 사진, 모래 거푸집으로 만든 원형 모형(prototype) 등을 활용해 알루미늄의 생산 주기와 알루미늄을 둘러싼 각종 정치적 관계를 소개하고, 알루미늄의 새로운 활용 가능성을 연구한다.

PLACES OF PRODUCTION, ALUMINIUM

Noura Al-Sayeh and Anne Holtrop

T35

The first aluminum smelter in the Gulf region was inaugurated in 1968 in Bahrain and is today the fourth largest single-site smelter in the world, continuing a history of metal trade that finds its roots in the third millennium BC. The presence of the smelter developed a local economy of aluminum- both formal and informal. Alongside, large locally based international companies producing typical by-products of aluminum, smaller workshops have developed with a focus on a smaller-scale production of aluminum. Today, aluminum stands as one of the strongest expressions of the contemporary evolution of the city, both in its application on the facades of new buildings, and in the re-cladding of older buildings, becoming synonymous with a projected image of modernity. Through an investigation of the gestures in the production processes of aluminum, the installation, using film, photography and a sand-casted aluminum prototype, is an attempt to understand the politics and production cycle of the material and to extract a different potential of its use.

Studio Anne Holtrop, Aluminum sand cast, 2019

국경 접경지대 공유정류장
에스투디오 테디 크루즈 +
포나 포르만

T36

지구적 경계에서 동네의 경계로 지리적인
스케일이 작아지는, 본 설치작업은 세계에서
가장 큰 국경도시이자 서반구 접경지대
기준으로 일일 교통량이 가장 많다고 알려진
샌디에고와 티후아나에 초점을 맞추고 있다.
이 프로젝트는 새로운 공생과 협력 방안을
모색하고자 국경도시들을 조사 대상으로
삼고 있다. 그 결과 연구에 기반한 정치적이고
건축적 산물을 이 지역에 배출할 뿐 아니라
UCSD 국경 접경지대 공유 정류장 건설을
통해 분리된 공동체 사이의 공존을 도모할
수 있었다. 공유 정류장은 두 국가가 접하는
국경선 인근 이민자 마을에 위치하고 있다.
이곳에서 대학 연구진, 학생, 공동체 지도부가
공동으로 연구, 교육, 도시화 활동을 하고
있다.

CROSS-BORDER COMMUNITY STATIONS
Estudio Teddy Cruz + Fonna Forman

T36

Descending geographically in scale from the global border to the border neighborhood, this installation focuses on the San Diego-Tijuana conurbation, among the largest binational metropolitan regions in the world, with more border crossings daily than any other check point in the western hemisphere. We investigate border cities as laboratories for new strategies of interdependence and collective action. Our research-based political and architectural practice is embedded in this region. We have developed the UCSD Cross-Border Community Stations as an infrastructure for co-existence between divided communities. The Community Stations are field hubs located in immigrant neighborhoods on both sides of the border where research, teaching and urban advocacy are done collaboratively among university researchers, students and community leaders.

T36

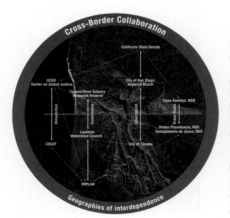

The Political Equator, Estudio Teddy Cruz + Fonna Forman, 2019. A research project that links border checkpoints across the globe, between the 30-38 degrees north parallel. When mapped alongside the climatic equator, the diagram reveals that this zone contains the most violent zones of conflict, poverty, migration, and disproportionate climate change impacts.

집합 공간의 변신: 공공 공간, 민주주의의 반영

아르키우르바노 + 타부:
존 오르티스, 이반 아세베도

T37

상호 창조적인 지역개발 전략에는 개별 공동체 주민과의 고차원적인 대화를 통한 요구사항 경청, 시민 참여, 그리고 집합건설의 품질 보장이라는 과정이 포함된다. 본 팀은 건축과 도시계획을 책임감 있는 사회 변혁의 수단으로 활용하여 공공 공간의 인간화와 취약 부문에 대한 사회적 접근을 우선 과제로 삼아왔다. 리오네그로, 우리 역사의 색채는 콜롬비아 도시 리오네그로의 역사적 도심을 되살리기 위해 구성된 시민 참여형 민관 협력체다. 외관의 전면을 시민들의 자발적인 참여를 통한 변화로 도심을 재생하겠다는 취지다. 본 프로젝트의 원동력은 공동 건설이다. 참여 시민들의 자긍심을 고취하고 소속감을 되살릴 뿐 아니라 보행 가능한 거리를 만들고 공공 공간을 활성화하는 것이 궁극적 목표이기 때문이다.

타부우는 부카라망가의 공공 공간을 복원하고 활성화했다. 수십 년 동안 방치되고 부패와 후견주의(clientelism)로 극심한 사회적 타격을 입은 곳이기도 하다. 우리 프로젝트는 수준 높은 설비를 갖춘 공공 공간, 환경과 도시의 조화, 지속 가능한 이동성, 통합 계획 등이라는 네 가지 분야에서의 변화를 감안하여 계획되었고 이에 대한 원칙에 따라 건설되었다.

TRANSFORMATION OF COLLECTIVE SPACES: PUBLIC SPACES, THE REFLECTION OF DEMOCRACY

ARQUIURBANO + TABUÚ: John Ortiz and Iván Acevedo

T37

The co-creative territorial strategy incorporates a high level of conversation with the inhabitants of each community, listening to their needs, guaranteeing the quality of citizen participation and collective construction. We have prioritized the humanization of the public space and social approach to vulnerable sectors, implementing architecture and urban planning as a tool for responsible social transformation.

Rionegro the Color of our History is a public-private alliance between the Mayor's office of Rionegro 2016-2019 and the Pintuco Foundation for the revitalization of the city's historical center, with citizen participation integrated throughout the transformation of its facades, generating a recovery that is voluntary and participative. This collective approach to construction is the motor of the project since its primary purpose is to generate pride in the citizens, contributing to building back a sense of belonging; along with the pedestrianization of the streets and the recovery of the public space.

TABUÚ reactivated and recovered public spaces in Bucaramanga that had been abandoned for decades and where deep social damage has been produced by corruption and clientelism (our methodology 'Building New Social Landscapes', Iván Acevedo G., Mayerly Medina Ch.). This project considered mainly four areas of transformation in planning and building the public spaces, in line with the following principles: 1) high quality equipment in the public spaces, 2) harmony between the environment and the city, 3) sustainable mobility and 4) integrated planning, following the guidelines from the New Urban Agenda for sustainable urban development.

T37

Parque de los Suen

푸른 혁명: 초투명 유리의 환경 파괴적 도시화에 대한 개입 조치

안드레스 하케 / 오피스
포 폴리티컬 이노베이션
(OFFPOLINN)

T38

신자유주의 사회는 파란색에 중독되었다. 초투명 유리(ultra clear glass)의 세계 소비량은 2008년 이후 두 배 증가하였다. 초투명 유리는 부유한 도시들의 업무공간, 아파트, 상업용 공간 등에서 흔히 보이는 재료가 되었다. 판유리 생산 과정에서 철분을 제거한 것이 초투명 유리다. 그렇게 유리는 초록 빛깔을 잃고 투명한 색을 띠게 되며 여기에 코팅까지 결합되면 자연 광선 중 자외선, 적외선, 주황색 가시광선의 투과를 막을 수 있다. 결과적으로 고급스러운 건물의 실내에서 보는 바깥 하늘을 원래보다 더 파랗게 보이게 한다. 2015년 초투명 유리의 소비가 증가함에 따라 유리 업계가 장기간 유지해온 배출 감축 추세에 변화가 일어나기 시작했다. 초투명 유리를 사용하면 유리창을 통해서 실내 온도의 상승을 막을 수 있다. 그러나 이 유리를 생산하는 지역의 대기 중 이산화탄소와 질소 산화물 농도가 전에 없이 상승 하였고 초투명 유리를 생산하기 위한 수압 파쇄 공법(hydraulic fracturing)이 가스 추출을 촉진시켰다. 파란색의 우월성을 추구하고 숭배하는 블룸버그 자선 재단(Bloomberg Philanthropic)의 C40 도시나 록펠러 재단의 100대 회복 탄력적 도시(100 Resilient Cities) 등과의 연계에 따른 지속적인 도시화로 인해 질소 산화물로 가득한 노란색 하늘의 도시들이 생겨난것이다. '파란색 반대'는 전 세계 도시들이 투명한 파란색 중독 현상에 개입할 것을 촉구한다. 전 세계 도시의 공조를 통해 특정 도시들의 소외현상과 노란색 하늘의 도시들의 상쇄에서 해방된 기술 사회를 구현 하자는 것이다.

BLUE REBELLION: AN INTERVENTION ON THE TOXIC URBANISM OF ULTRA CLEAR GLASS

Andrés Jaque / Office
for Political Innovation
(OFFPOLINN)

T38

Neoliberal societies have become addicted to BLUE. Since 2008, the consumption of ultra clear glass, UCG, in the world has doubled. UCG has become the material pervasively present in office, apartment and commercial buildings in wealthy global urban settings. UCG eliminates the use of iron in the composition of floated glass. The glass loses its green color to become clear, with a capacity to block the transmittance of the ultraviolet, infrared and orange spectrum of natural light, when combined with coatings. As a result, the outer sky looks bluer when seen from wealthy global interiors. In 2015, the increase of the consumption of UCG changed the long-sustained tendency in the glass industry of reducing its emissions. Though UCG can reduce the window-caused heat gainance in building interiors; in the sites of glass production it brought an unprecedented increase in CO_2 and NO_x in the air and promoted the practice of hydraulic fracturing gas extraction throughout the world. A realm of yellowish skies loaded with NO_x segregated from a continuous urbanity networked by intercity initiatives — such as Bloomberg Philanthropies' C40 Cities or Rockefeller Foundation's 100 Resilient Cities — where BLUE hegemony is pursued and worshipped. Blue Rebellion promotes an intervention on the global urban addiction to clear BLUE; a combined action to envision techno-societies emancipated from segregation and from the offsetting the airy yellow.

생각할(먹) 거리
페르난도 드 멜로 프랑코
T39

현 상파울루의 토지를 생산하고 운용하는
방식에는 여러 결정적인 단계들이 작용을
한다. 사회경제적 불평등과 공간적 격리로
인해 상당수의 사람들이 사각지대로 밀려난
것이다. 이들의 거처는 환경보호구역까지
뻗어나가기 시작했으며 이에 따라 도시는 수질
공급 위기에 직면하고 있다. 도시가 변두리
지역까지 팽창 하면서 저소득층의 주거환경은
더욱 열악해지고 공공시설과 관련 서비스
제공에 비상이 걸렸다.

생각할 거리 프로젝트는 상파울로와 도심 외곽
농경지대를 연구대상으로 한다. 환경보존,
식품, 영양보존, 지역사회경제개발 등을
통합적으로 살펴보며 이를 통해 지역 농업을
강화하고 사회의 환경지속가능성을 꾀하고자
하였다. 아울러 정책설계에 따라 도시의
위기를 완화하고 복합적인 문제 역시 해결할
수 있음을 제언하고자 하였다.

FOOD FOR THOUGHT
Fernando de Mello Franco
T39

There are multiple crises that affect
the ways of producing and using the
territory of São Paulo. The context of
socioeconomic inequality and spatial
segregation has pushed a significant
share of people to live in informal
settlements. This process continues
advancing on environmentally sensitive
areas, this way accentuating the water
supply crisis that affects the metropolitan
population. Urban sprawl in peripheral
areas intensifies the vulnerability of the
low-income population and difficults the
provision of infrastructure and services by
public authorities.

Food for Thought selects urban and
periurban agriculture in São Paulo as
an object of study in its integrative
dimension of aspects of environmental
preservation, food and nutritional security
and local socioeconomic development.
It aims to promote social environmental
sustainability of rural territory through
strengthening local agriculture.

It shows the extent to which design for
policy can offer answers to complex
problems in order to mitigate the crises of
the metropolis.

The project was with the collaboration of
Anna Kaiser, Carolina Passos, Leandro
Lima, Marcela Ferreira and Marta Bogea

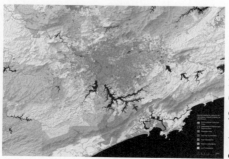

Overview of Sao Paulo's rural area

기록의 재정립
오픈워크숍
T40

오늘날 정보화시대를 맞이하여 기록보관실은 시대별 정보를 소장할 목적으로 건축되었다는 점에서 특별하며, 보관실 존재의 목적이 권력과 밀접한 연관성을 지닌다는 점에서도 흥미롭다. 고대사회에서는 정보의 수집 가능성이 곧 영토 지배와 확장 가능성과 직결 되었기 때문이다. 오늘날에는 다수 이해관계자의 의견을 수렴해야만 기록물을 재배치하거나 정보를 선별 혹은 분류할 수 있다. 한때는 권력의 집결지였던 기록보관실이 이제는 그 권력을 공유하고 타인에게 이양할 목적으로 변모하였고, 그 자체로 공적인 담론의 장이 되었다. 이를 통해 기록을 새롭게 정리하고 역사와 정보를 재구축할 뿐 아니라 현대적인 요소도 가미할 수 있게 되었다. 정보는 지식의 산물로 기록될 뿐 아니라 이의 제기나 비평의 대상이 될 수 있는 것이다.

RE-ASSEMBLING THE ARCHIVE
The Open Workshop
T40

Within our current Information Age, the archive holds a unique role in society as the building constructed to house information through the ages. The formation of the archive was closely tied to power, the archive being a critical instrument to rule territories by gathering records and compiling information. Cloaked as neutral repositories, the assembling of information within an archive was not innocent—it reflected how different societies structured their world and acted within it.

To reassemble the archive today asks how the selection and ordering of information can be situated as a discourse that is inclusive to a range of voices and their corresponding forms of information. Going beyond the archive as a static container of information and transitioning the building from a site of power to a place of empowerment, Re-Assembling the Archive situates the archive as a place of discourse and public assembly. By doing so, it brings this space to the foreground of the public sphere and allows for the reordering of information to construct new histories and knowledge. Inserting the contemporary subject as a participant in assembling the archive, information can now be questioned, critiqued and reordered as an act of knowledge production.

Data Garden: The Data Server becomes a public space of information, with its excess heat gathered to curate a botanical garden.

숲의 모양들

LCLA 오피스, 데일 위브

T 41

보레알 숲은 오슬로 공공지대의 정수를
보여준다. 토요일 오후 이곳 도시에 머문다는
것은 숲에 머문다는 뜻이기 때문이다. 이
삼림지대는 한때 나무를 기계로 베어 목재를
생산하는 곳이었다. 하지만 이러한 용도의
공간은 수십 년에 걸쳐 수많은 담론과 정책을
양산해 냈다. 그 결과 베어져 나간 목재와
공터를 눈에 띄게 하지 않기 위한 일련의
규칙이 만들어지기 시작했으며, 그 규칙은
다음과 같다.

1   평지에 수평으로 나무를 심지 않는다.
2   솔길과 호수를 너머로 숲이 자연스레
        우거지도록 한다.
3   공터의 크기를 제한하거나 아예
        제거한다.
4   인위적으로 길을 내지 않는다. 본래의
        지형을 따른다.

본 프로젝트는 노르웨이 삼림 관리법의 허점을
짚어보고 이에 대한 대응책을 제시하고자
하였다. 팀은 노르웨이 삼림을 일종의 거대한
카페트로 재해석하였으며, 벌목을 디자인
도구로서 보고 삼림 속 공터가 공공을 위한
공간으로 거듭날 수 있는지 살펴보고자
하였다. 대자연이 어느 정도의 인위성을
수용할 때 도시의 공간 자원은 확장될
가능성을 갖는다. 본 팀은 프로젝트를 통해 숲
속 공터가 열린 공간을 창출하는 매개체가 될
것으로 기대하며 이를 통해 숲이 아름다워지고
인류는 보다 역동적인 도시 기획 담론을
이끌어낼 수 있기를 기대한다.

SHAPES OF THE FOREST

LCLA Office with Dale Wiebe

T 41

The boreal forest is the quintessential
public space in Oslo. On a Saturday
afternoon to be in the city is to be in
the forest. This space is a machine to
produce timber, however its use as a
backdrop for all to enjoy, led to decades
of discussion and policymaking. The
outcome is a sophisticated set of rules
that hide timber extraction from sight.
This is done in Norway using the following
rules:

1   No planting trees in neat rows.
2   Let the forest grow through to
        succession where paths and
        lakes are.
3   Limit the size of clearings or clear
        cuts.
4   No orthogonal cuts, follow
        topography.

The project plays with the loopholes of
these Norwegian forestry rules and then
interprets the Norwegian forest as a large
wool carpet, imaging timber harvesting
as a design technique to create even
better collective spaces in the forest. By
embracing the artificiality of forestry, new
spatial opportunities are possible for this
urban asset. With this in mind, the project
aims to use the extraction of timber as a
design tool to create open-air rooms. New
shapes amplify the aesthetic possibilities
of the clearings, as well as playfully
establish a dialogue with the dynamic
forms that resist clear definition.

T41

Tuby, Jean-Baptiste. Plan du bosquet
de l'Arc de Triomphe. (1635-1700)

리오 세코: 그린 네트워크
빌딩 소사이어티 포 아키텍처 /
베라미노 산토스 "키우라"

T42

본 프로젝트는 배수로 개선과 도시재생 방안에
대해 다루고자 한다. 우리는 루안다 도심의
낙후된 배수로를 개선함으로써 이 지역의
도시화 문제를 일부 해결할 수 있을 것으로
기대한다. 이를 위해 본 프로젝트는 배수로
터널 건설과 도시의 주요 수로별 녹지 공간을
조성하여 수자원을 재활용한 도시재생방안을
모색하고자 한다.
우리는 보다 포용적인 도시, 혁신과 탄력성을
갖춘 역동적인 경제도시를 기대한다. 이를
위해 열린 광장, 조화로운 녹지 공간, 도로,
실내상업공간을 만들고, 리오 세코의 강
유역을 중심으로 주거지를 조성해 볼 수 있다.
본 프로젝트가 루안다와 유사한 난제에 직면한
도시들에 영감과 혜안이 되기를 기대한다.

RIO SECO: GREEN NETWORK
Building Society for Architecture / Belarmino Santos "Kiwla"

T42

This project intends to recover a
drainage ditch, reutilizing it from an
urban standpoint. This drainage is
located within the city of Luanda;
as a partial solution for some of the
urbanistic challenges that Luanda faces
nowadays, we have proposed a process
of urban regeneration that goes from
the infrastructure of a drainage canal
and its reuse through the construction
of several green spaces of collective use
that run through central arteries of the
city. The Rio Seco canal is 2.7 km long,
it starts in Zona Verde and ends in the
Kinanga neighborhood. The architectural
project has divided the canal into 5
different sections, each underwent an
environmental, social and economic
diagnosis, allowing the development of
proposes related to architectural design,
landscaping, infrastructures as well
as social and economic dynamization.
The proposed solution is to recover the
drainage ditch, intending to improve
rainwater management, control the air
pollution that spreads diseases, minimize
the environmental impact of drainage
on water quality, improve the urban
landscape and promote public safety. We
defend a vision that meets fundamental
conditions for a sustainable urban
development; we propose an intervention
that seeks to recover an integrated,
harmonious, clean and inclusive sense of
the city.

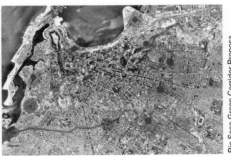

Rio Seco Green Corridor Proposa

밤섬 당인리 라이브

매스스터디스 / 조민석

T43

한강의 밤섬을 마주하는 당인리 발전소 부지의 현재 도시적 성격은 지휘자가 없는 오케스트라가 연주하는 교향곡과 같다. 이곳은 기계로서의 도시 비전을 통해 에너지 생산, 홍수 방지, 차량 이동에 제각기 초점을 맞추게 되면서 초래된 독립적 기능들의 엔트로피로서의 집적체이다. 90여 년 전 서울이 현대 거대도시로 공학화되면서 시작된 이곳은 역사 없는 연대기를 가진, 장소 없는 광대한 도시 지역이다.

그러나, 대부분 높은 건물들의 장벽으로 규정되는 한강 변의 장소들과 달리, 이곳은 화합적인 문화와 생태학적 야망을 위한 수변 공공 활동을 수용할 수 있는 큰 스케일의 마지막 공공 지역이 될 잠재력을 가진다. 총합이 집적된 부분들보다 거의 모든 경우 작은, 초 구획화된 세계에서 '집합도시'를 상상하는 것이 더욱 불가능해 보이는 지금, 복잡하고 공존 불가능한 것들로 가득한 이 도시 영역을 새로운 땅으로 만들 수 있겠는가? 어떠한 집합적 노력이 이 우발적으로 발견된 부지를 대체적 유토피아로 만들 수 있겠는가?

BAMSEOM DANGINRI LIVE

Mass Studies / Minsuk Cho

T43

The current urban quality of the area around the Danginri Power Plant site on the Han River, facing Bamseom Island, is that of a symphony played by an orchestra without a conductor. It's where the vision of the city as a machine resulted in an entropic agglomeration of independent performances focused on energy production, flood control, and vehicular movement. It's a vast urban territory without place, with a history-free chronology. This all started 90 years ago when Seoul re-engineered itself as a modern metropolis.

However, unlike most locations along the Han River walled off by tall buildings, there is potential for this to become the last public area of this scale that can accommodate public activities along the waterfront and with a larger cohesive cultural and ecological ambition. Now it seems beyond impossible to imagine a 'collective city' in a hyper compartmentalized world, where the sum is almost always less than the collected parts. How can this urban territory, with these complex incompatibles, become a new ground? What kind of collective effort does it take for the accidentally found site to create a substitute utopia?

Danginri Podium and Promenade from the Han River at Night (Image Courtesy of Mass Studies)

저항의 지도
포렌식 아키텍처
T 4 4

2014년 8월, 이슬람 무장조직 이슬람국가(ISIL)는 소수 종파·민족인 야지디(Yazidi)족을 몰살하기 위해 이들이 12세기부터 삶의 터전으로 삼은 이라크 북부 신자르(Sinjar) 지역을 습격했다. 당시 공격으로 야지디족 대상의 조직적 살해, 고문, 성노예화가 이루어졌고, 이들의 거주지와 농경지, 예배공간이 대대적으로 파괴되었다. 2014년 습격 이후 미국에 설립된 야지디족 국제 비영리 구호단체 야즈다(Yazda)는 ISIL 대원을 상대로 제소된 국제법적 사건을 지원하기 위해 2018년 포렌식 아키텍처와 함께 당시 범죄 현장의 증거를 모아 공식적으로 문서화하기 시작했다. 포렌식 아키텍처는 자체적으로 고안한 교육 프로그램을 통해 야지디족이 직접 설문 문항을 만들거나 사진을 활용한 설문조사를 진행할 수 있도록 지원해 왔다. 본 전시는 프로젝트 초기 단계의 결과물로, 민족의 중요한 문화들인 야지디 사원 과 영묘(mausoleum)들의 파괴를 분석한다. 야지디족 문화유산 중 가장 중요한 유적지들이 조직적이고 대대적으로 말살되었다는 점이 증명된다면 당시 공격이 집단학살(genocide)이었다는 주장이 인정받는 데 큰 힘이 실리며 이는 ISIL의 목표가 야지디족 문화 근절이었다는 증거가 뒷받침되어야 한다.

MAPS OF DEFIANCE
Forensic Architecture
T 4 4

In August 2014, the Islamic State (ISIL) invaded the Sinjar region of northern Iraq, intent on exterminating the Yazidi, an ethnic and religious minority who have lived there since the 12th century. Their campaign included the systematic killing, torture, and sexual enslavement of the Yazidi people, and the widespread destruction of their homes, agriculture, and places of worship.

Four years later, Forensic Architecture and Yazda, an international Yazidi NGO founded in the US in response to the 2014 attack, partnered on an initiative to document the sites of these crimes. They did this to support legal cases brought in international forums against members of ISIL.

A training program devised by Forensic Architecture allowed members of the Yazidi community to carry out a photographic and modelling survey themselves. This exhibit shows the initial stages of the project: an analysis of the destruction of culturally specific Yazidi shrines and mausoleums. Demonstrating that the eradication of the most important sites of Yazidi heritage was systematic and widespread is crucial in bringing about genocide claims, which demand proof of ISIL's intent to stamp out the Yazidi people and their culture.

Testing a 'community satellite' kite rig in Istanbul, Turkey (Image: Forensic Architecture)

프로젝트 발자취
사미프 파도라 건축연구소 (sP+a/sPare)

T45

본 프로젝트는 수년에 걸쳐 연구, 실습, 공동, 협력이라는 네 가지 세부항목에 따라 진행되었다. 우리는 상호 연계성과 영향력을 바탕으로 세부항목의 단순 응집체가 아닌 상호 유의미한 집합체를 도출해낼 수 있었다. 어쩌면 우리가 수많은 맥락과 변수에 대응할 수 있었던 핵심은 혼돈과도 같았던 이 접근법이었는지도 모른다. 세부항목을 통해 다룬 내용은 다음과 같다.

1. 주거라는 이름으로(연구)

이 연구는 뭄바이의 도시구성과 역사에 깊이 뿌리내린 저가공공주택의 유형학적 형태에 관한 것이다. 본 연구를 통해 최근 건립된 공공저가주택이자 생기 없는 주거공간으로도 알려진 주거양식의 대안을 제시하고자 하였다.

2. UDAAN (실행)

나비 뭄바이 지역 UDAAN 주거설계프로젝트는 사전연구자료를 토대로 한 실전역량강화를 위한 학습의 장이 실시간으로 마련되었다.

3. 반드라 공동 프로젝트 (공동)

무료봉사활동이 반드시 무에서 시작될 필요는 없다. 우리는 이미 계약이 체결된 공공프로젝트에 디자인 요소를 가미하여 부가가치를 창출하는 방식으로 참여하였다.

4. 우리가 지은 집(협력)

저가주택건립 프로젝트로 도시연구협력체인 URBZ, 지역주민 및 시공사와의 협력으로 진행되었다.

PROJECTIVE HISTORIES
Sameep Padora Architecture and Research (sP+a/sPare)

T45

The practice has evolved over the years into four distinct formats - practice, research, collaboration and collectives. These are not siloed but are shaped to constantly feed into each other. This almost schizophrenic format of practice is perhaps a way to address the vast multitudes of context and variations. A few manifestations of these models are as follows:

1. In the Name of Housing (Research):

This research documents the typological configurations of affordable collective housing in Mumbai sutured deep within the city's fabric and history, and presents this as an alternative to the apathetic apartment format of formal affordable housing being built in the city currently.

2. Jetavan (Practice):

Evolved through interactions with the local community and built by master craftsman and by local masons trained by the master craftsman, Jetavan is a community centre with skill & spiritual development programs. Collaborating with Hunnarshala we evolved construction techniques created to leverages waste and recycled materials

3. Bandra Collective (The Collective Project):

Working pro bono, the collective doesn't necessarily plan new ventures from scratch, but picks up on existing public projects directly assigned to contractors, attempting to add value through design.

4. A House We Built (Collaborative Practice):

This affordable house was built in collaboration with URBZ, an urban research and action collective, along with local actors in the form of residents and contractors.

이미지와 건축 #11: 팔만대장경

바스 프린센

T46

복제물, 사진, 예술품 등은 건축물이나 표현하고자 하는 공간을 얼마나 근접하게 묘사할 수 있는가? 한 공간의 물리적 이동은 이미지로 대체할 수 없다. 그럼에도 관객과 재현된 사물 간의 상관관계는 친밀하고도 직접적이라고 할 수 있다. 본 프로젝트는 새로운 사진 현상법을 수년에 걸쳐 시도하였다. 구체적으로는 화선지를 고화질 디지털 파일과 결합하여 실제 사이즈로 현상하였다. 이로써 건축에 활용된 재료와 재료의 특성을 이미지화하고 예술적 건축물을 놀라운 섬세함과 깊이로 재현하였다. 그 결과 유서 깊은 공간이 새로운 시각으로 재탄생되고, 관객에게는 엄선된 재현물로 직접적이고도 직관적인 경험을 선사하게 되었다.

이미지와 건축 시리즈의 Work #11는 1237년에 81,258의 목판으로 만들어진 세계적으로 위대한 대장경판인 팔만대장경을 다룬다. 대장경판은 이 유명세보다는 덜 알려진 해인사라는 불교 사원의 네 개의 건물에 나누어 보관되어 있다. 대장경판이 보관되어 있는 장경판각은 재생산되고, 복제되며, 전파되어야 할 지식의 보고이다. 또한 기발하게 구성되어 있는 자연 공기 순환 체계를 통해 장경판이 수백년을 걸쳐 보존될 수 있게 했다는 점에서, 장경판각을 일종의 기술적 상징으로 생각할 수도 있다.

IMAGE AND ARCHITECTURE #11: TRIPITAKA KOREANA

Bas Princen

T46

How close can a copy, an artistic depiction, or a photographic print come to an architectural object or space it depicts? The experience of moving through a space cannot be substituted with an image, but the relation between a viewer and the printed object can be intimate and direct. In the past several years, I have worked with a new photographic printing process, combining rice paper and high resolution digital files, resulting in life size image-objects, able to capture and transfer 'material' and 'materiality' of architectural spaces and artistic artefacts with marvelous fragility and depth. This allows for a new view of well-known spaces to be created, nearly as direct, one-to-one experiences of deliberately chosen details.

Work #11 in the series Image and Architecture captures The Tripitaka Koreana, the world's greatest collection of Buddhist scriptures consisting of 81,258 wooden printing blocks, dating back to 1237. The collection, housed in four buildings at the Haein Temple, is a sacred space, albeit a space which is not monumental. It is an archive of knowledge that is meant to be reproduced, duplicated, disseminated. It is also a monument of crafts — the buildings in which the collection is kept are equipped with an ingeniously designed natural air-circulation system which has been able to preserve the blocks over millennia.

도시와 농촌의 교류
아틀리에 바우와우

T 4 7

20세기에 들어 효율의 극대화를 추구하는 산업사회가 발전하면서 세상의 구조가 크게 달라졌다. 대도시의 삶은 편리해졌지만, 산업사회가 제공하는 서비스에 전적으로 의존해야 하는 도시의 삶에 숨 막혀 하는 사람들도 늘었다. 그 결과, 많은 도시 거주자들이 농업으로 눈을 돌려 새로운 가치를 찾고 있다. 한편 농촌은 인구 감소와 고령화 문제에 직면해 있다. 농가의 1차 산업 소득은 넉넉하지 않고 농촌의 자연이 위협을 받고 있으며 장애인들의 부족한 취업 기회도 심각한 문제다. 이러한 상황에서 출범한 '도시와 농촌 교류' 프로그램은 다양한 배경의 사람들이 농업에 도전할 기회를 제공하여 기존과는 다른 새로운 농업인을 배출하고자 한다.

오늘날까지 두 개의 프로젝트가 진행되었고, 기존의 지역 자원에 대한 접근성을 개선하고 신규 자원 발굴이라는 목표의 다른 프로젝트도 진행 중이다.

1.　코이스루부타 연구소는 햄·소시지 공장이자 지역에서 생산된 돼지고기 전문 식당이다. 연구소는 전원주택을 재해석해 도시인과 농업인이 만나는 공간으로 변신시킨 결과물이다.

2.　쿠리모토 다이이치 장작공급소는 코이스루부타 연구소 인근에 방치됐던 삼나무림과 농장을 정비 및 관리하여 장작과 고구마를 생산한다. 이곳은 버려진 공간을 수많은 방문객들에게 쉼터가 되는 대형 공간으로 재해석한 결과다.

3.　인근 숲에 위치한 다양성 캠핑장 프로젝트는 해당 지역에서 숙박과 야외활동을 즐길 수 있는 공간이다. 모든 시설에서 장애인과 출소자들이 일할 수 있으며, 시설들은 주변 풍경과 어우러지며 방문객이 이 곳을 여타 사회복지시설처럼 느끼지 않도록 설계되었다.

URBAN RURAL EXCHANGE
Atelier Bow-Wow

T 4 7

During the 20th century, the development of industry, guided by values of optimum efficiency, changed the world structure a lot. Life in the big city has become convenient but people also feel suffocated by being fully dependent on services provided by industrial networks One of the reactions of urban inhabitants is the rediscovering of new value in farming. On the other hand, rural areas are confronting shrinking and aging populations. The lack of a breadwinner, in the form of a primary industry, threatens the village's landscape. Meanwhile, there are very few opportunities for disabled people to find jobs. Urban Rural Exchange is a challenge to open the farming process and allow people with various backgrounds to participate, helping to generate a new type of membership. Two facilities have been realized and another project is underway, following the concept of discovering local resources and making them more accessible.

1.　Koisurubuta Laboratory is a ham-sausage factory and restaurant, for tasting local pork. The villa typology is reinterpreted, where people from both the city and the farm encounter each other.

2.　Kurimoto Daiichi Firewood Supply Station maintains an abandoned cedar forest and farm next to the Koisurubuta Laboratory, producing firewood and sweet potato on site. Here the large space of the barn typology is reinterpreted as a shelter in order to welcome large numbers of people.

3.　Diversity Camp Site Project is in the adjacent forest, providing accommodation and activities in the area. All the facilities create working opportunities for disabled people and for those released from prison, yet they do not appear as social welfare facilities, in the landscape nor in the eyes of visitors.

Kurimoto First Firewood Supply Station

독창성의 도시
알렉산더 아이젠슈미트 / 비져너리 시티스 프로젝트
T48

현대 도시의 잠재력을 발견하는 능력을 키우는 것은 극단적인 도시화의 세계에서 무엇보다 중요하다. 따라서 비져너리 시티스 프로젝트는 공간적, 조직적, 도시의 힘과 압력에 의해 발생한 물질적 독창성을 탐구한다. 이 독창성은 건축 비전문가들에 의해 만들어지고 모든 사람에 의해 사용된다. 여러 해 동안 프로젝트 팀은 세계 여러 곳에서 이 세상이 건축되는 방식에 매우 중요한 요소이며, 건축이 동원할 수 있는 역량에 영향을 미치고, 그 자체의 기이함에 매료되기도 하지만, 통상적 건축적 영역 바깥에 자리하는 다양한 사례를 수집해왔다. 전 세계의 이러한 도시 발명과 전술은 종종 터무니없고, 때로는 재치있으며, 항상 여기, 그리고 지금 이 순간에 내재되어 있다. 무엇보다 중요한 것은 이들에는 새로운 종류의 건축적 도시주의가 강하게 잠재되어 있다는 점이다. 따라서, 도시적 독창성을 지닌 집합도시의 밑그림은 개별적이고 혼성화된 도시 지능으로 구성된 도시화를 투영하기 위해 이러한 마이크로 어바니즘(미시적 도시주의)을 활용한다. 이 새로운 종류의 도시는 분열되어 있으면서도 공유된 가치들에 있어서는 집합적 성격을 띈다. 그 집합성은 세계적이면서도 지역적인 독창성으로 구성되어 있고, 연속적인 기반구조를 통해 제한적이면서도 끝없이 이어지며, 시대를 초월하면서도 현재에 가장 충실하다. 본 프로젝트에서 말하고자 하는 '독창성의 도시'는 건축이 도시화의 강력한 힘으로부터 배우게 된 개념이다.

CITY OF URBAN INVENTIONS
Alexander Eisenschmidt / Visionary Cities Project
T48

Cultivating an ability to detect potentials in the contemporary city is paramount in our world of extreme urbanization. The Visionary Cities Project, therefore, explores spatial, organizational, and material ingenuities born out of the forces and pressures of the city, answered by the architectural amateur, and used by everyone. Over multiple years, our team has collected examples from around the world that are important to the way the world is built, influential in its capacities to mobilize, mesmerizing in its strangeness, and yet outside the architectural radar. These urban inventions and tactics from across the globe are often outrageous, sometimes humorous, but always embedded in the here and now. Most importantly, these conditions are saturated with potential for a new kind of architectural urbanism. The drawing of a Collective City of Urban Inventions, therefore, utilizes these micro urbanisms to project an urbanism entirely composed of these individual and hybridized urban intelligences. This new kind of city is fragmented and yet collective in its shared qualities; global and yet composed of local ingenuities; limited and yet endless through its continuous infrastructure; timeless and yet committed to the present. It is a city where architecture has learned from the forces of urbanization.

Alexander Eisenschmidt / Visionary Cities Project, City of Urban Inventions, drawing composed of hundreds of urban ingenuities from across the world that together form a collective urbanism.

도시전
CITIES
EXHIBITION

INTRO

도시전에서는 전세계 약 80여 개 도시를 초대하며,
비엔날레의 전체 주제인 '집합도시'를 바탕으로,
각각의 도시에서 가장 중요하게 생각하고 있는 주제와
이슈들을 다루도록 요청하였으며, 이 과정에서 우리는
도시 간의 연결과 집합의 결과물로서의 도시에 대해
이해할 수 있도록 한다.

'집합적 결과물로서의 도시'는 우리의 도시가
공간적, 시간적, 그리고 사회적 환경의 집합체임을
보여줌과 동시에, 경우에 따라서는 의도하지 않은
혹은 계획되지 않은 요소들의 개입으로 끊임없이
변화하는 생물임을 보여준다. 도시전이 열리는
돈의문박물관마을이라는 파편화된 공간에서
이루어지는 전시는 전세계 도시의 상황을 대변해준다.
물리적으로 분리가 되어있더라도 여러가지 요소들로
다시 서로간에 연결되고 있는 도시들의 상황은
돈의문박물관마을의 공간과 중첩된다.

2019 서울도시건축비엔날레 도시전은 집합적
결과물로서의 도시가 어떻게 서로간에 분리되고 또
연결되는지 보여주는 전시가 될 예정이다.
특히 이번 도시전에서는 전세계 80여개 도시에서
다루는 다양한 주제는 10개의 소주제로 분류되어
전시된다. 이들 소주제들은 현대 도시들이 직면에
있는 도시적 화두가 무엇인지 분명하게 보여준다.

21세기 산업도시	강과 수변
도시의 유산, 시장	인프라스트럭처	참여형 어바니즘
도시밀도	집합적 유형	도시공간의 점용
상상의 도시와 분석방법	집합적인 도시기억과 레이어

큐레이터.˙.˙.˙임동우, 라파엘 루나
협력큐레이터.˙.˙.˙김유빈
코디네이터.˙.˙.˙조웅희
관람경험 디자인.˙.˙.˙NOLGONG

Under the overarching theme of the "Collective City," the Seoul Biennale of Architecture and Urbanism (SBAU) 2019 Cities Exhibition is a platform for discussing core issues and pressing themes raised by approximately eighty participating cities. The goal of this year's discussion is to better understand the collective consequences of connections and meetings between cities.

This concept of "cities as a collective consequence" highlights cities not only as an aggregated environment, but also as living, dynamic systems that are constantly changing. The fragmented nature of Donuimun Museum Village, reflects the contemporary condition of many cities around the world, highlighting that independent of the geographical divides, cultural differences, and physical boundaries that separate cities, there is a multitude of common questions, experience and overlap, spaces of intersection and co-existence, that is shared territory for speculation and debate.

It is against this backdrop that the SBAU 2019 aims to demonstrate how cities as a collective consequence are simultaneously separated and connected with each other at any given time. Especially, in the Cities Exhibition, ten sub-themes are introduced based on core issues that approximately 80 invited cities are presenting. In deed, they show what urban issues that modern cities are facing globally at the present.

INTRO

| Industry in the 21st Century | Rivers and Waterfronts |

| Market Inheritance | Infrastructure | Participatory Urbanism |

| Density | Collective Typologies | Appropriations |

| Methods and Speculations | Layers and Collective Memory |

Curator.∴.Dongwoo Yim, Rafael Luna
Associate Curator.∴.YouBeen Kim
Coordinator.∴.Tony Woonghee Cho
Visitor Experience Design.∴.NOLGONG

MAP

MAP

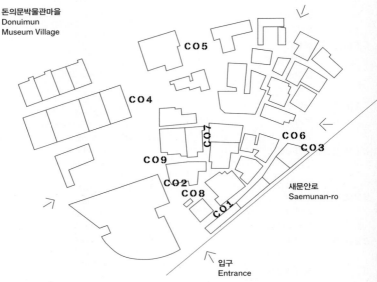

돈의문박물관마을
Donuimun
Museum Village

CO5
CO4
CO7
CO6
CO3
CO9
CO2
CO8
CO1

새문안로
Saemunan-ro

입구
Entrance

MAP

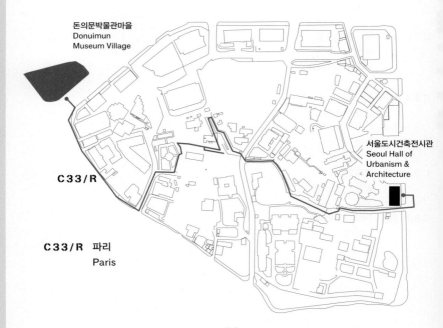

돈의문박물관마을
Donuimun
Museum Village

서울도시건축전시관
Seoul Hall of
Urbanism &
Architecture

C33/R

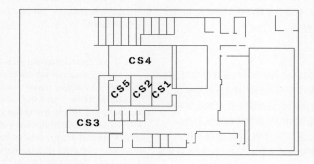

서울도시건축전시관 지하 1층

Seoul Hall of Urbanism & Architecture B1F

MAP

도쿄, 일본
도시 야생 생태학

후미노리 노우사쿠, 미오 쓰네야마 (큐레이터) / 아틀리에 로쿠요샤, 하기 스튜디오, 모쿠친 아키텍처, 준페이 노우사쿠 아케텍츠, 타카다 조우엔 (참여)
C 4

우리는 2011년 동일본에서 발생한 대지진과 그에 따른 후쿠시마 원전 사태로 인류의 일상을 지탱하는 요소들이 전 세계에 얼마나 큰 영향을 미치는지 깨닫게 되었다. 이로써 우리는 일상 용품, 주택, 음식, 쓰레기 등 매일의 생활과 관련된 물질과 에너지의 관계를 들여다보기 시작했다. 산업사회가 제공하는 인프라에 의존하던 생활 양식에서 벗어나 지구와 인간의 새로운 관계를 모색하기 시작한 것이다. 기존의 전력망 대신 태양 에너지를 사용하는 주택, 토양의 미생물을 활용하는 정원, 빈 건물을 이용한 공유 주택, 버려진 자재를 활용한 자유로운 건축 등이 그 사례가 될 수 있다. 지금껏 활용되지 않았던 자원을 활용함으로써 생태적 생활도시가 탄생했다. 도시 속에서 살아남은 야생(wildness)은 자연의 파괴와 오염을 견뎌내고, 생동적 삶을 누리며 온갖 어려움을 극복해낸다. 이번 전시는 도시의 야생 생태학 연구 활동과 더불어 이와 관련된 여러 키워드를 제시한다.

인프라스트럭처

Infrastructure

TOKYO, JAPAN
Urban Wild Ecology

Fuminori Nousaku, Mio Tsuneyama (Curator) / Atelier Rokuyosha, Hagi Studio, Mokuchin Architecture, Junpei Nousaku Architects, Takada Zouen (Participant)
C 4

The Great East Japan Earthquake and the consequent accident at the Fukushima nuclear power plant made us aware of how much the support for our daily life affects the entire planet. We began to look at the link between materials and energy related to daily life, such as household goods, homes, food and waste. We started to try to reinvent life from one that was dependent on infrastructure and industrial society to another relationship: off-grid homes that generate solar power, gardens that utilize the power of microbes in the soil, share-houses that use vacant buildings, and voluntary construction using discarded materials. By utilizing the resources of the city that have not been utilized until now, the ecological life of the city has been created. The wildness that survives the city does not yield to the environment that is being polluted and destroyed, but enjoys the fluctuating life and overcomes inconvenience. We introduce the collectives of these Urban Wild Ecology activities along with the keywords of its resources.

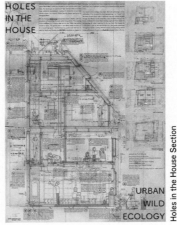

Holes in the House Section

리마, 페루
집합성 구현과 도시 생활 혁신을 위한 리마의 건축 전략

샤리프 카핫, 마르타 모렐리
C 3

현재 리마는 스스로의 일에만 몰두하는 모습(self-absorption)과 '타인'(the "other")에 대한 공포라는 두 가지 특성을 보인다. 공격적인 분위기가 가득한 거리, 곳곳에 높이 쌓여가는 벽과 담장, 호신 장치, 이웃에 대한 불신, 이 모두가 리마의 '공포'를 상징한다. 공포가 만연한 도시에서 집합적 교류를 위한 활동은 경멸의 대상이 될 뿐이다. 다시 말해, '공공의 문화'가 없기 때문에 리마에는 '시민'이 존재하지 않는다. 이번 서울도시건축비엔날레에서는 '건축을 통해 어떻게 집합성을 구현할 것인가'라는 주제를 연구한다. '건축을 통한 집합성 구현'이라는 주제는 도시 인구밀도 증가와 급격한 도시화 등의 문제에 대응하는 동시에 지가 상승, 도시에서 발생하는 시간과 에너지 낭비, 불평등, 자원 고갈 등 공공 문제와 더불어 사적인 영역에서 발생하는 고유의 문제 해결과도 연관되어 있다. 집합성 구현과 도시 생활 혁신을 위한 리마의 건축 전략은 지난 5년간 출품인이 프로젝트 더 나은 도시 리마를 위해(Taller Urban Lima)를 통해 진행한 연구와 성과를 모은 작품이다. 본 작품은 리마에서 가장 인구밀도가 높고 가장 활기찬 세 지역, 가마라(Gamarra), 메사 레돈다(Messa Redonda), 리마 구시가지의 '집합적 공공'(collective and public) 공간을 위한 새로운 도시 건축 청사진을 모색한다. 또한 리마 시민들의 사회적 교류를 촉진하는 새로운 유형의 건물과 공간을 제안한다.

LIMA, PERU
LIMA: Framing the Collective (Architectural Strategies to Reinvent Urban Life)

Sharif Kahatt, Marta Morelli
C 3

Lima's contemporary society lives in a self-absorption and with fear of the "other," which translates into aggressiveness in the streets, proliferation of walls and fences and security devices, distrust among neighbors and the contempt for any activity that promotes collective interaction. Therefore, contributing to the construction of "the collective" is the most effective "possibility" of starting a culture of the "public." This exhibition for the Seoul Biennale looks at the question of how to promote collectivity through architecture while responding to the process of densification and rapid urbanization of the city, along with other factors such as the rise of land value, the need for reducing time and energy consumption in urban mobility, urban inequality, and resource depletion, all combined with the vernacular phenomenon of the informal. The exhibition gathers research and projects developed by Taller Urban Lima at PUCP School of Architecture led by Sharif Kahatt y Marta Morelli during the last five years and shows architecture explorations of new urban scenarios for "collective and public" spaces and places in Lima, located in the most dense and vibrant neighborhoods of Lima: Gamarra, Mesa Redonda, and the historic Downtown. These projects propose new buildings and spatial typologies that promote the social interaction that Lima needs.

Lima © Evelyn Merino-Reyna

암만, 요르단
(또)다시 찾은 요르단 계곡
파디 마수드, 빅토르 페레스-아마도
C39-2F

AMMAN, JORDAN
(Re) Revisiting the Valley Section
Fadi Masoud, Victor Perez-Amado
C39-2F

본 작품은 도시화와 농업 발전을 위한 요르단 강 국경 지역의 물줄기 방향 전환과 그에 따른 사막화 현상을 조명한다. 요르단과 이스라엘이 접하는 이 지역은 세계에서 강을 둘러싼 국경 분쟁이 가장 첨예한 상징적인 곳으로 주목받고 있지만, 강의 방향 전환과 그로 인한 환경 변화의 상관관계는 국제적 관심을 끌지 못했다. 한가지 분명한 것은 요르단 강 유역에서 물의 분배를 둘러싼 논란은 물 부족 문제가 아닌 '물을 어떻게 할당하고 통제해야 하는가'에 있다. 이미 갈등이 첨예한 이 지역은 요르단 강 주변 농업과 도시 주거지가 새로운 현실에 적응하고 변화하지 않는다면 긴장이 더욱 고조될 것으로 시급한 변화가 필요하다. 본 작품은 요르단 강 유역이 발전하고 이에 따른 계획을 수립하기 위해서는 지역의 한계와 잠재력을 명확히 인식해야 한다는 점을 역설한다. 또한 변화의 대안으로 '요르단 계곡을 위한 새로운 선형적 농업-도시 체제' 형성을 제안한다. 정확한 지역 현실을 반영하고, 지나친 인공적 개발을 자제하는 탈중앙화(decentralized) 인프라가 새로운 체제의 바탕이 될 것이며 이 인프라는 새로운 토지 사용 양식과 주택 유형을 제시하여, 그동안 주목받지 못하던 요르단 계곡 지역을 재조명할 것이다.

This project uncovers the underplayed link between water-diversion for urbanization and agriculture and the resulting desertification in one of the world's most disputed and symbolic border-river regions. In the Jordan River Basin, it has been proven that the most controversial aspect of water distribution is not scarcity but allocation and control. The premise for expedited change is that both agriculture and urban settlement along the Jordan River in their current format will increase the contention in this already sensitive region if they do not evolve and adapt to new realities.
This is a proposal for a regional landscape planning and design scheme that necessitate an understanding of the landscape's limitations and potentials. The result is a new linear agro-urbanization regime for the Jordan Valley — one that is based on more localized, de-engineered, and decentralized infrastructures involving transboundary regions, new land use patterns, novel housing typologies, and the revisiting of the "Valley Section."

C39

2F

바쿠, 아제르바이잔

바쿠: 석유와 도시주의

예브 블라우, 이반 루프니크
C39-1F

한때 러시아 제국과 소련에 속했던 아제르바이잔의 수도 바쿠는 최초의 석유 도시로서 석유와 경제, 정치, 물리적으로 긴밀하게 얽혀 있는 곳이다. 바쿠의 첫 번째 석유 호황은 19세기 말에 찾아왔다. 지역의 석유재벌들이 새롭게 거머쥔 막대한 재산을 쏟아 부어 이른바 '카스피의 파리'라 일컬어지는 대도시를 건설하고, 노벨 가족 기업의 러시아 지부가 바쿠 지역의 유전을 현대화한 덕분이었다. 구소련 시기에는 '사회주의자의 석유 도시'를 형성하려는 도시 실험의 현장이 되기도 하였다. 실험 대상 도시로는 네프트 다실라리도 있었다. 카스피해의 거대한 다리 위에 건설된 네프트 다실라리는 2천 개의 석유 굴착기, 송유관, 집유소와 그 곳에서 일하는 수천 명의 노동자들과 이들을 위한 학교, 상점, 정원, 진료소, 심지어 영화관까지 있는 도시였다. 오늘날 바쿠는 세계 석유와 가스의 주요 생산지일 뿐 아니라 가장 빠르게 변모하는 도시 가운데 하나이다. 이 전시는 참여작가가 쓴 동명의 저서(Park Books, 2018)에 수록된 내용을 담고 있다. 이 책은 150년이 넘는 세월에 걸친 바쿠의 석유 산업과 도시화 사이의 밀접한 상호작용에 대한 최초의 종합적 연구를 담고 있으며, 전세계의 도시화라는 오늘날의 근본적인 변화를 이해하는 데 있어 바쿠의 역사가 얼마나 중요한지를 여실히 보여준다. 동 서적에는 네덜란드 사진가 이반 반의 포토 에세이도 수록되어 있다.

BAKU, AZERBAIJAN

Baku: Oil and Urbanism

Eve Blau, Ivan Rupnik
C39-1F

Baku, the capital of Azerbaijan and formerly part of the Russian Empire and Soviet Union, is the original oil city, with oil and urbanism thoroughly intertwined — economically, politically, and physically — in the city's fabric. Baku saw its first oil boom in the late nineteenth century, driven by the Russian branch of the Nobel family modernizing the oil fields around Baku as local oil barons poured their new wealth into building a cosmopolitan city that became known as the "Paris of the Caspian." During the Soviet period, Baku became the site of an urban experiment: the shaping of an "oil city of socialist man." That project included Neft Dashlari, a city built on trestles in the Caspian Sea and designed to house thousands of workers, schools, shops, gardens, clinics, and cinemas as well as 2,000 oil rigs, pipelines, and collecting stations. Today, Baku is a major oil and gas producer on the global energy market and one of the most rapidly changing urban territories in the world. The exhibition is drawn from the book Baku: Oil and Urbanism, by Eve Blau, with Ivan Rupnik (Park Books, 2018), the first comprehensive study of the close interplay of the oil industry and urbanism in Baku over a period of more than 150 years and reveals the critical significance of that history for our understanding of the global urbanization processes that are reshaping the contemporary world. It also includes a photo essay by Iwan Baan.

Infrastructure

1F

C39

인프라스트럭처

Black Town blocks, BC 1900

바르셀로나, 스페인
빅데이터 도시화

카탈로니아 고등건축원
C39-1F

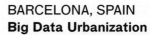

BARCELONA, SPAIN
Big Data Urbanization

Institute for Advanced Architecture of
Catalonia
C39-1F

본 전시는 1878년 바르셀로나의 일데폰스 세르다가 최초로 제안한 도시화 이론과 더불어 데이터 분석, 관리, 예측 정보시스템 등 새로운 디지털 시스템에 대해서도 소개한다. 영문으로 가장 먼저 번역된 알데폰스 세르다의 도시화 이론에서, 세르다는 도시개혁과 확장이라는 측면에서 도시주의라는 개념을 만들었으며 심도 깊은 연구를 위해 통계학, 위생학, 법학, 경제학, 건축학, 종교학 등을 공부했다.

그는 두 번째 저서에서 바르셀로나를 통계와 데이터에 근거해 분석하였으며 이를 통해 당대의 도시가 주민들의 생활공간으로 부적절함을 증명해냈다. 이로써 새로운 도시 모델이 제시되어야 했다. 그의 이론은 오늘날 영어로 번역되었고 통계자료가 도표 및 지도 등으로 시각화되어 19세기 당시의 상황을 잘 묘사하고 있다. 그는 도시주의 개념의 창시자일 뿐 아니라 정보 및 데이터의 중요성을 강조한 혁신가이기도 했다. 150년이 지난 오늘날, 다수의 도시에서 디지털 혁명이 일어나고 있는 것은 어쩌면 오래전 그의 비전이 실현된 것인지도 모른다. 오늘날의 우리는 도시분석지도, 도시관리, 참여절차, 비즈니스 플랫폼, 글로벌오픈데이터시스템 등을 활용하고 있다. 현대의 도시주의는 스페인 동북부에 위치한 카타로니아의 지역도시인 바르셀로나에서 싹티웠으며 우리는 지금 보다 더 살기 좋은 세상을 만들기 위하여 더 나은 도시와 지역을 위해 노력하고 있다.

How Big Data was used to coin Urbanism in 1867 and how today every city is trying to make art out of it.
This exhibition presents the first General Theory of Urbanization, written in 1867 in Barcelona by Ildefonso Cerda, together with new digital systems that allow for analyzing, managing, and projecting data and information systems.
In his General Theory of Urbanization, which has now been translated for the first time into English, Cerda coined the concept of urbanism, as the science to reform and expand cities. And for this, he studied disciplines such as statistics, hygienism, law, economics, architecture, and religion.
In its second volume, he analyzed the city of Barcelona using statistics and data in order to demonstrate that people's living conditions were not adequate. Thus a new urban model had to be invented.
Now, with the general theory translated into English, the statistical pages visualized through graphics and maps, making them compressible create the first representation of nineteenth-century urban big data. Cerda invented the concept of urbanism and also pointed out the need to use information and data to understand cities better.
Coinciding with the 150th anniversary of this important project, and amid a digital revolution where multiple initiatives in cities around the world are aiming to represent the functioning of cities through data, the work presents his vision into a reality.
The selection of work presented includes map projects used for urban analytical maps, city management, participatory processes, business platforms, or open data systems from around the world.
In Barcelona, capital of Catalonia, modern urbanism was invented. Now, we work to build better cities and territories to make a more livable world.

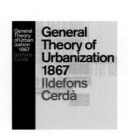

General Theory of Urbanization 1867 Ildefons Cerdà

베를린, 독일

연방 통계청 건물

퀘스트 팀 (크리스티안 부르크하르트,
플로리안 쾰)

C39-5F

지난 10년 동안 베를린 도심에 갈수록
거세어지는 상업적 압력에 따라 1970년대
이후로 베를린의 성격을 규정해온 행동주의와
시민 참여 전통은 도시 계획과 건축을 위한
새로운 실험 기반을 형성했다. 시민사회
단체와 지방정부는 변화무쌍한 도심 지역에서
장기적으로 비용이 덜 드는 장소를 확보하기
위해 협력에 나섰고 새로운 참여 계획 방식,
자금 조달 모형, 운영 체계를 검토했다.
지금까지 가장 정교하고 시대를 앞선 참여
개발 사례로는 '연방 통계청(Haus der
Stastistik)'이 꼽힌다. 이곳은 베를린 도심에
있으며 구 동독의 행정부 청사였다. 애당초
시민사회 단체의 계획은 현재는 비어있는 연방
통계청 건물에 예술가와 난민 전용 공간을
창조하려는 구상을 추진하는 것이었다.
이에 베를린시는 해당 건물을 매입하고 동등한
권리를 지닌 다섯 개 파트너 기관과 공동 개발
과정에 착수했다. 다섯 개 파트너 기관은
계획 협동조합으로 발전한 원래의 시민사회
단체, 베를린 시 산하 주택 건설 회사, 시
산하 부동산 업체, 베를린 시와 미테 지구의
도시 계획 당국이다. 결과적으로 해당 도시
개발 구상은 공공기관의 폭넓은 참여에 따라
의뢰 및 진행되었다. 현재 파트너 기관들은
프로젝트 개발에 필요한 관리 체계를 조사하고
있다.

BERLIN, GERMANY

House of Statistics

Quest (Christian Burkhard, Florian Köhl)

C39-5F

When the city centre of Berlin became
subject to increasing commercial
pressures during the last decade,
the tradition of activism and civic
engagement, which has defined the city
since the 1970s, created the basis for
new experiments in urban planning and
architecture. To secure low cost areas for
diversified neighbourhoods in the long
term, civil society initiatives and local
governments started to work together
to probe new participatory planning
instruments, financing models, and
operating schemes.
The most elaborate and advanced
participatory development to date is
the "House of Statistics," a former East
German administrative complex in the
heart of Berlin. Initially, it was a civil
society initiative, which promoted the
idea of creating space for artists and
refugees in the now empty buildings.
The City of Berlin bought the property
and started a joint development process
with five equal partners. They comprise
the original initiative, which evolved into
a planning cooperative, a municipal
housing company, the municipal real
estate company and the urban planning
authorities of the city and borough. As a
result, the urban development concept
was commissioned and developed with
broad public involvement. Currently
the five partners are conducting an
investigation into governance structures
to develop the project.

브리즈번, 호주

야누스의 얼굴을 한 도시
브리즈번: 88 엑스포부터 퀸즈 워프 카지노 리조트까지

샌드라 카지-오그래디, 실비아 미켈리
C39-2F

본 프로젝트에서는 호주에서 가장 빠르게 성장하는 도시 중 하나인 브리즈번을 가로지르는 강의 남쪽(사우스뱅크)과 북쪽 지역(노스뱅크)의 도시 개발을 분석한다. 사우스뱅크가 브리즈번의 문화 중심지로 발전한 것은 1982년 퀸즐랜드 미술관의 신규 건물이 이곳에 들어서면서 부터였다. 아울러, 1988년 브리즈번 엑스포를 계기로 다양한 야외 레저 시설이 더해지면서 사우스뱅크는 더욱 매력적인 지역으로 변모했다. 그 이후 사우스뱅크가 화려한 집합 인프라와 독특한 공공 공간이 있는 지역으로 자리잡게 되면서, 다양한 배경, 계층, 연령대의 사람들이 몰려들었다. 2015년 브리즈번 시정부는 노스뱅크 지역의 퀸즈 워프 거리에 민영 리조트와 카지노를 건설하겠다는 계획을 발표했다. 호주 퀸즐랜드 주지사는 사우크 뱅크 지역이 88 엑스포 이후 장족의 발전을 이뤄냈듯, 노스뱅크 개발 계획을 통해 브리즈번의 중심 업무 지구가 온전히 새롭게 변할 것이라고 역설했다. 88 엑스포와 노스뱅크 개발 계획 모두 대규모 사업으로, 국내외 기업인과 관광객을 유치하겠다는 목표를 갖는다. 그러나 노스뱅크의 공공 장소들은 사우스뱅크와는 사뭇 다르다. 두 지역 개발 프로젝트의 정치적 맥락, 목표, 그리고 이를 대하는 대중들의 변화에 따라 전세계의 사회·경제적 변화를 이해할 수 있다. 이러한 정치적 변화가 도시와 같은 인공적으로 조성된 환경에 미치는 영향은 무척 흥미로운 주제이다. 이번 전시회에서는 노스·사우스뱅크가 두 지역을 연결하는 도보 다리 건설이 예정된 상황에서 상호 영향을 미칠 수 있는 요소들을 면밀히 살펴보고자 한다.

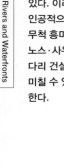

C39

2F

BRISBANE, AUSTRALIA
The Janus-faced City: Brisbane from Expo 88 to the Queen's Wharf Casino Resort

Sandra Kaji-O'Grady, Silvia Micheli
C39-2F

This project analyses the urban development of the south and north banks of the river that dissects one of Australia's fastest-growing cities, Brisbane. The South Bank's development as the city's cultural precinct began with the construction of the state's Queensland Art Gallery in 1982 and its appeal broadened after Expo 88, when a wide range of facilities for outdoor leisure were added. South Bank has since become a spectacular collective infrastructure and a unique set of public and institutional spaces, attracting people of diverse ethnicity, class, and age. In 2015, the State Government announced the development of a private casino and resort at Queen's Wharf, on the north bank of the Brisbane River. The Premier of Queensland claimed that the complex would transform Brisbane's Central Business District in the same way the South Bank development did after Expo 88. Both projects are grandiose in scale and aim to attract international and domestic business and tourists. Yet, the north bank has a fundamentally different relationship to public space. The shifting political context of, ambitions around, and responses to the two projects are symptomatic of global social and economic change. Our interest is in the impact of these political changes on the built environment. With the two precincts set to be connected by a pedestrian bridge, our study investigates the potential each bank has to influence the other.

Aerial photograph of the Expo 88
© Queensland State Archives

케이프타운, 남아프리카공화국
분산적 협동

알프레도 브릴렘버그, 후베르트 클룸프너,
스콧 로이드, 앤디 볼닉
C39-4F

본 전시는 '권익 증진과 통합을 위한 개발
프로젝트'에 대한 기록을 담고 있다.
프로젝트는 소박한 2층 주택, 주민 참여에
따른 공간 기획, 통합생활 조성 계획이라는
세 가지 핵심 요소를 개발하고 실행 및
평가함으로써 무허가 거주지를 개선하기 위한
혁신적 설계·조직 모델을 활용한다.
본 전시는 '도시를 형성해 나가는 집합적
참여 역량 동원'에 대해 소개한다. 전시의
핵심 작품은 거대한 집합 지도이며 이는 지난
28년 동안 케이프타운의 특정 지역을 함께
일구고 관리해 왔던 주민들의 손으로 작성한
지역 최초의 지도이다. 주민들의 노력에 더해
도시-싱크탱크, 취리히 연방 공과대학교와
건축대학 학생들 및 지역 NGO 이카발라미도
참여하였다. 지역 단체가 줄자, 연필, 종이를
들고 지역을 누비며 작성한 지도가 한데
모여 주민들의 거주를 증명하는 최초의 공식
기록물이 되었으며 이는 케이프타운 시
당국으로 부터 '체류권'을 확보하는데 중요한
역할을 했다.

CAPE TOWN, SOUTH AFRICA
The Distributed Cooperative

Alfredo Brillembourg, Hubert Klumpner,
Scott Lloyd, Andy Bolnick
C39-4F

The exhibition documents the Empower
Integrated Development Project. The
project leverages innovative design
and organisational models to upgrade
informal settlements through the
development, implementation, and
evaluation of three core components:
a two-story housing prototype; a
participatory spatial planning process;
and integrated livelihoods programming.
The installation explores the theme of the
mobilizing force of collective participation
in shaping the city. The focal piece is a
full-scale collation of the first mapping
of an occupied site by its residents, who
have collectively built and managed this
fragment of the city for the past twenty-
eight years. The mapping was conducted
by the residents with the Urban-Think
Tank, ETHZ, ETH architecture students,
and local NGO, Ikhayalami. Community
groups were tasked to survey their area
of the site with the aid of a tape measure,
graphite, and paper. The drawings were
then collated together to became the first
official document of occupation, and a
powerful tool to secure "the right to stay"
upon presentation to municipal officials.

Participatory Urbanism

4F

C39

참여형 어버니즘

The first collective mapping of BT Section
Site C (U-TT, ETH, Scott Lloyd, 2017)

엘 엘리코이데: 쇼핑몰에서 감옥으로

셀레스테 올랄퀴아가
C39-1F

1950년대 후반에 착공한 베네수엘라 카라카스의 최신식 드라이브인 쇼핑몰인 엘 엘리코이데 데 라 로카 타르페야는 결국 완공되지 못했다. 본래 계획에 따르면 건물을 오르내리는 이중 나선 형태의 차량 통행만 4킬로미터가 넘는 이 미래형 철근 콘크리트 쇼핑몰은 베네수엘라 현대화의 상징이 되어, 전세계의 관심을 끄는 명소가 될 예정이었다. 오늘날 기준으로 1,000만 달러에 육박하는 건설 예산으로 착공된 엘 엘리코이데는 혁신적인 민간 사업이기도 했다. 공사 자금을 충당하기 위해 300개의 상업 공간 대부분을 사전 분양했으나 마르코스 페레스 히메네스의 군부 독재 정권(1952-1958) 붕괴 이후, 사업은 정치적인 이유로 은행 대출에 난항을 겪으며 건설에도 차질이 빚어졌고, 1961년에는 공사가 전면 중단되었다. 엘 엘리코이데를 공공 목적으로 이용하려는 간헐적 시도가 여러 차례 있었지만, 성공한 사례는 두차례 뿐이었다. 1979년에는 산사태 이재민 수용소로 활용되어 1만 명까지 늘어난 이재민은 이 건물에서 3년 간 머물렀다. 1985년부터는 베네수엘라 정보경찰 본부로 사용되고 있다. 건물 내 대부분의 구역이 여전히 방치된 상태이지만, 엘 엘레코이데는 정보경찰 본부로, 그리고 악명 높은 정치범 수용소로 사용되며 라틴 아메리카 현대화의 어두운 이면을 단적으로 보여준다.

CARACAS, VENEZUELA
El Helicoide: From Mall to Prison

Celeste Olalquiaga
C39-1F

Built in the late 1950s in Caracas, Venezuela, El Helicoide de la Roca Tarpeya, a state-of-the-art drive-in mall, was never completed. Featuring a two-and-a-half-mile double helix of ascending and descending vehicular ramps, this futuristic structure of reinforced concrete was to become a flagship of Venezuela's entry to modernity and attracted worldwide attention. Initially budgeted at the modern equivalent of $10 million, El Helicoide was also an innovative private initiative — most of its 300 commercial spaces were pre-sold to help finance the construction. After the military dictatorship of Marcos Pérez Jiménez (1952-1958) collapsed, the project's bank loans became politically entangled and construction was interrupted, coming to a full stop in 1961. Despite numerous stop-start projects to put the vast complex to public use, the structure has only known two uses: as a refuge for landslide victims in 1979, which quickly extended to a 10,000-person, three-year occupation, and as a headquarters for the Venezuelan intelligence police since 1985. While large parts of the structure remain in abandon, El Helicoide continues to be used as a police headquarters and notorious prison for political detainees to this day, representing the dark side of modernity in Latin America.

El Helicoide. Photo Nelson Garrido, 2012. Cortesía Proyecto Helicoide

카사블랑카, 모로코
절충적 도시, 카사블랑카

오우아라로우 + 초이
C39-5F

일찍이 현대화를 수용한 카사블랑카는
특히 20세기 건축과 도시설계에 관한
실험적인 도시로 꼽힌다. 카사블랑카는
급진적으로 변모하는 과정에서 도시가
건설·물질·공공·건축학·자국·사회적인
면모를 갖출 수 있도록 노력하였다.
우리는 일련의 행동계획을 통해 도시가
건축 유산을 보존하고 계승할 수 있도록
심혈을 기울였다. 본 설치물은 카사블랑카
현대주의자의 여정을 담은 것이다. 우리는
가두풍경을 진화와 변혁의 토대로 삼아,
1900년대 초 옛 터전을 포함해 1930년대
유럽계획과 1940년대 회교도 도시,
1950년과 60년대의 위성도시와 유대인
도시를 다루었다.

이를 통해 우리는 도시의 공간적 변화를
탐색하고자 하였으며 공유지와 사유지, 열린
공간과 사적인 공간, 틈새 등을 살펴보고,
카사블랑카의 다양한 이해관계자들과
거주민들이 도시의 팽창과 수축 과정에서
교집합을 형성해가는 과정을 담아냈다. 이처럼
본 작품은 중립지대로서의 역할을 수행하는
카사블랑카를 소개한다.

CASABLANCA, MOROCCO
Negotiated Territories: Experiences in Adapted Habitats in Casablanca

OUALALOU + CHOI
C39-5F

Welcoming modernity from its inception,
Casablanca, Morocco was above all
a laboratory for the modern project in
architecture and urban design in the
twentieth century. The city allowed
for — perhaps even called for — radical
investigations into the urban form that
pushed boundaries: constructive and
material, formal and architectonic,
domestic and social.

As the city becomes increasingly invested
in the preservation and promotion of this
significant built heritage, led in large
part by OUALALOU + CHOI's devising of
a Plan of action for the Urban Agency,
this installation presents key moments
in Casablanca's 'modernist adventure',
positing the streetscape as a negotiated
territory existing in a state of evolution
and transgression. Investigations include
the medina of the early 1900s, the
'European plan' of the 1930s, the Cité
musulmane d'ain chock of the 1940s,
through to the 'satellite city' (carrières
centrales) and the Cité israélite el hank of
the 1950s and 60s.

A series of intricate models explore
these spatial transformations within the
city, from the street to the enclosure,
from public to private and in-between.
Instances of sometimes-rapid
multiplication, contraction, dilatation,
and intersection portray Casablanca
as a territory continuously in search of
a terrain d'entente between its various
stakeholders and inhabitants.

도시의 유산, 시장

Dar Lamane_photo B. Taylor-Aga Khan Trust for Culture

도시재생

한국토지주택공사(LH)

C39-2F

CHEON-AN, SOUTH KOREA
URBAN REGENERATION

LH (Korea Land & Housing Corporation)

C39-2F

천안 동남구청 도시재생 사업의 목표는
쇠락하는 동남구청 주변의 원도심을 새로운
지역 경제 중심지로 되살리는 것이다.
이번 도시재생 사업은 천안시,
한국토지주택공사, 주택도시기금, 현대건설이
민관 합동으로 진행한다. 주택도시기금법 시행
이후 주택도시기금 도시계정이 지원하는 첫
번째 사업이며 도시재생 사업의 모범 사례로
평가되고 있다.
공공보건센터, 지식산업센터, 동남구청,
어린이회관, (대학생을 위한) 행복기숙사, 공용
주차장, 주거 시설(현대 힐스테이트 천안)이
동남구청 부지에 건설된다.
동남구청, 행복기숙사, 어린이회관 등
공공시설은 2019년 12월까지 완공되어 운영
할 예정이다. 지식산업센터는 올해 하반기에
리모델링 공사를 착수하며, 이미 15개 기업과
단체가 입주를 확정한 상태이다. 2021년
완공 및 입주 예정인 주상 복합단지는 천안
원도심에서는 보기 어려웠던 47층 건물로
조성되어 도시의 랜드마크 역할을 톡톡히 할
것으로 보인다.

The urban regeneration project of Dongnam-gu Office in Cheonan is aimed at creating a new economic hub to revitalize the declining downtown area around the Dongnam-gu Office in Cheonan.
It is a private participation urban regeneration project jointly promoted by the city of Cheonan, LH, the Housing and Urban Fund, and Hyundai Engineering & Construction. It is the No. 1 project supported by the Urban Account of the Housing and Urban Fund since the implementation of the Housing and Urban Fund Act, and is considered as the best example of urban regeneration projects. Public facilities such as public health centers, knowledge industry centers, Dongnam-gu office, a children's hall, a happy dormitory (for college students), public parking lots, commercial facilities, and residential facilities (Hillstate Cheonan) will be installed on the site of the Dongnam-gu office.
In the future, public facilities such as the Dongnam-gu Office, dormitories and the children's center will be completed and fully operational in December 2019. The Knowledge Industry Center will begin construction on remodeling later this year, with fifteen companies and organizations confirmed to move in. The residential and commercial complex apartment is expected to become a landmark building with forty-seven floors which was difficult to see in the old downtown area and will be completed and moved in by March 2021.

청주, 대한민국
도시재생

한국토지주택공사(LH)
C39-2F

청주시 옛 연초제조창 일원 도시재생 사업은
청주시, 한국토지주택공사, 주택도시기금이
공동 투자하고 중앙정부, 지자체,
공공기관과의 협력으로 진행되는 사업
모델이다.
문화 상업 지역이라는 비전 아래 진행되는
이번 도시재생 사업은 청주시가 소유한
옛 연초제조창 부지를 활용한다. 문화와
도시재생을 결합하는 새로운 유형의 경제
기반 도시재생 사업을 통해 낡은 담배 공장이
복합문화공간으로 탈바꿈하게 될 것이다.
리모델링 이후 1층과 2층에는 상업 및 문화
체험 등의 용도로 민간 임대 시설이 들어서고,
3층과 4층은 공예전시관과 창고로, 5층은
복합 ICT 센터와 열린 도서관으로 꾸며질
예정이다.
이번 청주 도시재생 사업을 통해 민간 투자가
활성화되고 공예·문화상품 판매 시설, 전시관,
공연장, 새로운 문화 체험 시설과 결합된 상업
시설이 들어설 것이다. 더 나아가 도시 일자리
창출을 위한 기반을 넓히고 시민들에게 신선한
자극을 제공하여 지역경제 활성화에 기여할
것으로 기대된다.

CHEONG-JU, SOUTH KOREA
URBAN REGENERATION

LH (Korea Land & Housing Corporation)
C39-2F

The Urban Regeneration Project of the
former tobacco manufacturing plant in
Cheongju is a business model jointly
invested in by Cheongju City, LH, and the
Housing and Urban Fund in cooperation
with the government, local governments,
and public institutions.
Under the vision of a "Cultural Business
Park" on the site of the old manufacturing
plant owned by Cheongju City, a new
business model for an economic-based
urban regeneration project that combines
culture and urban regeneration was
applied to the old tobacco factory as a
complex cultural space.
After remodeling, the first and second
floors of the building will be private rental
facilities, including sales facilities and
cultural experience facilities. The third
and fourth floors will be used as craft
cluster exhibition rooms and storage
rooms, and the fifth floor will be used as a
multipurpose room, ICT center, and open
library.
It is expected to contribute to the
revitalization of the local economy
through the creation of an employment
base and new incentives for the
urban population by attracting private
investment and creating commercial
facilities that combine crafts, cultural
sales facilities, exhibition halls,
performance halls and new cultural
experience facilities

Industry in the 21st Century

5F

C39

21세기 산업

크라이스트처치, 뉴질랜드

기쁨을 조직하다: 도시 트라우마에 대한 집합적 대응

바나비 베넷, 김동세
C39-1F

2011년 크라이스트처치를 강타한 지진으로 185명이 목숨을 잃었으며 시내의 약 70%가 파괴되었다. 2019년에는 시내 부근 두 곳의 이슬람 사원에서 발생한 테러 공격으로 51명이 사망했다. 이 두 사건은 도시에 커다란 반향을 일으켰다. 크라이스트처치는 인명 피해와 직접적인 고통을 겪었을 뿐 아니라, 도시의 정체성에 대해 보다 넓은 시각으로 다시 바라보는 계기가 되었다.

정치적 생태주의자 제인 베넷의 저작에서 영감을 얻은 이번 출품작은 크라이스트처치에서 일어난 충격적인 사건들에 대한 집합적 대응을 보여주고 그 과정에서 생겨난 큰 기쁨과 대중의 힘을 연구한다. 아울러 다양한 프로젝트와 조직이 서로를 변화시키는 이야기와 이들이 창조해 낸 환상적인 경험을 공유한다. 크라이스트처치 시민들이 함께 애도하고, 힘들어하고, 웃고, 울고, 즐기던 시기에 콘서트, 추모 행사, 예술 행사가 여러 차례 열렸다. 이 행사는 시민들의 다양한 열망과 요구를 포착하고 표현하였으며 시민들 역시 능동적으로 참여했다. 또한, 행사를 위해 정부, 대학, 의회, 예술인 단체, 연극인 단체, 디자인 회사, 새로운 NGO 등 여러 단체들이 집합적 지성과 물리적 노력으로 함께 기여하였다.

이번 전시회를 통해 우리는 고통스러운 사건 이후 트라우마를 겪는 도시인의 경험이 (흔히 생각하는 바와는 반대로) 창의적 정신과 즐거움이 있는 큰 기쁨을 자아내고, 이를 통해 도시에 영속적인 변화를 가져올 수 있다는 점을 시사하고자 한다. 빠르든 느리든 변화를 겪고 있는 전세계 다른 도시에서도 크라이스트처치의 사례를 참고할 수 있기를 바란다.

CHRISTCHURCH, NEW ZEALAND

Organizing Enchantment: Collective Responses to Urban Trauma

Barnaby Bennett, Dongsei Kim
C39-1F

In 2011, a massive earthquake hit Christchurch killing 185 people and leading to the demolition of around 70% of the central city. In 2019, a terrorist attacked two mosques near the center of Christchurch, killing 51 people. Both events had a massive impact on the city: from direct suffering and death to broader conceptualisations about its identity. Following the writing of political ecologist Jane Bennett the exhibition examines the sense of enchantment and the public agency that was generated by the collective responses to these traumatic events. It tells the story of the interplay between projects and organisations and the experiences of enchantment they generated. Concerts, memorial services, and art projects are discrete entities that represent and capture public desires and demands and were captured by the people of the city as it mourned, struggled, laughed, cried, and played together. Behind these projects, organisations such as governments, universities, councils, artist collectives, theatre groups, design firms, and new NGOs organised the collective intellectual and physical labour.

This exhibition suggests that experiences of post-traumatic urbanism, perhaps counter-intuitively, offer creative, joyful experiences of enchantment that can affect the permanent state of the city, offering lessons for other cities globally undergoing slow or rapid changes.

FESTA 2012 - Luxcity - By Bridgit Anderson

디트로이트, 미국
버려진 디트로이트, 도시에 남은 사람들의 도전

마리아 아르쿠에로 데 알라르콘, 마틴 머리
올라이아 치비테 아미고, 미시간대학교 터브먼
건축대학
연구 협력 조교.·.·.·천쉐웨이, 쇼리아 제인,
먀오이신, 그웬겔, 한쿤헝, 다라 미탈, 니샨스
미탈, 마이클 애미던

C39-2F

María Arquero de Alarcón, Martin Murray
and Olaia Chivite Amigo, Taubman
College University of Michigan
Collaborators Graduate Research
Assistants.·.·.·Xuewei Chen, Shourya
Jain, Yixin Miao, Gwen Gell, Kunheng
Han, Dhara and Nishant Mittal, Michael
Amidon

C39-2F

지난 수십 년 동안, 산업시설이 사라지고 쇠락하는 도시를 주제로 한 학문적 연구가 활발하게 진행되어 왔다. 산업의 쇠퇴로 고통받고 방치된 세계 여러 도시에 대한 많은 논문이 발표되었으나 도시의 '유휴공간에 남은 사람들이 어떻게 살아가는지'에 관한 이해는 아직 턱없이 부족하다. 디트로이트는 이렇게 쇠락하는 도시의 전형적인 사례로 꼽힌다. 사라진 투자와 일자리, 도시 외곽 주거지역으로 '탈출'하는 백인 중심 중산층 가족, 점점 악화되는 공공 서비스, 무관심 속에 방치된 여러 지역. 바로 누더기처럼 버려진 디트로이트의 현주소다.

본 작품은 디트로이트의 쇠락 이후 남겨진 공간의 현실을 비판적 시각으로 다루며, 특히 거주 가구의 반 이상이 이주해버린 세 주거지역을 집중 조명한다. 디트로이트 시 당국은 여러 지역에서 부흥을 위한 노력을 펼치고 있지만, 이 세 곳은 공식적인 계획에도 포함되지 못해 미래가 불확실하다. 남아있는 주민들에 대한 현장 조사와 시각·문헌 자료 연구 결과를 담은 이번 작품은 해당 지역 주민들이 자신을 둘러싼 환경을 바꾸고, 그 과정에서 지역의 새로운 미래에 따라 개인의 비전을 이루기 위한 노력을 보여준다.

Over the past several decades, a vibrant scholarly literature has addressed struggling post-industrial cities in decline as objects of inquiry in their own right. Yet while much has been written about abandonment in deindustrializing cities around the world, very little is understood about the "after-life" of these "leftover" spaces. Detroit is perhaps the quintessential exemplar of a struggling deindustrialized city in decline. The impact of disinvestment and job loss, the exodus of largely white middle-class families to the residential suburbs, shrinking municipal services, and widespread abandonment have left much of the urban landscape in tatters. This project critically explores what happens to leftover spaces after abandonment. We focus on three distinct Detroit's neighborhoods with high vacancy rates, where over half the homes have completely disappeared. Unlike other struggling neighborhoods where city managers hope for recovery, these areas have remained in institutional limbo, with no official plans for rehabilitation. Through a combination of ethnographic fieldwork and an examination of visual and written records, our project uncovers how ordinary residents have taken it upon themselves to reshape the environment in which they live and, in the process, to imprint their own vision for the future of their neighborhoods.

Industry in the 21st Century

2F

C39

21세기 산업

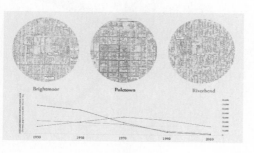

Brightmoor   Poletown   Riverbend

1930   1950   1970   1990   2010

갠지스강, 인도
갠지스강, 거대한 물의 세계

앤서니 아치아바티, 레이철 르페브르, 맥스
오웰릿 호위츠
C39-2F

1820년대 이래 갠지스강 유역은 굉장한 변화를 겪으며, '강과 연결되는 방대한 물의 세계'로 변모해 왔다. 이 변화를 이해하려면 깊이 있는 연구가 필요하며, 우리는 갠지스강 유역의 역동적인 변화를 담은 지도를 제작하였다. 이 유역이 지난 200년간 어떻게 변화되어 왔는지, 매해 우기에 어떤 변화를 겪는지를 새로운 방식으로 시각적으로 설명하고자 한다. 단순히 지도 제작을 위해 이러한 작업을 진행한 것은 아니다. 갠지스강 유역의 변화는 수백 년 동안 진행되어 왔다. 기존의 강에서 뻗어 나가는 새로운 강을 만들어 낸 과거의 방식이 강 내부에 새로운 강을 만드는 방식으로 변하면서 이 지역 사람들의 생각과 일상도 바뀌었다. 이를 바탕으로, 이번 작품에서는 갠지스강 유역이라는 '방대한 물의 세계'에 나타난 급격한 변화에 대해 연구하고 그 의미를 해석한다. 때로는 상공에서, 때로는 지상에서 바라본 강 유역의 모습을 담으며, 갠지스강이 어떻게, 그리고 왜 (전세계가 놀라워하며 모방하고자 했던) 새로운 수자원 관리 방식의 시험 무대가 되었는지를 들여다본다. 또한, 해마다 찾아오는 변덕스러운 우기(雨期) 때의 모습부터, 거대한 갠지스강 운하, 주민들이 직접 판 수많은 우물, 매년 히말라야 산맥에서 쏟아져 내리는 토사, 임시로 지어진 텐트촌, 끝없이 엇갈려 펼쳐지는 농촌과 도시의 모자이크 같은 정경까지 담은 역동적인 시각 자료를 제시한다. 이번 작품을 통해 우리는 무질서해 보이는 갠지스강 유역의 숨겨진 질서를 드러내고, 이 지역 발전을 위한 지능적 설계를 제안하며, 강과 연결된 거대한 물의 세계가 어떻게 형성되었는지에 대해 이야기한다.

GANGES, INDIA
The Ganges Machine: A River Without Banks

Anthony Acciavatti, Rachel Lefevre, Max Ouellette-Howitz
C39-2F

Since the 1820s the Ganges River basin has been transformed into a vast water machine. Such a transformation demands scrutiny. It is for this reason that the world's most hyper-engineered river basin merits a dynamic atlas: a new form of visualizing how this landscape has changed over the last two hundred years and how it changes every year with the arrival of the monsoons. A purely cartographic concern? If only. Centuries in the making — the shift from creating new rivers from existing rivers to building rivers within rivers reoriented people's thinking and daily lives, forming a vast and varied infrastructural landscape. With these changes in mind, the exhibition investigates and interprets the "Ganges Water Machine," that radical transformation of the sacred Ganga River basin. Moving between the terrestrial and celestial, this exhibition examines how and why the Ganges became a testing ground for a new culture of water management — one that inspired awe and copycat projects across the globe. From the capriciousness of the monsoons to the great Ganges Canal system, from the proliferation of millions of private borewells to annually shedding over one billion tons of silt from the Himalayas, from the temporality of tent cities to an endless agro-urban mosaic, a dynamic atlas of the Ganges Water Machine reveals a hidden order, a design intelligence, and the formation of a vast hydrological supersurface.

이스탄불, 터키
설계지침서
네메스튜디오
C39-1F

> "일상적이고 시시한, 하찮은 감각을 —
> '섬광처럼 번뜩이는 감각'의 반대말이
> 아니라 — 위대한 단순함으로, 감각이
> 스스로를 초월한 상태로 받아들일
> 수 있을까?" 장 뤼크 낭시, <u>세계의
> 감각</u>(1997)

<u>설계지침서</u>는 거대 건축물의 '지루함',
'시시함'한 특성을 소개한다. 설계 단계에
있거나, 현재 건설 중이거나, 혹은 이미 완공된
다양한 대형 건축 사업을 조명하는 이번
작품은 일상적인 건축·건설 작업의 시시함과
주변 환경을 변화시키는 장대한 건설 사업의
원대한 상상력의 관계를 연구한다. 대규모
건설 사업과 일상에서의 건축 작업, 눈에
보이는 작업과 보이지 않는 작업, 그리고
건축과 마감 작업을 병렬 배치함으로써, 건축
작업과 노동의 절차적이고 관료주의적인
성격을 짚어본다. 작품은 건축 도면과 모형을
전시하며, 전시장 자체가 건축 현장이 되며,
건축가가 수행하는 설계, 건설, 해체 작업을
하나의 퍼포먼스로 인식하며 그에 대한 담론이
이루어지는 공간이다.

ISTANBUL, TURKEY
Manual of Instructions
NEMESTUDIO
C39-1F

> "Can we think of triviality of sense
> — a quotidianness, a banality, not
> as the dull opposite of scintillation,
> but as the grandeur of the
> simplicity in which sense exceeds
> itself?" Jean-Luc Nancy, <u>The Sense
> of the World</u> (1997)

<u>Manual of Instructions</u> is a project that
speculates on the mundane and banal
aspects of large-scale construction.
Focusing on various large-scale projects
in Istanbul that are currently in design
or construction phases, or complete,
the project sets a dialogue between
environmental imaginations of the
large-scale and the quotidian day-to-day
workings of architecture and construction.
By juxtaposing the planetary and the
quotidian, the foreground and the
background as well as the construction
and the finish, the project speculates
on the bureaucracy and procedures of
architectural work and labor. Through the
presentation of drawings and models,
the exhibition space itself becomes a
construction site and a place to prompt
a discourse around architecture's own
making, assembly and unmaking as
performance.

Infrastructure

1F

C39

인프라스트럭처

Jeff Wall, Restoration, 1993.
@ The artist.

가오슝, 타이완
시민의 집2.0

스튜디오 허우 x 린

C39-4F

참여형 어버니즘

C39

4F

Participatory Urbanism

시민의 집 2.0은 유서 깊은 항구 도시 가오슝을 보다 살기 좋고, 기업 친화적이며, 다양한 기능의 도시로 만들기 위한 도시재생 계획이다. 가오슝이 지속가능한 열대 도시의 모범 사례가 되기 위해 극복해야 하는 난관이 하나 있다. 바로 인구밀도를 제곱 킬로미터 당 15,000명까지 높이는 동시에 도시 고유의 건축 양식을 보존하는 것이다.

타이완의 제2 도시 가오슝은 한국의 부산처럼 중공업 단지가 있는 항구 도시다. 거대한 쓰레기 매립장과 함께 탄생한 초기 도시는 제대로 된 도시 계획 조차 없었다. 1900년부터는 1제곱 킬로미터 면적의 구시가지에 주택, 상점, 사무실, 놀이 공간, 다용도 건물이 지어지면서 자급자족하는 이상향을 이뤄냈다. 하지만 1980년대부터 컨테이너를 이용한 혁신적인 해상 화물 운송이 확산되면서 가오슝의 구시가지는 불경기에 신음해 왔다. 이에 따라, 신규 공동체, 신흥기업, 그리고 많은 사람들이 가오슝의 옛 시가지를 되살리겠다는 목표로 풀뿌리부터 시작하는 시범적 도시재생 사업을 진행하며 오래된 집을 보존하고 도시에 새로운 생명을 불어넣고 있다. 이러한 지역 사회 차원의 운동은 타이완 정부의 지역 인프라 개선 투자 계획과 맞물려, 인구밀도가 높은 도시의 지속가능한 미래에 대한 가능성을 제시한다.

시민의 집 2.0은 새롭게 부상하는 몰입기술을 도입하였으며, 풀뿌리 도시재생 운동이라는 혁신적인 프로젝트와 시민과 함께 상호작용하는 발전 모델을 공유하여 관람자의 상상력을 일깨운다. 가오슝은 무질서한 발전이 아닌, 도시 고유의 DNA를 보존한 21세기형 발전을 추구하고 있다.

KAOHSIUNG, TAIWAN
House of Citizens 2.0

studio HOU x LIN

C39-4F

House of Citizens 2.0 is an urban intervention that envisions a livable, entrepreneur-friendly, mixed-use urban regeneration along the old harbor of Kaohsiung. Being a tropical sustainable model, the challenge is to raise up the urban density to 15,000 people per km2, yet with the neighborhood's original building typology.

Kaohsiung, the second largest city in Taiwan, is a harbor city with heavy industries, more or less like Busan. Started with a large portion of landfill, Kaohsiung was born with a clear tabula rasa plan. Within 1km2, the old city center contains housing, retail, offices, and recreation, a mixed-use, self-contained utopia from 1900. The old city center, with the revolutionary container shipping from the 80's, has suffered from depression since then. New collectives, start-ups, and people who are devoted to this community experiment with micro urban regeneration by preserving and bringing new lives to the old houses. With the governmental investment of upgrading the local infrastructure, the local grass-roots movement shows a possible alternative for a high-density yet sustainable urban future.

Through the emerging immersive technology, House of Citizens 2.0 triggers the imagination by sharing pioneer projects and interactive models of the bottom-up movement. Instead of sprawl, Kaohsiung moves towards the 21st century with its own DNA.

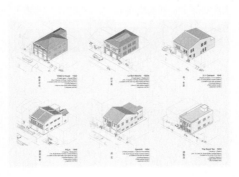

쿠알라룸푸르, 말레이시아
벽으로 이뤄진 유토피아

MIIM 건축사무소
C39-1F

최근 쿠알라룸푸르의 부동산 시장이 급성장하면서 다른 개발도상국 도시들과는 전혀 다른 통합적 복합개발 열풍이 불고 있다. 그러나 쿠알라룸푸르만의 진정한 특징은 도시의 대동맥이 되는 (고속)도로를 경계로 도시가 여러 넓은 구역으로 나뉘어 있다는 점이다. 쿠알라룸푸르 부동산 시장의 시 외곽 여러 지역으로 분산되는 과정과 지속적인 도로 인프라 확장이 맞물리면서 도시 곳곳에 10만 제곱미터 이상의 대규모 개발 지구가 속속 생겨나게 되었다. 수많은 개발 사업은 도시의 형태를 점점 바꾸어 나가고 새로운 구역이 출현하면서 구역을 서로 연결하는 도로가 건설되었다. 이 도로 연결망은 타 개발 구역, 역사적으로 유지되어 온 경계선, 자연적 지형, 그리고 최근 조성중인 지하철 부지와 경계를 이루며 역설적으로 물리적·심리적 장벽을 형성하게 되었다.
 이번 작품은 쿠알라룸푸르의 특징인 고도의 탈중앙적 도시 형태, 경계 바깥의 불쾌한 요소를 차단하고 자신만의 유토피아를 추구하는 도시 형태의 바탕에 깔린 역사적·정치적 요소를 짚어본다. 관람자들은 도시에 대한 교과서적인 관점으로는 설명할 수 없는 쿠알라룸푸르의 탈중앙적 형태를 통해 도시의 연결성과 괴리성이라는, 동시에 나타나지만 상반되는 특징에 대해 다시 생각해 보게 된다. 두 특징 사이의 모순은 어쩌면 자동차 중심의 도시가 당면하는 자연스러운 결과일지도 모른다.

KUALA LUMPUR, MALAYSIA
Forting Utopia

MIIM Office for Architecture
C39-1F

In recent years, KL's resurgence as a real-estate engine has resulted in a boom in integrated mixed-use developments that are unlike other similarly developing cities. However, what sets the city apart is its unusual organization into large districts that are bounded by major arterial roads or highways. The decentralization of real estate from the city core coincided with an ongoing expansion of road infrastructure and has resulted in a proliferation of large-scaled developments in the order of at least 100,000m2. The litany of such projects increasingly defines the city's morphology, becoming subdistricts whose peripheral roads and connective networks are also, ironically, physical and psychological barriers that coincide with local sub-districts, historical demarcations, natural topography, and recently the basis for the city's metro rail network. We explore the historical and cultural basis for this highly localized urban form, where utopia is defined by the blockade against undesirable elements beyond its boundaries. It disentangles itself from textbook understandings of what cities ought to be, and instead re-examines the juxtaposition of the ideas of connectivity and disjunction as a natural outcome of an automobile city.

Infrastructure

1F

C39

인프라스트럭처

KL Sentral Transportation Interchange Development Area dominating the Brickfields district

로스앤젤레스, 미국

로스앤젤레스: 자율주행 교통 천국을 향해

시티 폼 랩, 하버드대학교 디자인대학원
(안드레스 세브추크, 에번 시에)
C39-1F

로스앤젤레스(LA)라고 하면 흔히 "세계 자동차 수도"라는 별명을 떠올리게 되지만, 사실 LA는 대중 교통 인프라 확충을 위해 또 다시 유례없는 투자를 하고 있다. 한편, 이 도시는 우버, 주문형 마이크로 교통수단, 다양한 유형의 공유 자전거, 전동 스쿠터 등 민간 모빌리티 기술이 치열한 경쟁을 벌이는 곳이기도 하다. 더 나아가 LA는 '교통 중심 도시'라는 비전에 큰 변화를 가져올 자율주행차량의 도래에 대비하고 있다. 교통 전문가들은, 적절한 규제 없이 시장에만 교통 정책을 맡긴다면 자율주행 택시로 인해 자동차 여행이 다시 크게 늘어나고 그 결과 교통체증 증가, 자동차 우선 인프라 증가, 무분별한 도시 확장, 에너지 소비 증가로 이어질 것이라고 경고한다.

이번 작품에서는 LA의 새로운 대안, 즉 기술에 종속되지 않으며 더 나은 도시를 만들기 위해 기술을 전략적으로 규제하는 청사진을 모색한다. 이 새로운 도시에서 자율주행차량은 개인 이동수단 확산이 아닌 대중 교통, 특히 버스를 혁신적으로 개선하는 데 활용된다. 또한 그래픽 노블을 활용하여 관람자들에게 청사진에 따라 LA 시민들이 일상에서 자율주행 대중 교통을 이용하는 가상의 미래를 제시한다. 이러한 미래 대중교통은 양질의 서비스를 제공할 뿐 아니라, 도시라는 인공 환경을 크게 변화시키고, 교통 수단의 다양성, 사회적 포용성, 환경의 지속가능성을 극대화하는 방식으로 운영될 것이다.

LOS ANGELES, USA

Los Angeles: Towards Automated Transitopia.

City Form Lab, Harvard GSD
(Andres Sevtsuk, Evan Shieh)
C39-1F

Despite its popular stereotype as the car-capital of the world, Los Angeles is once again making unprecedented investments into public transport infrastructure. Yet, L.A. has also become a hot battleground for private-sector mobility technologies, such as Uber, on-demand micro-transit services, as well as a number of dockless and docked bikes and e-scooters. Furthermore, L.A. is bracing for a looming arrival of automated vehicles that are bound to challenge the transit-oriented vision of the metropolis. Transportation scholars have warned that if left to market forces alone, automated taxis are likely to once again boost car-based travel, leading to more congestion, vehicular infrastructure, sprawl and energy consumption.

This exhibit explores an alternative future for L.A. — one that is not dictated by technology, but where technology is strategically regulated to produce a better city. Instead of expanding private mobility, automated vehicle technology is used to revolutionize public transportation, particularly bus service. The exhibit uses a graphic novel as a medium to depict how everyday Angelenos could experience their city on automated public transport in ways that not only provide a high-level of service, but also reshape the built environment of the metropolis, maximizing multi-modal, socially inclusive, and environmentally sustainable outcomes.

인프라스트럭처

C39

1F

Infrastructure

로스앤젤레스, 미국 /
베이징, 중국 / 호치민 시, 베트남
이주민과 함께 만들어 나가는 도시 공간

애넷 미애 킴
C39-2F

현대 인류는 최초의 '도시 시대'를 살아 가고 있다. 수많은 사람들이 도시로 이동하면서 세계 대격변에 따른 광범위한 파급효과를 낳았다. 전세계에서 정체성과 소속감, 공공·사적 공간에 대한 권리라는 기존의 관념이 그 뿌리부터 의문과 변화의 대상이 되고 있다.

공간분석 연구소 SLAB은 이번 전시를 통해 베트남, 중국, 미국에서 거주하는 이주민들의 도시 생활, 즉 평범한 사람들이 새로운 가능성을 써 나가는 참신한 공간적 실천을 주제로 한 세 가지 연구 프로젝트를 제시한다. 또한 다양한 배경의 사람들을 분석한 지도를 한 편의 영화처럼 펼쳐내면서, 도시가 보다 아름답고 인간적이며 포용적으로 변모하도록 도시 공간을 재정의하고 함께 공유할 수 있는 방안을 모색한다.

보행의 도시는 베트남 호치민 시의 공공 장소가 지난 10년간 주민, 노점상, 경찰의 상호작용 속에서 겪은 변화를 보여준다. 땅 밑의 도시는 당국 몰래 백만 명의 이주노동자가 거주하는 중국 베이징 시의 방공호와 지하 시설을 조명한다. 다양성의 도시는 미국 로스엔젤레스 시의 복잡한 언어 지형과 도시라는 인공 환경에서 나타나는 다양한 문화들의 자유로운 표현을 소개한다. 또한 시선을 옮겨 미국 흑인들에게 특별한 장소인 Bill's Taco 가게의 내부를 들여다본다. 이 가게는 멕시코계, 한국계 미국인을 비롯한 모두가 소속감을 공유하는 공간이다.

LOS ANGELES , USA /
BEIJING, CHINA /
HO CHI MINH CITY, VIETNAM
Migrant Urbanism and the Co-Production of Places

Annette Miae Kim
C39-2F

For the first time in history, the human race has become urban. Human migration to cities has set off a global upheaval with far-reaching implications. Our fundamental notions and institutions of identity and belonging, and of entitlements to public and private space are being challenged and re-written all over the world.

SLAB presents three research projects of im(migrant) urbanisms, the emergent spatial practices of everyday peoples carving out new possibilities in Vietnam, China, and the United States. Our ethnographic, cinematic maps explore how we might reconfigure and share urban space to be more humanistic, inclusive, and beautiful. Sidewalk City maps how public space in Ho Chi Minh City was re-negotiated over a ten-year period between neighbors, street vendors and police. Subterranean City makes visible Beijing's underground layer of bomb shelters and basements which privately housed a million migrant workers. ethniCITY presents the complex linguistic landscape of Los Angeles, a city that allows the expression of multiple cultures in the built environment. The exhibition then moves beyond building exteriors into Bill's Taco restaurant, an important African American space that is co-produced by Mexican- and Korean-Americans, all sharing a stake in creating places of belonging.

집합적인 도시기억과 레이어

집합적 유형

마닐라, 필리핀

상황 대응형 인프라, 도시 마닐라를 위한 새로운 대안

디트마어 오펜후버, 카차 쉐츠너
C39-1F

디트마어 오펜후버, 카차 쉐츠너
C39-1F

전세계 많은 개발도상국의 역동적인 도시들처럼 마닐라는 시민들에게 충분한 인프라와 공공 서비스를 제공하는 데 애를 먹고 있다. 이러한 환경에서의 인프라 문제는 개인 차원에서, 혹은 지방 의회, 주민, 민간 기업의 협력으로 해결된다. 최근 등장한 '상황 대응형 인프라' 개념에 따르면, 인프라 거버넌스는 다양한 사람들이 상황에 맞게 "필요할 때 도움을 주고받는" 과정이다. 마닐라 시 당국과 공공 서비스 기업이 진행하는 현대화 사업에도 이 관점이 적용될 수 있다. 마닐라의 가로등과 전력망을 조명하는 이번 작품은 '어떻게 시민들이 상황에 맞게 대응하는 과정을 통해 도시의 인프라 시스템을 구성해 가는가'를 연구한다. '상황 대응형 인프라'는 다양한 사회적 의미를 갖는다. 이는 시민들이 자신들의 방식으로 도시 인프라를 활용하며 기존 체제에 도전하는 행동이면서, 동시에 인프라 수리와 유지 보수에 대한 창조적인 접근 방식이다. 또한 상황에 따라 적응하는 공동 생산 양식이기도 하다. '주어진 시점에서 구할 수 있는 자원을 최대한 활용해 계획하고 만들어 나가는 행동 양식'이라고 정의할 수 있는 '상황 대응형 인프라'는 시민들이 살아 가면서 급박한 일을 겪거나 예기치 않은 기회에 직면했을 때의 적절한 대응 과정에서 탄생했다. 도시 인프라는 '상황 대응적' 시각에서 소통을 위한 의미 있는 매개체가 된다.

마닐라의 가로등을 벗어나 세계 어느 도시를 가더라도 '상황 대응형 인프라'를 만날 수 있다. 이와 같은 인프라에 대한 새로운 시각은 특히 개발 도상국의 스마트 시티 개발 계획에 적용될 수 있다. IT 기업들이 제시하는 천편일률적인 해결책을 넘어 지역의 문화, 사회, 환경적 맥락을 고려하여 문제에 기민하게 대응하는 스마트 시티가 가능해지는 것이다.

C39

1F

MANILA, PHILIPPINES
Manila Improstructure

Dietmar Offenhuber, Katja Schechtner
C39-1F

Dietmar Offenhuber, Katja Schechtner
C39-1F

Like many dynamic megacities of the Global South, Manila is struggling to provide adequate infrastructure and public services to its constituents. In this environment, many infrastructural issues are collaboratively resolved through personal interventions and collaborations involving council members, residents, and companies. The concept of improstructure describes infrastructure governance as an improvisational process of "call and response" among a diverse set of actors. We apply this perspective to ongoing modernization efforts by the city of Manila and its utility companies. Focusing on social practices in Manila's streetlight and electricity grid, the project investigates how actors shape the infrastructural system through an improvisational process.

Improstructure encompasses subversive practices of appropriation, creative approaches to repair and maintenance, and ad-hoc models of co-production. Defined as the coincidence of planning and doing with the resources available at the moment, improvisation emerges in response to an urgent need or an unforeseen opportunity. From an improvisational perspective, infrastructure becomes a material medium of communication.

Beyond the management of Manila's streetlight infrastructure, improvisational governance can be found in every city of the world. The improstructure perspective becomes especially relevant in the context of smart city initiatives in the Global South: to move beyond generic solutions prescribed by IT companies, towards a nimbler approach that takes the local cultural, social, and environmental context into account.

메데인, 콜롬비아

메데인: 환경친화적 어바니즘, 사회, 교육, 문화

페데리코 구티에레스 줄루아가, 카를로스 파르도 보테로, 니콜라스 에르멜린 브라보, 에드가 마조 자파타, 세바스티안 메히아 알바레스, 알레한드로 레스토포-몬토야
C39-2F

메데인: 환경친화적 어바니즘, 사회, 교육, 문화는 지속가능한 도시화 전략을 위한 메데인의 자연존중 도시 계획을 들여다본다. 우리는 이번 작품을 통해 강 유역의 자연을 복원하는 도시 계획의 핵심 개념, 즉 도시의 '자연적' 구조를 조명한다. 또한 자연을 공공 공간과 연계하여 시민이 다가가기 쉬운 환경 친화적 도시로 만들고 시민들이 서로 교류하는 새로운 공간을 조성하고자 한다.

우리는 가로 170센티미터, 세로 300센티미터 크기의 메데인 도시 모형을 제작하고 그 위에 네 개의 도시 발전 역사 단계를 담은 이미지를 투영할 것이다. 네 가지 단계란 (1) 자연적 구조 (2) 도시 생성, 초기 확장 과정 (3) 도시 건설과 무계획적 팽창으로 사라져 버린 강 유역과 주변 자연 (4) 자연 복원을 목표로 하는 오늘날의 도시 계획을 말한다. 현재의 도시 계획을 통해 우리는 강 유역, 자연 지형과 풍경을 복원하고 이를 도시의 공공 공간과 연계하여 기존의 도시 구조 속에서 새로운 공공 공간을 만들어 내고자 한다.

전시 공간 한쪽 벽면에 상영되는 영상은 메데인을 바꿔 나가기 위해 진행중인 여러 활동들과 점차 늘어나는 환경 친화적 공공 공간을 조명한다.

MEDELLIN, COLOMBIA

Medellín: Environmental Urbanism, Society, Education, and Culture

Federico Gutiérrez Zuluaga, Carlos Pardo Botero, Nicolás Hermelín Bravo, Édgar Mazo Zapata, Sebastián Mejía Álvarez, Alejandro Restrepo-Montoya
C39-2F

This exhibition explains Medellín's urban planning process from its natural components as strategies for the sustainable occupation of the territory. The aim is to highlight the natural structure of the city as a fundamental concept for urban planning that re-densifies the natural components in its hydrographic basins and articulates them with the public space to qualify it in spatial and environmental terms, and generate new places for citizen gatherings.

For the exhibition, we propose the construction of a 170 cm x 300 cm topographical model, on which images of four historical moments of the city are projected: (1) Natural structure; (2) First occupations of the territory and first expansion processes; (3) Disappearance of the natural structure of the hydrographic basins and their natural components due to the construction of the city and its unplanned expansions; and (4) Current territorial planning based on its natural components: watersheds, topography and natural landscape, as instruments to articulate the structural natural components of the city with the existing public space and for the generation of new public spaces in the existing urban structure.

On one of the walls, another video projection will show the current urban and landscape interventions in the city, with the increase of public space and its environmental qualification.

Infrastructure

C39 2F

인프라스트럭처

밀란: 건축적 맥락

스튜디오 디 마우리치오 카로네스 (마우리치오
카로네스, 조항준, 루카 스칼린지)
C39-2F

밀라노는 현대도시와 건축이 겪은
우여곡절이 가장 잘 드러나는 도시이다.
자연·사회·문화·경제적 환경에 더해
밀라노를 뒤흔들었던 20세기의 여러 역사적
사건을 통해, 도시에 다양하고 때로는
실험적인 아이디어가 19세기 후반부터
20세기 말까지 꽃필 수 있었다. 이번 작품은
도시를 묘사하는 몇 가지 핵심 요소를
짚어본다. 도시를 묘사한다는 것은 결국
도시를 그 자체로서 만이 아니라 역사와도
연관 지어 바라본다는 의미이다. 이런
시각으로 도시를 바라보면 어떤 해석도
의미를 가지게 된다. 이미 다양한 지도 제작,
연구, 설계 사례를 통해 도시에 대한 여러
가지 시각이 제시된 바 있으며, 본 작품도 그
중 일부를 담고 있다. 이번 작품에서 묘사된
도시의 모습을 통해 밀라노를 '현대성'의
상징적인 도시로 바라보는 시각을 접할 수
있다. 밀라노는 언뜻 보면 획일적인 도시
같지만, 사실 다층적인 복잡성이 존재하는
곳이다. 그 덕분에 다양한 파편적 요소가
계속해서 중첩되고 병렬 배치되어 현대 도시의
탄생 과정을 묘사하는 방식을 실험하기에
더 없이 좋은 곳이기도 하다. 또한, 밀라노를
구성하는 다양한 요소가 현실에서 구현되는
과정이 이 도시를 하나의 단일한 대상으로
바라보는 도시 계획 및 발전 프로젝트와는
어떤 점에서 다른가를 살펴보아도 좋을
것이다.

MILANO, ITALY
Milan: Building Contexts
Studio di Architettura Maurizio Carones
(Maurizio Carones, Hang-Joon Gio, Luca
Scalingi)
C39-2F

Milan arguably exemplifies several
elements of modern urban and
architectural vicissitudes. The
natural, social, cultural, and economic
environment — as well as the dramatic
historical events of the 20th century —
has promoted the implementation, even
an experimental one, of ideas whose span
stretches from the second half of the
19th century through the end of the 20th.
This work suggests some key elements
that describe the city. To describe a city
entails not only looking at the modern city,
but also taking into account its history.
When approaching the city from this
perspective, any interpretation becomes
potentially meaningful. A number
of views have already been provided
through various mapping, research, and
design initiatives, some of which are
also included in this project. The views
of the city depicted in this project offer
an approach to the city as the symbolic
embodiment of "modernity." Despite
its seeming uniformity, there is a multi-
layered complexity to Milan. Because of
this, it is a great city for experimenting
with techniques in describing how a city
arises from the continuous overlapping
and juxtaposition of its various fragments.
At the same time, it is worth examining in
what ways the processes by which these
various elements that make up Milan have
differed from the urban planning and
development projects that approach the
city as a single entity.

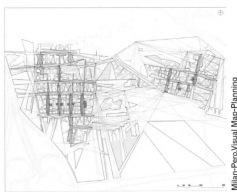

Milan-Pero.Visual Map-Planning

나이로비, 케냐

움직이는 나이로비: 도시 기동성에 관한 이야기

시빅 데이터 디자인 랩, MIT (프로젝트 리더: 새라 윌리엄스 / 리서처: 카르멜로 이그나콜로, 딜런 할펀)

C39-1F

전시작 움직이는 나이로비은 각각 도보, 자전거, 오토바이, 버스를 이용해 출퇴근하는 네 사람의 시선에서 바라본 나이로비의 모습을 소개한다. 또한 우버 운전사를 호출하여 나이로비 부유층 지역에서 서민 지역으로 이동하면서 보이는 도시의 모습을 조명하고, 급격히 발전하는 나이로비의 변화상을 저속촬영 영상을 통해 관람자가 눈과 귀로 느낄 수 있도록 한다.

나이로비의 하루 유동인구는 350만이나 된다. 그러나 조직화된 도시 교통 계획이 부족하기 때문에 도시의 교통은 종종 마비되고 통근자들은 몇 시간씩 차 안에 갇히곤 한다. 본 작품은 우버, 휴대전화, 구글을 통해 수집한 시민 이동 동선 데이터를 애니메이션으로 상영하고, 벽면에는 나이로비의 숨막히는 교통 체증에 대한 자료를 전시한다. 이러한 시각화 자료를 나이로비에 사는 통근자 네 사람의 이동 경로와 결합하여 제시하고, 이들 개인의 경험과 도시 전반 교통 문제 사이의 연관성을 보여준다.

나이로비 시민들은 이러한 유동성의 흐름을 낳은 케냐의 정치상황에 대해 종종 논쟁을 벌이곤 한다. 본 작품 역시 케냐의 신문을 통해 이 이야기를 관람자들에게 소개한다. 신문 기사에 소개되는 여러 사실들과 수치가 나이로비 통근자들의 모습과 함께 전시장 벽면에 전시되며 관람자들도(역자: 나이로비의 기동성, 더 나아가 도시의 기동성에 대해) 의견을 제시할 수 있다.

NAIROBI, KENYA

Moving Nairobi: Stories of Urban Mobility

Civic Data Design Lab, MIT (Project Leader: Sarah Williams / Researchers: Carmelo Ignaccolo, Dylan Halpern)

C39-1F

Moving in Nairobi is an exhibition that explores Nairobi from the eyes of four commuters as they walk, bicycle, ride motorcycles (boda boda), take buses (matatu), and hire Ubers from the wealthy neighborhoods of this East African capital to its informal neighborhoods. The sounds and sights of this rapidly developing city come to life through their time-lapse videos.

3.5 million people move through Nairobi every day. However due to the lack of coordinated transportation planning, urban traffic often grinds to a halt, leaving commuters stuck in hours of gridlock. Human movement data acquired from Uber, cell phones, and Google will be animated and the walls of the exhibition will showcase materials on Nairobi's stifling congestion. The paths of our four commuters will be embedded in the visualization to help connect their experience to the experience on the road. The politics that have created Nairobi's mobility flows are often debated amongst the city's residents and these stories will be illustrated in a newspaper given away at the exhibition. The facts and figures discussed in the newspaper will also be displayed on the wall alongside portraits of Nairobi commuters. Those who come to the exhibition can start their own debate.

프로비던스, 미국
도시의 빈 공간, 그 여백의 실험

울트라모던

C39-2F

빈 공간은 어느 도시에나 있다. 산업시설이 사라진 미국 동부 지역에서는 빈 공간이 위기의 동의어로 여겨진다. 도시에 텅 빈, 혹은 제대로 개발되지 않은 지역이 늘어나면, 기본적인 인프라 공급과 유지마저도 주민의 세금으로는 감당하기 어려워지는 상황이 오게 되기 때문이다.

따라서 빈 공간을 위기로 바라보는 인식은 매우 보편적이다. 현재 진행중인 많은 도시 계획들 역시 도시의 부동산을 효율적으로 관리하여 빈 공간을 최대한 줄이는 데 초점을 두고 있다. 이런 관점에서 보자면 빈 공간은 그저 부정적인 상태, 즉 토지와 건물이 경제적인 잠재력을 충분히 발휘하지 못하고 있는 상태이다. 그러므로 경제학자와 도시 계획 담당자들이 한 목소리로 비난하는 대상이 된다. 이들의 눈에 빈 공간이란 부동산 가격과 세금 정책, 다양한 인센티브와 부동산 시장 부양을 통해 없애야 하는 사회문제일 뿐이다.

하지만 이런 접근법이 간과하고 있는 사실이 있다. 도시의 빈 공간이 모두 세수 창출 용도(revenue-generating uses)로 개발되고 채워진다면, 그로 인해 도시의 거대한 잠재력이 빛을 잃게 된다는 점이다. 미국 로드아일랜드 주 프로비던스시는 빈 공간을 활용해 새로운 형태의 집합성을 창조하는 대안적 도시 모델을 제시한다. 물론 이 도시도 부동산 투자를 마다하지는 않지만, 다른 한편에서는 건축가, 도시 계획 전문가, 문화 단체들이 지난 수십 년간 비어 있던 프로비던스의 빈 공간에서 새로운 주택 양식과 문화를 실험하고 있다. 낙후되었지만 많은 가능성을 품은 도시의 빈 공간, 그리고 이 여백의 공간(margin)에서 진행되는 실험은 때로 여백이야말로 도시 풍경의 진정한 특징일 수 있다는 점을 시사한다. 결국 프로비던스의 대안적 도시 모델 실험은 일시적인 비효율 상태를 활용하여 도시의 참신한 미래를 만들어가고자 하는 것이다.

PROVIDENCE, USA
Vacant: Experiments at the Margins

Ultramoderne

C39-2F

Vacancy is a condition common to all cities. In the post-industrial cities of the American northeast it is often seen as a crisis. Empty and underdeveloped lands bring the cities below a critical density at which point the cost of delivery and maintenance of infrastructural services no longer matches the population's ability to pay.

The economic challenges of vacancy are thus well established, and much current thought in city planning is indeed focused on the efficient economic management of urban real estate as a means of minimizing vacancy. In this context, vacancy is understood as a purely negative condition. It denotes a site (land or building) that is not living up to its economic potential. Economists and planners alike see vacancy as a pejorative condition: a problem to be eradicated through pricing, taxation, incentives, and market props.

What this approach neglects is the enormous potential left behind when all the empty space is efficiently filled by revenue-generating uses. The city of Providence, Rhode Island, proposes an alternative model where vacancy is seen as a generator of new forms of collectivity. While the city by no means discourages real estate speculation, architects, planners, and cultural groups have used the economic fallows of the last several decades as an opportunity to experiment with new housing types and cultural development in these vacant spaces. Simultaneously scrappy and potent, these experiments in the margins of the city — margins that at times have appeared to be the dominant feature of the urban landscape — have leveraged a condition of temporary inefficiency to reflect on alternative futures for urban life.

Providence Big Top Parking Lot

로마, 이탈리아
로마의 동쪽: 집합적 의례의 도시

2A+P/A (잔프랑코 봄바치, 마테오 코스탄초),
다비데 사코니, 가브리엘레 마스트리글리
C39-4F

로마는 고요한 도시이다. 하나로 합쳐지지
않은 수많은 조각들이 쌓여 만들어진
도시다. 원형 극장, 공연장, 광장, 빌라,
수도관 등 수많은 고대와 현대의 구조물이
서로 간에, 그리고 도시의 경관과 조화를
이루고 있지만, 로마 전체를 아우르는 질서는
존재하지 않는다. 이런 관점에서 로마는
현대의 도시화 패러다임에 새로운 대안을
제시한다. 일반적으로 도시란 개성없이
늘어선 촘촘하고 반듯한 구역 배치를 통해
끝없이 확장되는 경제적 관계와 경영 활동을
구현하는 것으로 인식된다. 그러나 이와 달리
로마는 유한하고 개별적인 요소들로 구성된
도시이며 이곳의 건축은 집합성이라는 정치적
아이디어를 재구성하는 핵심 요소이다.
도시화의 새로운 대안을 제시하는 이번
작품은 주로 로마의 동쪽 구역을 조명한다.
예전부터 로마에서 가장 빈곤하고 관심의
대상에서 벗어난 지역이었으며, 오늘날에는
도시구조상의 선명한 대조, 유동적이며
불안정한 인구구성을 보이고 있다. 생활
조건은 열악하지만, 한편으로는 판타지와
팍팍한 생존을 오가는 삶에서 나오는
원초적인 에너지가 있는 곳이기도 하다.
이번 작품은 도시 문제에 대한 구체적인
해결책을 보여주기보다는 로마의 새로운
비전을 제시하는 데 초점을 둔다. 이를 위해
건축이라는 형식을 통해 도시의 원초적인
에너지를 끌어내고, 정형화된 계획이나
프로그램 대신 사람들의 집합적 행동을
활용하는 방식을 제시한다. 이런 관점에서,
작품이 제안하는 로마를 위한 새로운
도시 비전의 목적은 즉각적인 행동주의나
규범적이고 총괄적인 도시 계획을 기반으로
하는 건축을 넘어, 건축의 원초적인 힘을
수용하여 사람들이 함께 살아가고 행동할 수
있도록 하는 건축의 새로운 지평을 만들어
가는 것이라고 할 수 있다.

ROME, ITALY
Rome East: A City of Collective Rituals

2A+P/A (Gianfranco Bombaci and
Matteo Costanzo) and Davide Sacconi in
collaboration with Gabriele Mastrigli
C39-4F

Rome is a still life, a city made by
the accumulation of pieces, of parts
without a whole. Circuses, theatres,
forums, villas, aqueducts and all sorts
of ancient and modern structures are
juxtaposed to relate one another and
with the landscape, without an overall
predefined order. As such, Rome offers an
alternative to the contemporary paradigm
of urbanization. Where urbanization
embodies the endless field of economic
relationships and managerial practices
through the neutral and scaleless
dimension of the grid, Rome is a
composition of finite and discrete forms
that claims the crucial role of architecture
in reinventing a political idea of the
collective.
The proposal focuses on the east sector
of Rome, traditionally the poorest
and most neglected part of the city,
characterized by stark contrasts in the
urban fabric and by a fluid and unstable
population mix. The area produces
extreme living conditions but also a
savage energy which suspends life
between fantasy and survival. Rather
than providing specific solutions, the
project proposes a vision for the city of
Rome through a series of interventions
that aim to mobilize such energy through
form, substituting function or program
with the collective action of a subject. As
such the proposal intends to construct
a horizon for the architectural process
that goes beyond both the immediacy of
activism and the normative character of
the masterplan to instead embrace the
savage power of architecture of bringing
people to act and live together.

Collective Typologies

4F

C39

집합적 유형

사마르칸트, 우즈베키스탄
도시와 도시를 잇다, 사마르칸트
문화유산기술연구소
C39-4F

유라시아 대륙의 한 중간에 위치한
사마르칸트는 동서 문명이 충돌하는 곳이자
교류하는 허브로서 기능해왔다. 수많은
문화와 인종들이 뒤섞이며 3천여년간
누적되어 온 도시의 층위들은 주인이 수없이
바뀌었을지언정 허브로서의 역할과 기능은
마치 태초부터 부여된 숙명처럼 유지되어
왔다는 것을 보여준다.
1965년, 사마르칸트 동북쪽 언덕에서 도로
공사 중 발견된 7세기 아프라시압 궁전벽화는
수천년간 사마르칸트가 어떤 도시였는지
함축적으로 표현한다. 벽화에는 이 지역과
교류한 세계 각국의 사절들이 그려져 있고
그 중에서는 놀랍게도 극동에서 온 한국인도
등장한다. 이 벽화궁전과 도시를 둘러싼
견고한 성벽에는 케슈문, 나우베차르문,
부하라문, 중국문 등 세계 각지로 연결되는
성문이 별도로 나 있다.
사마르칸트는 도시와 도시를 연결하고, 다시
그 도시들이 성장할 수 있도록 문명과 재화를
공급하는 도시들의 도시였다. 전시에서는
7세기 아프라시압 유적을 통해 허브 도시
사마르칸트를 함축적으로 조망해 본다.

SAMARKAND, UZBEKISTAN
The Hub of the World City Network, Samarkand
Technology Research Institute for Culture & Heritage
C39-4F

Located in the middle of the Eurasian
continent, Samarkand has served both
as a place where Eastern and Western
civilizations collide as well as a hub for
their interaction. A variety of strata of the
city, which have accumulated over 3,000
years along with the mixture of numerous
cultures and races, show that the city's
role and function as a hub, although its
owners may have changed numerous
times, has been maintained as if it were a
fate granted from the beginning.
The 7th century Afrasiab palace
murals, discovered in 1965 during
road construction on a hill northeast of
Samarkand, suggest what Samarkand
was like for thousands of years. In the
murals, there are envoys from around the
world who interacted with this region,
including, surprisingly, Koreans from the
Far East. The solid ramparts surrounding
the palace and the city have separate
gates that are connected to various parts
of the world, such as the Kesh gate, the
Naubechar gate, the Bukhara gate, and
the China gate.
Samarkand was a city for cities that
connected cities with cities and supplied
civilizations and goods for them to grow.
The exhibition offers a view of the hub
city of Samarkand through the ruins of
Afrasiab from the 7th century.

산호세, 코스타리카

건축, 도시 변화의 촉매제: 주민 회관을 사회적 기업의 안식처로 (산호세 '빛의 동굴'을 비롯한 다양한 활동을 중심으로)

엔테르 노스 아틸리에르 (마이클 스미스 마시스, 알레한드로 바예호 리바스, 에스테반 다르세)
C39-4F

새롭고, 대안적이며, 혁신적인 유형의 건축 활동을 꿈꾸는 건축가가 되려면, 복잡한 도시 구조속에서 꾸준한 영향력을 위한 건축 방법론을 새로운 방식으로 생각해야 한다. 다양한 사회·경제적 역학관계로 빚어진 정치적이고 공간에 따른 복잡한 상황을 겪으면서 사람들은 회복력을 키웠고, 또한 사회적 맥락을 반영하는 건축 공간을 만들어냈다. 이러한 변화는 다양한 지역에서 그 나름의 방식으로 무한한 역량을 발휘하여 주어진 자원으로 사람들의 삶을 개선해 나갈 때 더욱 증폭된다. 건축, 도시 변화의 촉매제: 주민 회관을 사회적 기업의 안식처로는 중앙 아메리카의 인공 환경(역주: 도시)에서 싹트는 참신한 생산 방식과 지역의 사회적 권익 향상을 위한 노력을 조명한다. 사회적 성격을 가지며 시민 권리 신장을 위해 노력하는 행동주의적 기업들이 주민 회관을 활용하여 모여서 융합하고 교류할 수 있는 안식처를 제공하는 것이다. 이러한 노력 위에 지역사회, 다양한 학계 전문가와 협력하는 과정을 더하고 참여를 바탕으로 지속가능한 환경을 위한 도시 설계 전략을 수립한다. 또한 본 작품은 다양한 매체를 활용하여, '정의롭지 못한 사회, 기회가 주어지지 않는 사회, 도시에서 나타나는 여러 가지 문제들'과 같은 인류 시대의 중요한 이슈를 탐구하고, 이런 문제들을 해결하기 위해 빛의 동굴에서 우리가 실제적으로 시작할 수 있는 노력을 조명한다. 다양한 사고 과정을 담은 도표, 지역의 모습을 담은 이미지를 전시하여 지역의 변화를 이끌어내고 상황을 개선하기 위한 길을 모색한다. 결국 이번 전시는 다양한 사람들과 관람자가 함께 만들어가는 공간이다. 전시장에 설치된 나무 소재로 제작한 실물 크기의 '만남의 장소' 모형이 이 점을 잘 보여준다.

SAN JOSÉ, COSTA RICA
Architecture as Excuse: Community Centers as Catalysts for Social Agency in Informal Cities (The Cave of Light in San José, Costa Rica and other Emergent Initiatives)

Entre Nos Atelier (Michael Smith Masis, Alejandro Vallejo Rivas, Esteban Darce)
C39-4F

Recognizing our agency as designers to envision emerging, alternative, and new radical forms of practice, it is critical to reconceive our methods to have a coherent impact within the complexity of our cities. Continuously exposed to challenging political and spatial contexts compounded by divergent socioeconomic dynamics has engendered a human resiliency and contextual sensible space of practice. This condition is enriched by a boundless capacity for social endeavors working with local processes, existing resources, and promoting local assets within diverse communities and geographies. This exhibition will expose unconventional modes of production and social empowerment within the built environment in Central America. Utilizing community centers as catalysts, these spaces serve as the first line of convergence, interaction, and shelter for alternative enterprises defined by social agency, advocacy, and activism. Through this undertaking strategies of participatory and sustainable environmental design endeavors emerge by working alongside communities and in collaboration with transdisciplinary teams. The exhibition will encompass narratives with diverse mediums of display undertaking the importance of: the lack of justice, opportunities and urban challenges; Cave of Light as a practical demonstration on how to start among all collateral efforts; Process and sequence diagrams, pictures over the territory to express change and recreate site conditions; Show main stockholders and users stories as project enablers; A timber spatial mock-up of space for gathering and interaction.

참여형 어버니즘

95

상파울루, 브라질
상파울루: 그래픽 바이오그래피
펠리페 코레아
C39-1F

브라질 상파울루의 역사는 450년도 더 되었지만, 대부분의 도시 발전은 주요 성장 동력이 농업에서 공업으로 옮겨간 2차 세계대전 이후에 이루어졌다. 에콰도르 건축가 펠리페 코레아는, 상파울루가 공업 도시에서 서비스 경제 허브 도시로 변모하고 있는 현 시점에서 산업시설이 떠난 광범위한 도심 구역을 현대 도시에 적합한 용도로 활용할 수 있는 보다 나은 방안을 면밀히 검토해야 한다고 역설한다. 코레아는 다양한 기록 자료와 사진, 자신이 그린 도면, 텍스트가 담긴 전시작 상파울루: 그래픽 바이오그래피를 통해 브라질 최대 도시 상파울루를 속속들이 보여줌과 동시에, 이 도시가 거쳤던 고속성장 이야기를 들려준다. 또한 관람자들은 본 작품에 참여한 조경, 생태, 거버넌스, 공공 보건 등 다양한 분야 전문가의 글을 통해 상파울루의 역사와 발전을 바라보는 간(間)학문적 시각을 접할 수 있다.

본 전시작은 파울리스타 지역이 최초로 도시의 형태를 갖추기 시작하던 시절, 그리고 상파울루가 브라질의 제조업 중심 도시로 성장해 나가는 과정을 보여주는 데 그치지 않고, 산업시설이 떠나간 구역을 도심의 저렴한 복합용도 주거지구로 바꿔 나갈 방법을 제안한다. 또한 도심의 빈 공간을 바라보는 새로운 관점을 제시함으로써, 도심과 교외 지역간 극심한 사회경제적 격차 완화에 꼭 필요한 도시 구조 개편의 비전을 설득력 있게 제시한다. 21세기 도시의 청사진을 소개하는 본 작품은 도시 개선에 관한 탁월한 연구 성과와 함께 도시 스스로가 미래를 그려갈 수 있는 방안에 대한 독특한 시각을 담고 있다. 디자인 그룹 서매틱 콜라보레이티브의 펠리페 코레아와 데빈 도브로볼스키가 기획한 상파울루: 그래픽 바이오그래피는 동명의 책 출간을 위해 진행되었던 작업에 바탕을 둔 작품으로, 전시 형식으로 상파울루의 다층적이고 복잡한 면모를 관람자들에게 제시한다

SÃO PAULO, BRAZIL
São Paulo: A Graphic Biography
Felipe Correa
C39-1F

While the history of São Paulo dates back more than 450 years, most of its growth took place after World War II as the city's major economic engine shifted from agriculture to industry. Today, as São Paulo evolves into a service economy hub, Felipe Correa argues, the city must carefully examine how to better integrate its extensive inner city post-industrial land into contemporary urban uses. In São Paulo: A Graphic Biography, Correa presents a comprehensive portrait of Brazil's largest city, narrating its fast-paced growth through archival material, photography, original drawings, and text. Additional essays from scholars in fields such as landscape architecture, ecology, governance, and public health offer a series of interdisciplinary perspectives on the city's history and development. Beyond presenting the first history of Paulista urban form and carefully detailing the formative processes that gave shape to this manufacturing capital, São Paulo shows how the city can transform its post-industrial lands into a series of inner city mixed-use affordable housing districts. By reorienting how we think about these spaces, the exhibition offers a compelling vision of a much-needed urban restructuring that can help alleviate the extreme socioeconomic divide between city center and periphery. This twenty-first century urban blueprint thus constitutes an impressive work of research and presents a unique perspective on how cities can imagine their future.

The exhibition São Paulo: A Graphic Biography curated by Felipe Correa and Devin Dobrowolski from Somatic Collaborative, builds on the work produced for the book of the same title, and presents in exhibition format, the layered complexity of this South American Metropolis.

도호쿠 + 구마모토, 일본
모두의 집
일본 국내외 건축가 52인과 건축 사무소
C39-1F

모두의 집은 2011년 동일본 대지진에 대한 건축가로서의 대응 방안을 모색하는 프로젝트이다. 모두의 집 프로젝트 팀은 획일적인 임시 주택에서 지내는 지진 피해자들의 고단한 삶을 관찰한 끝에 이들에게 다함께 모여 온기를 느끼고 먹고 마시며 대화할 수 있는 대안적 주거지를 제공하고자 한다.

토요 이토와 가즈요 세지마는 여러 건축가와의 협업과 세계 각지의 넉넉한 기부금의 중추 역할에 힘입어 도호쿠 지방에서 모두의 집 프로젝트를 착수할 수 있었다. 현재까지 모두의 집 16채가 지어졌다. 집들은 모두 서로 다른 특징을 갖으며 이재민을 위한 만남의 장소, 어민들을 위한 공간, 어린이 놀이방 등이 마련되어 있다.

센다이에 있는 최초의 모두의 집은 구마모토 현의 지원으로 건설되었다. 2012년 7월 구마모토 광역에 홍수 재해가 일어났을 때 센다이의 경험을 바탕으로 모두의 집 두 채가 추가로 지어졌다. 구마모토현 지사는 2016년 4월 구마모토 지진 기간 중에 공공부문과 민간부문의 자금을 활용하여 100채의 모두의 집을 목조로 건설하는 데 앞장섰다.

전시에서는 피해 지역에서 우리의 지원책 이상의 역할과 미래의 공공건축과 사회에 역동적인 본보기가 된 모두의 집 프로젝트를 소개할 예정이다.

TOHOKU + KUMAMOTO, JAPAN
Home-for-All
52 Architects and Architecture Practices inside/outside Japan
C39-1F

Home-for-All is a project that explores how architects can respond to the Great East Japan Earthquake in 2011. After observing the tough life of the victims living in standard temporary housing, Home-for-All aims to provide the affected people with an alternative home where they can gather, take warmth, drink, eat, and talk.

In Tohoku, Toyo Ito and Kazuyo Sejima play the central role to create "Home-for-All" projects in collaboration with many architects and supported by generous donations from all over the world. Sixteen Home-for-All houses have been built to date. They have variations in character and include gathering places for displaced residents, spaces for fishermen, playhouses for children, etc.

The first Home-for-All in Sendai was built with the support of Kumamoto Prefecture. Building upon this experience, another two Home-for-All houses were built after the Kumamoto Wide Area Flooding Disaster in July 2012. During the Kumamoto Earthquakes in April 2016, the Governor of Kumamoto Prefecture took the initiative in using funds from both public and private sectors as capital to build 100 Home-for-All houses constructed with timber structure. Through this exhibition panel, we will introduce Home-for-All projects that surpassed their intended support role for the afflicted areas, and became aspirational models for both tomorrow's public architecture and for the society to come.

Participatory Urbanism

1F

C39

참여형 어바니즘

"Home-for-All" in Komori 2, Nishihara
© Yousuke Harigane

통영, 대한민국
도시재생
한국토지주택공사(LH)
C39-2F

URBAN REGENERATION
LH (Korea Land & Housing Corporation)
C39-2F

한때 통영의 경제 성장을 견인하던 신아 SB 조선소는 2015년 11월 파산하고 말았다. 대규모 실업이 발생했고 주변 곳곳에 빈 집이 늘어갔다. 사람들이 빠져나가고 빈 집이 늘자 통영의 지역 경제 역시 빠르게 쇠퇴했다. 이에 대한 대책으로, 국토교통부는 2017년 12월 신아 SB 조선소와 주변지역을 (경제 기반) 도시재생 뉴딜사업지로 선정하고 국비 지원 계획을 수립했다. 사업 시행 주체인 한국토지주택공사는 2018년 3월 신아 SB 조선소를 매입했고, 통영시와 공동으로 지역개발 프로그램을 실행할 예정이다. 신아 SB 조선소 도시재생 사업의 목표는 폐조선소 부지를 문화관광 허브로 되살리고, 산업 전환을 통해 통영 지역을 세계적인 관광명소로 자리매김시키는 것이다. 통영의 폐조선소를 관광자원으로 활용함으로써, 국제 핵심 업무시설, 리조트 시설, 수변 휴양 공간, 최고급 주거 시설 등으로 구성된 종합 세계 수변 도시가 새로이 들어설 예정이다.

Shina sb dockyard, which was a major source of economic growth in Tongyeong, went bankrupt on November 2015. As a result, workers lost their jobs en masse and residential units in the surrounding areas became vacant. The vacant residential units resulted in rapid stagnation and a decline in the regional economy.
In response, the Ministry of Land, Infrastructure and Transport designated Shina sb shipyard and the surrounding area as a beneficiary for the Urban Renewal New Deal (economy-based) in December 2017 and plans to provide support for the area from the national budget; the implementer of the project, the Korea Land and Housing Corporation, purchased the Shina sb shipyard site in March 2018, and will carry out the project as a joint development program in tandem with Tongyeong city.
The purpose of the (former) Shina sb dockyard urban regeneration project is to form a cultural and tourist hub on the site of the closed dockyard, thereby establishing a global hub of tourism through industrial reorganization.
By using the former dockyard as a tourist attraction, a globally recognized model for a comprehensive waterside town consisting of international core facilities, resort facilities, areas for seaside leisure activities, high-end residential facilities, etc., will be established.

C39

2F

울란바토르, 몽골
일반적인 문제에 대한 남다른 해결책

제르허브
C39-5F

도시 거주민들은 공기 오염, 교통 혼잡, 도시 공간의 상실 등과 같은 난제에 직면하고 있다. 우리는 공동의 노력과 남다른 접근으로 이와 같은 난제를 해결할 수 있다고 믿는다. 본 프로젝트는 일반적인 문제들에 대해 남다른 해결책으로 대응하고자 한다. "일반적인 문제"는 유목민, 도시를 만나다라는 제목의 영상을 통해 제기된다. 이를 통해 울란바토르 뿐 아니라 수많은 도시들이 골머리를 앓고 있는 사회·환경적 고충에 대해 살펴본다."남다른 해결책"은 수 백년 전 몽골 유목민의 전통가옥인 게르에서 영감을 얻었다. 게르는 독특한 집단적 특성을 지닌다. 우리는 이 특성을 활용해 일반적인 도시문제에 도전장을 내민다. 본 프로젝트는 "일반적인 문제"를 매주 제시하고 게르를 찾은 관객들은 벽면에 혁신적인 아이디어를 기재할 수 있다. 폐막 무렵이 되면 게르는 모두를 위한 담론의 장이자 남다른 아이디어의 집합소로 자리매김할 것이다.

ULAANBAATAR, MONGOLIA
Uncommon Solutions to Common Problems

GerHub
C39-5F

As city dwellers, many of us experience similar urban problems such as air pollution, traffic congestion, and the loss of urban spaces. We believe that complex issues such as these can only be solved through collective and uncommon actions. Our exhibit explores the idea of providing uncommon solutions to common problems. "Common Problems" are shown through video series of the documentary, <u>Nomad Meets the City</u>. It elaborates on the social and environmental issues of Ulaanbaatar which is common in many other cities. "Uncommon Solutions" are shown in the Ger, a portable nomadic dwelling made with wood and felt covers, which is to be used as a canvas for idea submission and collaborative actions. Ger has been a collective gathering space for the nomadic people for centuries. As it has a unique collective nature, we aim to incorporate this and experiment how we can collect and convey public ideas to solve common urban issues. Throughout the exhibition, visitors are encouraged to provide their innovative ideas inside the walls of the Ger. A common problem would be provided each week and by the time the biennale ends, the Ger would house the collection of collaborative public ideas.

Density

C 39 5 F

도시리빙

Uncommon Solutions

울산, 대한민국
회복력 있는 도시를 위한 도시와 산업의 공생

조상용, 니얼 커크우드
C39-4F

대한민국 산업 지형은 지난 50년간 변화해 왔다. 대한민국은 농업 국가에서 전자, 반도체, 석유화학, 그리고 조선, 자동차 등의 중공업을 아우르는 제조업 강국으로 변모했다. 그러나 기후 변화, 환경 파괴, 인구 고령화가 현실로 다가오고 세계 경제 역시 불안정한 상황에서, 오늘날의 대한민국 산업 도시는 미래 성장을 위해 회복력 있는 사고 방식을 받아들여야 한다. 아울러, 세계 경제가 불황을 겪으며 전세계 경쟁이 격화되면서 한국의 다른 산업 도시들처럼 울산도 산업 시설이 빠져나가는 모습이 가시적으로 드러나고 있다. 어려운 경제 상황 속에서 지속가능성을 유지하는 만병통치약 같은 해법은 없지만, 지속가능한 개발을 위해 창의적인 방안을 모색할 필요가 있다. 울산은 순환경제를 통해 도시의 회복력을 높이고자 도시의 산업, 시민, 생태, 환경을 아우르는 협력 프로젝트를 '집합적'접근 방식으로 추진하고 있다. 본 전시는 울산의 지리, 거버넌스, 제도적 역량, 상품의 생산 흐름을 조명하고, 폐기물을 에너지로 전환하는 프로그램 개발을 위한 정부, 민간, 학계, 다자 기구간 파트너십 기반의 다부문 협력체계 구축 노력을 들여다보며, 상황에 적절하게 대응하는 거버넌스, 도시 계획·설계에 대한 포괄적인 접근을 통해 도시와 산업의 공생을 사회적 맥락 안에서 추구하고자 한다.

ULSAN, SOUTH KOREA
Co-opting Urban Industrial Symbiosis for Urban Resilience

Sang-Yong Cho, Niall Kirkwood
C39-4F

The landscapes of industry have been changing throughout the last fifty years in Korea. Industrial activity has transformed from agriculture to the agglomeration of manufacturing industries that comprised of electronic, semi-conductors, petrochemical and heavy industries such as shipbuilding and automobile manufacturing. Yet climate change, environmental degradation, ageing population, and unstable global economic forces require Korean industrial cities to engage in resilient attitudes to their future growth. Furthermore, like many other industrial cities in Korea, Ulsan has been showing visible signs of de-industrialization processes as a response to global economic decline and increasing pressure from overseas competition. While there are no 'one-size-fits-all' approaches to maintaining sustainability during a slowing economy, creative approaches in advocacy for sustainable development should be sought out. The City of Ulsan has demonstrated through a 'collective' approach to implement collaborative initiatives with its industry, people, ecology, and the environment to promote urban resiliency through the practice of a circular economy.

C39

4F

credits: Niall Kirkwood

우한, 중국
중국의 집합적 형태
샘 저코비, 징루(사이언) 청
C39-4F

중국의 집합적 형태는 사회적 프로젝트,
공간, 현실이 중국의 도시 설계와 계획을
이해하는 데 필수적인 세 가지 맥락을 어떻게
구성하는가에 대한 연구를 담고 있다. 세 가지
맥락이란 바로 (1) 과거의 집합적 형태·공간의
역사와 현대 공동체 건설 계획 간의 관계,
(2) 도시·농촌 개발의 사회-공간적 변화,
그리고 (3) 정부의 성격이다.
본 작품은 정부가 공간 설계를 수단으로
활용하여 집합적 형태, 집합적 공간, 집합적
주관성, 혹은 오늘날의 맥락에서는 공동체
자체를 구성하는 현상을 탐구한다. 과거
인민공사, 노동단위, 현대 도시 공동체의
사례를 살펴보면, 도시 구성원을 구체적인
사회활동과 관심사, 혜택을 공유하는 집단으로
아우르는 작업이 사회-정치적으로, 그리고
경제적으로 필요하다는 사실을 알 수 있다.
이러한 노력으로 사람들이 서로를 지원하고
돌보는 새로운 사회적 네트워크가 만들어진다.
마오쩌둥 시대에 싹튼 인구 구조의 근본적인
변화로 인해 진행되고 있는 중국 사회 집단의
변화, 그리고 이들의 요구와 가치관은 세계
여러 지역 도시 계획의 밑바탕이 되는 핵가족,
주거 단위, 또한 자유주의적 사회운영이라는
관점으로는 설명할 수 없다. 대중, 공공 공간,
혹은 공간의 구성이라는 관념에 기반한
서구중심의 도시 이론과 실천은 중국과는 다른
사회-공간적 역사와 난관 — 개념적으로나
실제적으로나 대개는 비생산적인 — 이라는
맥락 속에서 탄생한 것이기 때문이다.

WUHAN, CHINA
Collective Forms in China
Sam Jacoby and Jingru (Cyan) Cheng
C39-4F

Collective Forms in China is a study
of how social projects, spaces, and
realities shape three contexts critical
to understanding urban design and
planning in China: the history of collective
forms and spaces in relationship to
current community building agendas,
socio-spatial changes in urban and
rural development, and modes of
governmentality.
The study investigates the
instrumentalisation of spatial design
by government to shape collective
forms, collective spaces, and collective
subjectivities, or, in today's context, the
building of communities. Examining
both historical and contemporary case
studies — from the people's commune
and work unit to new urban communities
— a socio-political and economic need
for urban constituencies brought together
by concrete, shared social activities,
interests, and benefits emerge, in
order to provide new social networks of
support and care. Driven by fundamental
demographic changes rooted in the
Maoist era, the changing social groups
and their needs and values continue to
defy traditional beliefs in the nuclear
family, neighbourhood unit, and liberal
models of governance that underpin
much of global urban planning. Western-
centric urban theories and practices
based on notions of the public, public
space, or place-making are in a context
of different socio-spatial histories and
challenges largely unproductive, whether
conceptually or practically.

Participatory Urbanism

4F

C39

참여형 어버니즘

The service centre for party members
and the masses, Geguang Community,
Wuhan, Hubei Province, P. R. China

자카르타, 인도네시아
보이지 않는 집합성: 내면의 개혁 + 유동성의 결합

모하마드 카흐요 노비안토, 율리 칼슨 사갈라 (큐레이터), 아니사 드야하 라쥬아리니, 이네사 퍼르나마 사리 (어시스턴트 큐레이터)
C19-2F

인볼루션은 내면을 의미하는 "Inner"와 개혁을 뜻하는 "Revolution"의 합성어다. "도시 인볼루션"이란 인구는 증가하는 반면 도시의 공공시설과 제반여건은 이에 미치지 못하거나, 반대로 물리적 환경은 증가하는 반면 인구 성장률이 더딘 경우를 뜻한다.

컨플루이디티는 융합, 유동성을 뜻하는 "confluence"와 (특히, 물과 연관된) 유동성, 불안정성을 뜻하는 "fluidity"의 합성어다. 불안정성은 "불안정한 시각"에서 접근할 필요가 있다. 따라서 식수와 관련된 도시문제를 일컫는 "도시 컨플루이디티" 역시 유동적이고 불안정한 관점에서 대응할 필요가 있다.

카말 무아라는 자카르타 북부 해안지역에 위치한 다섯 도시 중 하나이다. 우리는 인볼루션과 컨플루이디티를 기반으로 자카르타 주민들이 지금껏 목도해 온 것과는 매우 다른 형태의 집합도시를 선보이고자 하였다. 이는 기존의 관점, 혹은 대지에 기반한 부동성의 패러다임에서는 포착 불가하기 때문이다.

JAKARTA, INDONESIA
Unseen Collectivity: Involution + Confluidity

Mohammad Cahyo Novianto, Yuli Kalson Sagala (Curator), Annisa Dyah Lazuardini, Inesa Purnama Sarii (Assistant Curator)
C19-2F

'Involution', can be ascribed to a combination of two words: 'Inner + Revolution', or it can be 'Inner + Evolution'. 'Urban Involution' is a condition when accelerating population growth is not accompanied by the growth of infrastructure and public facilities in the city. Or infrastructure and public facilities are increasing, but the increase is far less rapid compared to the extraordinary rate of population growth. 'Confluidity', the compound word of 'Confluence' (meeting) and 'Fluidity' (liquid state, instability), as a meeting between unstable conditions (especially those related to water). Unstable conditions that should also be responded to from the "unstable" point of view, too, not with the eyes of stability. 'Urban Confluidity', is a situation when a city eventually becomes a meeting of various problems related to water, which should be solved by "consciousness of fluidity", not "consciousness of stability".
Kamal Muara Village is one of five urban villages in the coastal area of North Jakarta. Through framing of 'involution' along with 'confluidity', we tried to offer a reading about the collective city which so far might have escaped the general observations of Jakarta residents, because this collectivity is not visible if we see it from the frame of a 'consciousness of stability' / land-based paradigm.

로테르담, 네덜란드
영구적인 일시성을 지닌 도시

ZUS [Zones Urbaines Sensibles]
C19-2F

지난 수십년 간, 로테르담 시는 신자유주의를
표방한 도시임을 몸소 증명해왔다. 투기,
공허, 타락과 재생을 거듭해온 것이다. 그러나
도시는 추진력으로 무장한 실행가들의
광범위하고도 뜻밖의 개입을 통해 새롭게
거듭나기 시작했고 색다른 도시 모델로서의
가능성을 내비쳤다. 예를 들어 시민과
활동가들이 공유지의 가치를 직접 결정하고
설계한다. 단기간의 프로젝트가 연속적으로
진행되어 도시에 영구적인 영향력을 끼치게
된다. 이를 통해 정부, 시민, 도시개발자의
역할이 재정의된다.

영구적인 일시성을 지닌 도시 프로젝트는
공적인 관점, 즉 로테르담 시를 중심으로
도시에 접근하며, 실제 사례를 토대로 설계와
개발 과정을 설명한다. 본 프로젝트는
18년이라는 장기간 동안 도시에 직접
현실적으로 개입하는 방식으로 진행되었으며
스히블록 보행교와 루흐트진겔 육교
프로젝트와 연계될 뿐 아니라 그간의 발자취와
미래의 청사진도 다루었다. 미완성 프로젝트는
도시에 깊이 뿌리내린 유산을 찾아가는 노력을
담았으며 이를 통해 지역사회가 고층건물과
다국적기업들과 공생하는 방법을 모색하고자
하였다. 아울러 고급주택가의 진입에 따른
기존주민의 이탈을 다룬 젠트리피케이션과
도시 단조로움이라는 난제를 넘어서는
역동적인 청사진을 제시한다.

ROTTERDAM, NETHERLANDS
City of Permanent Temporality

ZUS [Zones Urbaines Sensibles]
C19-2F

In recent decades, Rotterdam Central
District has proved to be a striking
illustration of urban development in the
neo-liberal city. Speculation, vacancy,
degradation, and revival have followed
one another in rapid succession.
But through a resurgence of urban
interventions by broad and sometimes
unexpected coalitions of actors, it has
been demonstrated that another urban
model is possible: a model in which
citizens and entrepreneurs have been able
to determine and design public values.
A model in which temporary projects
can permanently influence the course of
events. A model in which the division of
roles between government, citizens, and
developers are redefined.

City of Permanent Temporality
investigates, from the perspective of
the city's public capital, the design and
development of the city through the lens
of an actual case study: the Rotterdam
Central District. The project involves
eighteen years of intense intervention by
ZUS into the urban reality of this part
of the city, through projects such as
Schieblock and the Luchtsingel, and also
emphatically examines its history and
possible future. Incomplete & Unfinished
is a search for a rooted commons within
which the original urban fabric and a
local community coexist with high-rise
buildings and multinationals. Incomplete
& Unfinished is a living manifesto for a
dynamic blueprint that moves beyond
gentrification and monotony.

Luchtsingel Top Roundabout ZUS Ossip
van Duivenbode

타이후 호, 중국
타이후(太湖)호 XL
Circular Metropolis
C19-1F

타이후호 XL은 중국 항저우 타이후호(太湖) 주변 '원형 대도시'를 구상한 작품이다.. 타이후 호수는 인접한 거대도시 상하이의 대칭점으로 이해할 수 있다.

호수 주변의 7개 주요 도시 인구를 모두 합하면 상하이 인구와 맞먹으며, 이들 도시는 서로 빠르게 연결되어 네트워크를 구성해 나가고 있다. 본 작품은 세 부분으로 구성된다

1. 양쯔강 삼각주의 정경을 담은 영상. 관람자들은 본 작품의 배경을 이해할 수 있다.
2. 2050년을 바라보는 도시화를 소개하는 설계도. 관람자들이 앉아서 쉴 수 있는 양탄자 형태의 설계도이다.
3. 미래 도시 형성의 바탕이 되는 설계와 자원 구조를 소개하는 가상 의도시 모형.

본 전시에서 소개하는 원형 도시 모델을 통해 두 가지 새로운 연구가 가능하게 된다.

1. 새로운 도시 계획 모델
원형 대도시는 선형 도시와 네트워크 도시 사이의 흔치 않은 중간단계이다. 호수를 둘러싸는 원형의 도시에서 미래 성장 모델이 구체화된다.
2. 새로운 자원 모델
원형 대도시는 기존의 대지 사용 최적화 개념에서 벗어나 새로운 형태의 공유와 조정 모델을 제시한다. 또한, 전체적인 자원의 균형을 확보하기 위해 게임 기반 플랫폼을 활용한다.

TAIHU LAKE, CHINA
Taihu XL
Circular Metropolis
C19-1F

Taihu XL is a speculative proposal for a 'Circular Metropolis' around Lake Taihu. The site is a unique natural counterpoint to the global megacity Shanghai immediately adjacent. Around it seven major cities with a population comparable to Shanghai are rapidly being networked to each other. The installation is comprised of three parts:

1. A video projection that maps a journey through the Yangtze River Delta landscape that familiarises viewers with the context of the project.
2. A scaled urban plan that demonstrates a pattern of urbanisation towards 2050. This vision takes the form of a carpet that viewers can relax on.
3. A virtual data cityscape that demonstrates the underlying design and resource structures that support the formation of the city.

This emerging ring configuration presents two opportunities for emerging research:

1. A NEW PLANNING MODEL
The circular metropolis presents a rare intermediary stage between a linear and network city, a state of liminality in which future growth is consolidated around the urban ribbon enveloping the lake.
2. A NEW RESOURCE MODEL
By folding the city into the condition of a liminal ring, the proposal challenges the concept of land use optimization, and explores models of sharing and mediation needed to achieve it using a game-based platform through which a holistic resource balance can be approached.

토론토, 캐나다

수퍼 스트리트: 토론토의 "교외 스프롤현상" 에서 나타나는 집합도시 형태

마이클 파이퍼, 로베르토 다미아니, 폴 헤스
C19-1F

토론토의 도시화는 사방으로 확산되는 형태를 띠었지만, 사실 겉보기만큼 무질서하지는 않다. 다른 북미 도시들이 그렇듯, (역주: 영국의) 식민지 시기에 토론토에 형성된 반듯한 격자 구조가 무한하고 무정형적으로 확장되는 듯한 도시를 탄생시켰다. 개인 소유의 목적성 있는 낮은 건물들로 형성된 2차 대전 이후 토론토의 교외 지역이 '무질서하게 뻗어나간' 것처럼 보이는 반면, 토론토 주변지역은 (역주: 반듯한 격자 구조에 바탕을 둔) 식민지 시대 개인 부동산 소유 시스템의 논리에서 벗어나 사람의 형상을 한(figural) 도로들로 이루어져 있다. 이 균형 잡힌 도로들은 격자 모양의 도시를 가로질러, 고립된 것처럼 보였을지도 모르는 작은 도시 구역들을 한데 묶어준다. 토론토의 '수퍼 스트리트')는 북미 여러 도시 교외 지역의 특징이라고 할 수 있는 개인 토지 소유 문화 속에서 집합성을 창조해내는 곳이다.
본 작품은 토론토와 그 주변 지역을 담은 커다란 이미지를 통해 '수퍼 스트리트'와 식민지 시대에 형성된 격자형 구조 사이의 관계를 부각시킨다. 아울러, 사람 형상을 한 토론토 도로의 캐리커처를 표현한 여러 개의 인형을 비롯해, 도시의 확산에 따라 확장되는 개인의 공간 안에서 집합성의 가능성을 제시하는 사진, 도표, 설계 계획안이 수록된 서적도 전시한다.

TORONTO, CANADA

Super Streets: Emerging Collective Forms Within Toronto's "Suburban Sprawl"

Michael Piper, Roberto Damiani, Paul Hess
C19-1F

Dispersed urbanization in Toronto is not as shapeless as it seems. As with many cities in North America, the city's original, colonial property grid structures a seemingly endless and amorphous urban expansion. And while this landscape of low-slung, privately owned, object buildings makes Toronto's postwar suburbs seem "sprawled," the urban periphery here is structured by a series of figural roads that defy the logic of its colonial private property system. These shapely roads cross the grid and bring together neighborhood subdivisions that would otherwise seem isolated. Toronto's Super Streets produce a sense of collectivity within the cultures of private land-ownership that has come to define much of the North American suburban landscapes.
Our exhibition includes a very big drawing of the Greater Toronto Area that foregrounds the relationship between its Super Streets and the region's colonial property grid system. We will also show a series of stuffed figurines that provide caricatures of these figural roads, and a book which includes photographs, diagrams and design proposals about the possibility of collectivity within the private spaces of urban dispersal.

웰링턴, 뉴질랜드

전원에서 도시경관으로: 웰링턴 하타이타이의 도시화

샘 케벨, 김동세
C19-1F

본 작품은 웰링던의 교외지역인 하타이타이가 생산적 농장에서 집합적 도시 공간으로 서서히 변화하는 과정에 기여하기 위한 목표로, 실행되거나 예측적인 일련의 프로젝트들로 구성된다. 하타이타이 지역은 웰링턴 항에서 200미터 솟아오른 언덕의 동쪽에 위치한다. 1840년대 영국의 식민 도시 계획 전문가들이 '도시 외부' 지역이라 선언한 이곳은 20세기까지도 목축지로 사용되었다. 20세기가 한참 흘러 도시 접근성이 향상되고 나서야 주거지로 개발되었다. 이 지역의 복잡한 지형을 따라 도로가 등고선처럼 형성되었고, 그 결과 보행자 보다는 차량 친화적인 도로망이 되었다. 하타이타이는 웰링턴 중심가로부터 겨우 2킬로미터 떨어져 있지만 36%의 통근자가 차로 출퇴근하고, 평균 인구밀도는 헥타르 당 33명에 불과하다. 인구 밀도 증가와 보행자 친화적 공간 구성에 대한 요구가 높아지는 한편, 지역 공동체 차원의 활동과 공공 공간에 대한 관심도 다시 높아지고 있다. 본 작품에서 소개하는 각각의 프로젝트는 하타이타이가, 인류의 발길이 지구상에서 가장 늦게 닿은 지역 아오테아로아(뉴질랜드의 마오리어 이름)의 수도인 웰링턴으로 편입되는 과정에서 보다 보행자 친화적인 인프라 시설과 보다 많은 공공 공간을 갖춘 곳으로 발전할 수 있는 가능성을 모색한다.

WELLINGTON, NEW ZEALAND

Countryside to Cityscape: The Slow Urbanisation of a Wellington Hillside

Sam Kebbell, Dongsei Kim
C19-1F

This exhibit presents a series of projects, both commissioned and speculative, that contribute to the slow evolution of a Wellington suburb, Hataitai, from productive farm to collective urban environment. Hataitai is on the eastern side of a 200m high hill which rises out of Wellington Harbour. Colonial city planners in the 1840s declared it beyond the 'town belt' and it was used for grazing stock well into the twentieth century when improved access opened the way to residential development. Roads were formed along the contours of its complex terrain and the result is a road network that works better for cars than pedestrians. Hataitai is only 2km from the city centre, yet 36% of commuters travel by car and it has an average density of only 33 people / hectare. The increasing pressure to both intensify and pedestrianise comes with a renewed enthusiasm for community activities and public space. Each project in the exhibit explores the possibilities of an expanded pedestrian infrastructure and more public space which contributes to this shift from countryside to cityscape, in the capital city of the most recently inhabited landmass on earth, Aotearoa New Zealand.

암스테르담, 네덜란드
우리...도시

암스테르담 시 국제사무소,
도시계획·지속가능성부
C20-2F

암스테르담은 오랜 도시계획의 전통이
있는 도시이다. 암스테르담은 이번
서울도시건축비엔날레에서 '집합도시 구축'에
관한 생각을 공유하여 관람자의 상상력을
자극하고자 한다. '집합도시 구축'이란 디자인,
문화, 협력, 탐구, 스토리텔링으로 이뤄지는
탐구 작업을 뜻하며, 이러한 연구가 현재
진행중인 곳이 바로 바위크슬로테르함이다.
버려진 과거 산업단지가 포용성을 갖춘 현대적
도시경관으로 변모하고 있다.
바위크슬로테르함 지역은 암스테르담이
과거 도시들의 하향식 구조에서 탈피하려는
노력을 잘 보여준다. 지역의 새로운 공동체,
선구자들, 스스로 집 짓는 사람들, 새로 유입된
주민들은 시 당국과 함께 새로운 도시 유형과
설계 방식을 실험하고 있으며 암스테르담
시는 바위크슬로테르함 지역의 새로운 마스터
플랜을 구상 중에 있다. 산업사회의 유산,
시민의 집합성, 행동주의, 참여, 실험을 통해
형성된 억센 공간적 특성이라는 DNA를
바탕으로 더 나은 도시를 위해 노력하고 있다.
이번 서울도시건축비엔날레는 관람객들에게
암스테르담 시가 제공하는 도시의 선구적
프로젝트에 관한 정보, 인터뷰 영상,
애니메이션 자료와 더불어 체험 기회를
제공함으로써 도시의 설계, 밀도, 다목적
용도, 공동 창조, 순환 경제, 지속가능한
이동성, 그리고 생산적인 도시 모습에 대한
암스테르담의 원대한 꿈을 들여다보도록
한다. 우리...도시 전시는 우리에게 도시
환경과 특징에 관심을 갖고, 더 나아가 우리가
살아가는 공간에 대해 고민하고 반성적 사고를
갖는 창의적인 도시민이야말로 창의적 도시
설계의 출발점이라는 사실을 깨닫게 한다.

AMSTERDAM, NETHERLAND
We...City

International Office and Department
for Planning and Sustainability, City of
Amsterdam
C20-2F

Amsterdam is a city with a strong
tradition in urban planning. For the
Seoul Biennale of Architecture and
Urbanism, Amsterdam wants to trigger
the imagination by sharing ideas about
collective city making. Collective city
making means research by design,
research by culture and collaboration,
exploration and storytelling. This research
is shown through an ongoing project:
the continuous development of the
Buiksloterham, where former industrial
brownfields are being transformed into a
modern inclusive cityscape.
In this area the city tries to break with its
strong tradition of a more formal system
of top-down planning. New collectives,
pioneers, self-builders and new arrivals
together with city-officials experiment
with new urban typologies and design
methods. The city is currently working on
a new masterplan for the Buiksloterham,
where it seeks to build upon this
DNA of rough spatial quality through
industrial heritage, collectivity, activism,
participation, and experiment.
Through project-information on
pioneer projects, filmed interviews,
animations and an interactive element
for the biennale-visitor, Amsterdam
gives insight into its current ambitions
on urban design, density and mixed-
use, co-creation, circular economy,
sustainable mobility and productive
urban landscapes. The We...City entry
aspires to tickle the visitor, to show that
innovative urban design approaches start
with creative urbanites who care about
the city's environment and its diverse
characteristics, and thus, to think and
reflect on their own habitat.

Layers and Collective Memory

2F

C20

집합적인 도시기억과 레이어

Aerial picture of sustainable floating
village 'Schoonschip' © David de Bruijn,
May 2019

하노이, 베트남
도시와 기억

응우옌 태 손
C20-1F

대도시 하노이는 여러 작은 구역들로
이루어졌다. 도시 위에 또 다른 도시가
지어지듯, 새로운 구역이 옛 구역을 덮는다.
옛 구역의 도로, 거리, 시장도 새로운 이름으로
바뀌면서 그곳에 간직된 추억과 지역은 과거로
사라져 버린다. 도시에 대한 기억은 시간과
끊임없이 이 도시로 몰려드는 사람들의 물결과
함께 소멸해 간다.

어쩌면 도시는 기억 상실의 공간일 수도 있다.
아니, 도시의 기억 같은 건 애초에 존재하지
않을지도 모른다. 전쟁과 같은 특수한 사태가
없다면, 도시에 대한 기억을 파괴하는 주범은
바로 비인간적인 도시 계획이다. 도시는 점점
비대해지고, 도로는 넓어지며, 보행자들이
걷는 인도는 서로 멀어져간다. 비인간적 도시
계획이 늘어가면서 사람들 사이의 거리도
멀어진 것이다.

어쩌면 우리에게 필요한 것은 도시의 '인간성
지수'일지도 모른다. 시민들이 하루에 몇 보를
걷는지, 사람들이 가까이 모여 교류할 수 있는
시원한 그늘이 공공장소에 얼마나 있는지
등으로 측정할 수 있을 것이다.

한때 하노이는 '인간성 지수'가 매우 높은
도시였으나 최근의 도시 계획은 전혀 다른
방향으로 가고 있다. 영혼 없는 콘크리트
건물이 들어서며 하노이의 높은 '인간성 지수',
도시의 기억을 간직한 공간은 과거의 산물이
되어 가고 있다.

도시와 기억은 시각 예술을 통해 도시의
과거를 발견하고 지켜 가기 위한 노력의
산물이다. 이번 전시를 통해 관람자들이
가까운 미래 도시에서 기억이 갖는 가치를
생각해 보는 계기가 되었으면 한다.

HANOI, VIETNAM
City and Memory

Nguyen The Son
C20-1F

Hanoi is a big city in which there are
small urban areas, it's like city upon city,
the latest covering up the earliest, the
urban areas are vanishing into the past,
vanishing with nostalgia along with the
roads, streets, and markets that already
have new names. Memories of the city
keep vanishing with time, with the flood
of people immigrating tirelessly into this
city.

A city might lose its memory or perhaps
it doesn't have any memory at all. If it is
not because of the war, it mostly comes
from the inhumane urban planning which
has led to the breakdown of memory. The
cities are bigger and bigger, the roads are
wider and wider and the pavements are
further and further from each other. There
are more and more inhumane planning
ratios and they have pushed people apart
from one other.

Perhaps, we should measure the
humanity ratio of the city by the steps of
people, by the shadows in public spaces
that make people move closer to each
other and have more interaction with one
another.

Hanoi once had perfect ratios of a
humane urban life like that, but in recent
years, urban planning is moving in a
different direction that creates more and
more soulless blocks of concrete and it
is breaking the humanity ratio that we
previously had. Moreover, it is breaking
the memory spaces of the city.

This exhibition is a collection of efforts
in the journey to seek and save pieces
and pieces of our city memory through
my visual art projects, with the purpose
of stimulating the audience to raise
questions about the value of the memory
of the imaginary city in our near future.

아바나의 이미지

메건 피터스

C20-1F

도시 #1 심미적 이미지

고풍스러운 자동차, 무너져 가는 건물, 굵은 시가, 짙은 빛깔의 럼주, 그리고 독재자. 아마 외부인들이 아바나라는 도시에 끌리게 되는 건 이런 이미지 덕분일 것이다. 도시 #1 심미적 이미지에서는 외부의 시선, 미학적 시각, 거리의 효과를 분석하고, 그 과정에서 사진의 역할을 살펴본다. 환상적인 사진 이미지들은 아바나의 현실을 넘어선 또 하나의 현실을 보여준다.

도시 #2 구체적 이미지

아바나의 이미지에는 애국주의, 선전·선동, 자기 표현, 문화적 정체성에 관한 복잡한 이야기가 녹아 있다. 이렇게 강력한 이야기를 접한 사람들은 상반된 반응을 보인다 — 쿠바를 우러러보거나, 쿠바의 현실에 보다 큰 의문을 품거나. 정부가 아무리 표현의 자유를 억압하고 검열하려 해도 쿠바 고유의 이미지는 끈질기게 살아남았고, 그 이미지를 더 효과적으로 발전시키기 위한 방안이 모색되고 있다. 이러한 현실은 아바나가 도시 스스로 정체성을 창조하고 그 실현을 위해 적극적인 노력을 기울이고 있다는 점을 잘 보여준다.

건축, 그리고 건축에 대한 우리의 이해는 심미적 경험을 통한 시각과 지식 정보를 바탕으로 한다. 건축가의 관점에서 제작된 이번 전시는 도시 재현 작업이 그저 수동적인 반영에 그치지 않는다는 사실을 보여주며, 도시의 재현과 도시의 새로운 모습을 적극적으로 창조하기 위한 노력 간의 상관관계를 제시한다.

The Image of Havana

Megan Peters

C20-1F

CITY #1 THE AESTHETIC IMAGE

Classic cars, crumbling architecture, fat cigars, dark rum, and despotic leaders. These are the images that Havana evokes to an outsider. Analyzing the effects of the external gaze, aesthetics, distance, and the role of photography, these images reveal a city that has become more real in its fantastical image than in its reality.

CITY #2 THE EMBODIED IMAGE

The images found within Havana tell complex stories of patriotism, propaganda, self-expression, and cultural identity. They are powerful in ways that either command their audience into deference or empower the questioning of the status quo. Try as a government might to muffle and censor free speech, Cuban-authored images persist and find a way to prevail using calculated and clever means, revealing a city that is actively pursuing an identity of its own creation.

Architecture, and our understanding of it, is deeply rooted in the visual and the information we attain from the aesthetic experience. Using the eye of an architect, this work begins to understand how representations are complicit in not only passively representing a city but also in actively creating one.

Layers and Collective Memory

1F

C20

집합적인 도시기억과 레이어

VIVA LA RESOLUTION!

상트페테르부르크, 러시아
상트페테르부르크를 통해 보는 구소련 도시의 다층 구조

MLA+

C20-1F

본 작품은 구소련 지역의 전형적인 도시 중에서도 상트페테르부르크의 단면을 살펴본다. 구소련 지역의 도시들은 모두 비슷한 개발 과정을 거쳐 그 도시들만의 특이한 다층 구조를 갖는다.

이러한 다층형 구조에 대한 연구는 도시의 가치를 높이고 단점을 보완하기 위해 꼭 필요한 과정이며, 우리는 또 다른 층을 만드는 것이 아니라 현재의 도시를 개선해야 한다.

본 작품은 상트페테르부르크를 구성하는 여덟 개의 층을 소개하고 도시 밀도를 증가시킬 수 있는 잠재 요소를 보여준다. 보다 적극적인 도시 개발 계획을 요구하는 목소리가 높은 것은 사실이지만, '철거 후 재건축'이 아닌 대안적 개발 모델을 고려할 필요가 있다.

대안적 개발 모델은 도시를 구성하는 각각의 층이 가지는 맥락과 특성을 반영한 다양한 실험프로젝트를 통해 검증된다. 우리가 진행한 도시설계 연구 결과에 따르면, (역주: 추가적인 개발이 없어도 도시는) 현재의 도시 영역에서 적어도 향후 20년 동안 예상되는 모든 성장을 감당할 수 있다.

우리는 본 작품을 통해 도시 내 미개발지역에 자원을 퍼붓는 것보다 기존 도시 구역 개선을 위해 투자하는 편이 훨씬 큰 혜택이 있다는 점을 역설한다. 보다 많은 사람들의 생활이 향상되고, 전체적인 도시구조가 개선될 것이다.

ST. PETERSBURG, RUSSIA
Undiscovered St. Petersburg: Layers of Post-Soviet City

MLA+

C20-1F

This exhibition is a cut through layers of a classical post-Soviet city, in this case St. Petersburg. Post-Soviet cities went through very similar patterns of development, leaving layers of specific built form behind.

Today we see that studying their layers is a necessary step to fix values and correct shortcomings. At the present time we do not want to create another layer, but instead improve the city we already have. The exhibition looks into eight different layers and illustrates potential for densification. We understand that there is demand for more program in the city, but we believe that instead of demolition and new construction alternative development models should be thought of.

Alternative development models are tested through experimental projects that are responding to the context and special character of each layer. Our research by design showed that the city of St. Petersburg can accommodate all its projected growth within existing territories at least for the next twenty years.

We want to show that it is very beneficial to invest in the existing city instead of pushing resources into greenfields. More people benefit and the whole urban fabric gets an upgrade.

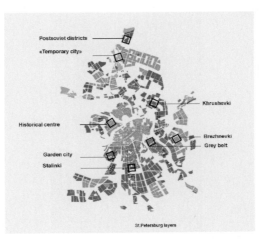

St.Petersburg layers

잔지바르, 탄자니아

응암보: 잔지바르 타운의 새로운 중심지

잔지바르 토지위원회, 아프리칸 아키텍처 매터스
C20-2F

탄자니아 잔지바르 주의 수도 잔지바르 타운은 2000년 유네스코 세계유산으로 지정된 훌륭한 건축 유산인 스톤타운으로 잘 알려져 있다. 그러나 스톤타운 동쪽에 위치한 응암보('반대편')라는 이름의 지역은 같은 잔지바르 타운에 있으면서도 과거에는 개울로, 요즘에는 붐비는 도로로 스톤타운과는 사실상 분리되었으며 이 지역을 아는 사람도 그다지 많지 않다. 2015~16년 잔지바르 주정부는 암스테르담 시와 비영리 단체인 아프리칸 아키텍처 매터스와 함께 응암보 지역의 지도를 제작했고, 이로써 지역의 풍부한 역사와 문화가 빛을 볼 수 있었다.
응암보 지역 발전 계획의 일환으로 진행된 지도제작 프로젝트는 유네스코의 역사적 도시 경관 프로젝트를 기반으로 한다. 2015년 수립된 잔지바르 구조적 개선 계획에 이어, 2017년에도 주 정부는 응암보 지역 발전 계획을 응암보의 재도약을 위한 핵심 전략으로 채택하고 이 지역을 미래 잔지바르 타운의 도심으로 지정했다.
본 작품은 잔지바르와 암스테르담의 집합적 실천, 응암보의 형성과 역사 이야기를 비롯해 이 지역을 구성하는 사회와 문화를 소개한다. 한때는 잔지바르 타운의 역동적인 문화 중심지였으며 '행복한 거리'라는 별명이 붙었던 응암보 지역은 스톤타운 만큼이나 역사적 의미가 있는 곳이다. 응암보 지역에 대한 지식을 보다 많이 연구하여 이곳을 새로운 도심으로 만들기 위한 미래 전략의 기초를 다질 수 있을 것이다.

ZANZIBAR, TANZANIA

The Ng'ambo Atlas and the New City Centre of Zanzibar Town

Zanzibar Commission of Lands, African Architecture Matters
C20-2F

Zanzibar Town, the capital city of Zanzibar, is known to many through the outstanding architectural heritage of Stone Town, a UNESCO World Heritage Site since the year 2000. The neighbourhoods commonly known as Ng'ambo, the 'other side', located east of Stone Town and separated historically by a creek and nowadays a busy road, are far less known parts of the city. In 2015–2016 the Government of Zanzibar carried out a mapping of Ng'ambo in collaboration with the City of Amsterdam and African Architecture Matters, which brought to the fore the historical and cultural richness of the area.
The mapping was part of the development of the Local Area Plan for Ng'ambo, based on the UNESCO Historic Urban Landscape approach. In 2017 the Government of Zanzibar adopted the Local Area Plan as the leading document in the redevelopment of the area, designated as the future city centre of Zanzibar Town, following the Zanzibar Structural Plan 2015.
This presentation gives insight into the collective action of Zanzibar and Amsterdam, the history and morphology of Ng'ambo, as well as its social and cultural make-up. Once a culturally vibrant heart of Zanzibar Town, known by some as 'the happy streets', Ng'ambo has proven to be as historically as significant as Stone Town. Knowledge that is now considered the basis for the future plans for the area, the new city centre.

공유지의 붕괴, 거리 101

ZUS [Zones Urbaines Sensibles]
C20-2F

거리는 누구의 소유인가? 누가 거리를 공공의 혹은 개인의 소유라 규정하는가? 건축이 거리의 쓰임에 미치는 영향은 무엇인가? 거리는 지난 3천 년의 역사 속에서 어떻게 변모해왔는가?

기존의 거리가 새롭게 단장되고, 이상적인 거리에 대한 우리의 관념이 달라진다는 사실은 그 자체로 아직 거리가 변화 중임을 암시한다. 우리는 다양한 거리의 유형들을 흑백 선 드로잉으로 담아냈다. 흑색은 열린 공간, 백색은 사적인 공간을 나타내며 이를 통해 거리의 건축적 특성을 부각했다.

거리가 어느 정권하에 만들어지고 언제부터 기능하게 되었는지 등 정치적 맥락은 차치하고서라도 거리에 관한 한 우리는 두 가지 중요한 결론을 도출해볼 수 있다. 하나는 거리가 교통 및 상업적 요소를 지니고, 인근 건축양식에 오락적 요소가 가미되어 다양한 공간을 수직적 혹은 수평적으로 연결한다는 것이다.

둘째는 공유지가 붕괴하기 시작했다는 것이다. 쇼핑몰, 실내형 공공시설, 테마공원, 사유화된 공유지 등에서 살펴볼 수 있듯이 한때는 탁 트인 형태로 모두에게 열린 공간이었으나 이제는 실내형, 사적인 공간으로 탈바꿈한 곳들이 있다. 공공성과 규율을 중시하는 거리와 그렇지 않은 거리 사이의 간극은 더욱 벌어지고 있다.

EINDHOVEN, NETHERLANDS
101 Streets, the Implosion of the Public Domain

ZUS [Zones Urbaines Sensibles]
C20-2F

Who owns the street? What makes it public or private? What is the influence of the architecture of our streets on its use? How have streets developed over the last 3000 years?

The evolution of existing and utopian streets shows the fragmentation of the concept of the street and indicates that the street is still transforming. The portraits of the different street typologies are shown as black and white line drawings: black indicates (semi) public space and white private space. Through this, the architecture of the street is made explicit.

Apart from the political context in which the streets were made or under which regime they function, there are two important conclusions to be drawn. First of all it becomes clear that the simple concept of the street as a linear element for transport, commerce, and entertainment has become a multi-faceted construction which connects various spaces horizontally but also vertically.

Secondly, an implosion of the public domain can be seen. The public domain is gradually introverting, as we can see in the examples of shopping malls, gated communities, theme parks and privately owned public spaces. The differences between formal, regulated streets on the one hand and informal streets on the other become more extreme.

베이루트, 레바논
창의적 집단

건축·도시생활 연구 플랫폼 플라타우 [산드라
프렘(프로젝트 리더), 보우로스 도우아히, 사빈
디나, 마리안 부가바, 비키 사미아, 린 타니르]
C21-1F

레바논 국민들의 공공영역과 집단 행동은
레바논 내전(1975~1990) 종료 이래
정치 양극화로 거의 사라져 버렸다. 수도
베이루트에는 모든 시민을 위한 공간이 턱없이
부족했고 그나마 남아 있는 공간 마저도
지나친 통제, 제한적 접근, 그리고 무관심으로
그 의미를 잃어갔다. 하지만 21세기 초반
이후 정보화 시대가 도래하고 개인이 기술을
활용하기 용이해지면서 과거에는 불가능했던
방식의 집단 표현이 가능하게 되었으며, 이
새로운 목소리는 기존의 정치·종교적 틀에
얽매이기를 거부했다.

이러한 맥락에서 우리는 창의적 집단를 통해
베이루트 공동체가 거쳐온 공간의 역사를
살펴보고, 2000년부터 오늘에 이르기까지 이
도시에서 나타났던 정치를 비롯해 사회·경제
분야를 아우르는 선구적인 집단행동과 변화의
조짐을 전부 되짚어 보고자 한다.

아울러, 이번 프로젝트에서는 베이루트의
창의성과 기업가 정신을 확인할 수 있는
지도와 화면 자료를 통해 새롭게 떠오르는
집단 행동 양식을 들여다보고자 한다. 개인과
공동체 행동이 교차하는 곳에서 이러한 새로운
행동 양식은 천천히 그리나 꾸준히 베이루트의
기존 건축물에 반영되어 집단 경험의 대안을
만들어 나갔다. 이러한 집단 경험은 통제와
검열로 인해 공공 영역이 조각난 섬처럼
흩어져 버린 도시에서 너무나 소중한 것이다.

창의적 집단은 베이루트의 집단성에 대한
다양한 견해가 공유되는 네트워크, 즉
창의적인 집단이 깃드는 도시로 발전할 수
있는 미래를 그리고자 한다. 또한, 취약한
도시 구조를 창의적으로 재정의하고, 대화를
통해 개인 차원의 행동 양식과 집단 경험
간의 긍정적인 조화를 모색하는 열린 공간을
창출한다.

BEIRUT, LEBANON
Creative Collectives

platau | platform for architecture and
urbanism [Sandra Frem (Project Leader),
Boulos Douaihy, Sabine Dina, Marianne
Boughaba, Vicky Samia, Lynn Tannir]
C21-1F

Since the end of the civil war, Lebanon's
polarized politics cultivated a deliberate
vacuum of the public realm and collective
practices, evident in the lack of public
spaces in Beirut, and their neutralization
through excessive control, filtered access
and neglect. Yet with the rise of the
information age and personal access
to technology since the early 2000s,
previously inaccessible ways of collective
expressions became possible, that broke
away from established political and
confessional definitions.

In such context, Creative Collectives looks
at the spatial history of the collective
in Beirut, re-charting all initiatives and
manifestations that can be dubbed as
communal — from political to social and
economic — and their expressions in the
city's urban space from the year 2000
until the present.

Secondly, the project investigates
emerging forms of collectives in the city,
looking through mappings and footages
at Beirut's creative and entrepreneurial
clusters. Operating at the intersection
of private and communal initiatives,
such practices in Beirut have slowly but
steadily inhabited existing buildings
and provided alternatives for collective
experiences, mostly invaluable in a city
where the public realm is fragmented into
filtered archipelagos.

The project imagines a speculative future
where Beirut can be overlaid by a network
of nodes — Creative Collectives — that
hold specific criteria for collectivity: the
creative re-appropriation of vulnerable
urban fabric and open spaces to
host a positive negotiation between
conservation, individual modes of
practice, and collective experience.

브뤼셀, 벨기에
브뤼셀은 작업중: 생산적 도시 건설하기

브뤼셀 시 총괄 건축가
C21-2F

21세기 산업

C21

2F

Industry in the 21st Century

산업 시설이 사라진 유럽 도시들의 폐허에 새롭고 매력적인 도시 거주구역이 들어왔다. 물론 좋은 일이지만, 이러한 변화로 제조업, 유지·보수 산업, 식품 공급 산업 등과 같은 생산적 경제가 도시 밖으로 떠밀려 나가고 있는 것도 사실이다.

그러나 생산은 도시의 중요한 부분이다. 우리는 도시의 전면에서 벌어지는 활동뿐 아니라, 눈에 띄지 않는 도시의 이면에서 벌어지는 활동 역시 중요한 일상으로 받아들이는 법을 배워야 한다. 도시가 지속가능한 유기체가 되려면, 그저 소비의 공간이기만 해서는 안된다. 생산이 도시에서 완전히 중단되기를 원하지 않는다면, 생산과 주거를 조화시킬 수 있는 새로운 도시 유형을 개발하고 이를 우리 도시의 공공 공간에 접목시켜야 있다.

이러한 아이디어는 이미 브뤼셀에서 정치적으로 논의되었지만, 현재 공간에 대한 명확한 해법이 있지는 않아 브뤼셀은 현재 혁신적이고 의미 있는 해법을 추구하는 실험의 장이 되고 있다. 브뤼셀의 생산 경제에서 가장 중요한 지역은 당연히 이 도시를 관통하는 운하 지역이다.

따라서, 이번 전시에서는 최근 브뤼셀 운하 지역에서 진행되는 신규 프로젝트를 조명한다. 도시의 생산 경제가 깃들 수 있는 공간을 제시하고 주변 환경과의 통합을 모색하는 다섯 가지 프로젝트가 전시된다.

— 브뤼셀 맥주 양조장 프로젝트(오피스 커스턴 기어즈 다비드 판 세베렌, 2018)
— 도시 인큐베이터 그린비즈 (아키텍테사속, 2016)
— 건축 자재 마을(테트라 아키텍츠, 2018)
— 혼합 개발의 장 노바시티 (보흐단 판 브룩 Bogdan Van Broeck/DDS, 2018)
— 식료품 시장과 농장의 만남 푸드멧 (ORG, 2015/2018)

전시품은 공간에 따라 두서없이 자율적으로 배치된다. 장황하지는 않지만, 중요한 문제에 대해서는 의문을 제기하는, 브뤼셀 스타일의 배치랄까.

BRUSSEL, BELGIUM
Brussels at Work: Building the Productive City

Bouwmeester Maitre Architecte (Brussels Chief Architect)
C21-2F

In many European cities post-industrial wastelands have been replaced by new attractive urban neighbourhoods. Of course this is a good thing, but at the same time we are seeing productive economy, manufacturing, maintenance and repair jobs, and food supply...pushed out of the city.

However, production belongs to the city. We must learn to accept the activities that take place behind the scenes of the city as an integral part of our daily urban lives, just as much as what is happening at the forefront. If we want the city to become a sustainable metabolism, then the city cannot just be the place of consumption. If we do not want production to disappear from the urban landscape, we must invent new urban typologies that make it compatible with housing and connect it with the public space of our cities.

In Brussels, these ideas are already on the political agenda, but this does not mean that there are obvious spatial solutions at hand. Brussels today is a laboratory searching for innovative and high-quality solutions. The most important area for the productive economy in the city is undoubtedly the Canal Area that crosses the urban fabric of Brussels.

The exhibition shows some recent projects in the Canal Area that give the productive economy a prominent place in the city and integrate it with its surroundings. These five projects are presented each in a different and appropriate manner:

— Brewery for Brussels Beer Project (OFFICE Kersten Geers David Van Severen, 2018)
— Urban Incubator Greenbizz (architectesassoc., 2016)
— Construction Materials Village (TETRA architects, 2018)
— Mixed Development Novacity (Bogdan Van Broeck/DDS, 2018)
— Food Market and Farm Foodmet/ BIGH (ORG, 2015/2018)

The exhibition layout is an autonomous constellation of representations meandering in the available room. Laconic but also questioning. A bit like Brussels itself.

인천, 대한민국
상향식나무도시 씨드

하태석
C21-1F

씨드는 디지털 바이오 하이브리드
도시장치이다. 이 장치는 블록체인, 센서,
사물인터넷과 가로등, 벤치, 정자 등이 나무와
융합하여 도시의 정보와 에너지를 먹으며
성장한다. 이 디지털 바이오 하이브리드
나무는 도시에 기생하며 매일, 매계절, 매년
다른 풍경을 연출하며 주민과의 인터액션으로
성장한다. 씨드 네트워크의 밀도가 올라가고
주민과의 인터액션이 올라가면 갈수록
도시에는 새로운 에너지들과 소속감이
창발한다.

INCHEON, SOUTH KOREA
Urban SEED (Smart Ecology & Energy Distribution)

Tesoc Hah
C21-1F

SEED is a digital bio hybrid urban
device. The device grows by eating
blocks of urban information and energy
by integrating blockchain, sensors, IoT,
street lamps, benches, and pavilion. This
hybrid bio-digital tree is parasitic on the
city, and it grows by interacting with the
residents creating a different scene every
day, year, and season. As the density of
the SEED network and the interaction
with the residents increases, new energy
and a sense of belonging emerges.

마드리드, 스페인

환형도시에서 구현되는 집합성: 산업화 이후의 생산적 도시를 꿈꾸는 마드리드

구티에레스 델라푸엔테 아르키텍토스 (훌리오 데 라 푸엔테, 나탈리아 구티에레스) (협력: Urban Reports), 어젠다 어반 에스파뇰라
C21-2F

환형도시에서 구현되는 집합성은 산업화 시대 이후의 생산적 도시의 새로운 흐름을 고찰하고 그 정치·생태·공간·사회·경제적 의미를 살펴본다. 또한 도시의 원형화 과정에 대한 다섯 가지 핵심 요소, 즉 정치적 문제, 지역적 분리, 도시의 지역블록화, 다양한 요소의 혼합과 집합적 상상력을 짚어본다

'환형 도시'의 바탕에는 생산-유통-소비 사이의 새로운 관계와 (일방 통행이 아닌 선형적) 순환이 있다. 이 새로운 순환의 고리는 얼마나 확장해 갈 수 있을까? 어떤 집합적 공간이 이 새로운 환경에서 출현할 수 있을까?

여러 문제를 안고 있는 취약 지역과 발전 가능성이 높은 유망 지역이 공존하는 마드리드, 특히 마드리드의 동남부 지역은 이러한 문제를 연구하는 최적의 도시이다. 이번 작품은 스페인 도시 어젠다(AEU2019) 에서 제시한 '인류 시대의 문제를 해결하기 위한 세부적 방안'이라는 주제를 담고 있다. 마드리드/AEU2019 행사는 '생산적 저택'에서 열린다. 이 저택은 벨기에 브뤼셀과 오스트리아 빈의 협력으로 조성된 공간으로서, 유럽의 순환적이고 생산적인 도시상(像)을 여실히 보여주며 저택과 전시 공간의 조화가 이뤄지는 곳이기도 하다. 마드리드/AEU2019 행사장에는 편안한 거실처럼 벽에 그림이 걸려 있고 큰 탁자가 놓여있다. 관람객들은 다른 도시와도 공유되는 이 공간에서 책을 읽고 비디오를 시청할 수 있다.

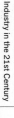

MADRID, SPAIN

Representing the Collective in the Circular City: Madrid, New Proximities in the Post-Industrial Productive City

Gutiérrez-delaFuente Arquitectos (Julio de la Fuente & Natalia Gutiérrez) with the collaboration of Urban Reports, Agenda Urban Española (AUE)
C21-2F

Representing the Collective in the Circular City is a reflection about the new proximities in the post-industrial productive city and its impacts on a political, ecological, spatial, social and economic level. The research is supported by five fundamental aspects about circularization and territorial metabolism: political scale, territorial segregation, urban armatures, hybridization, and the imaginary of the collective.

The circular city is anchored to the new distances and cycles, non-linear, between production-distribution-consumption. What are the scales of these new loops and which collective spaces can emerge from these environments?

The city of Madrid serves as a case-study, specifically the southeast area, which contains at the same time the most vulnerable and potential areas.

The research includes a reflection about the Spanish Urban Agenda 2019 (AUE2019), a transversal urban guide to address the challenges of the Anthropocene age.

The Madrid/AUE2019 space is located at the "Productive Villa" in cooperation with Brussels and Vienna, showcasing the circular and productive city in the European arena. A dialogue between villa-space and the display-concept. Madrid /AUE2019 is "the salon," with its pictures on the wall and the great table as common ground. A living room to read the book, watch the video and share with the other cities.

© Urban Reports

싱가포르, 싱가포르
사회적 스트라타

스페이셜 아나토미 (캘빈 추아)
C21-1F

스트라타 몰은 싱가포르에 위치한 집합적 소유 형태의 쇼핑몰들이다. 싱가포르 건국 초기에 사회적 집합 공간으로 기획된 이 쇼핑몰은 1970년대와 80년대 쇼핑 명소로 호황을 누렸지만 관리와 점포 구성이 훨씬 뛰어난 부동산 개발 회사의 쇼핑몰들이 등장하면서 상황이 바뀌었다. 스트라타 몰은 과거의 영광을 잃고 쇠락해갔으며, 황폐한 공간으로 변해 본래의 기능을 상실했다. 이미 여러 곳의 스트라타 몰은 재개발이 예정되어 있다. 한편, 몇몇 이익 단체는 이 쇼핑몰들을 일상의 유산으로 보존하자는 운동을 벌이고 있다. 이번 전시는 스트라타 쇼핑몰의 공간적·사회경제적 가치를 제시하고, 이 건물들이 어떻게 싱가포르의 미래상에 맞게 재활용될 수 있을지를 논의하는 장이 될 것이다. 전시공간은 스트라타 몰의 기본 건축 소재인 알루미늄 골조로 구성되며 스트라타 몰을 체험할 수 있는 축소판인것이다. 스트라타 몰의 설계 도면, 모형, 그리고 이곳에 남아 있는 세입자(상점 주인)와 건축가들의 인터뷰를 담은 전시는 스트라타 몰과 같은 집합적 소유 모델의 본질, 그리고 이러한 소유모델로 인해 생겨난 건축유형을 오늘날의 사회·문화적 맥락에서 살펴본다.

SINGAPORE, SINGAPORE
Social Strata

Spatial Anatomy (Calvin Chua)
C21-1F

Strata malls are collectively-owned malls in Singapore. Designed as social condensers during early years of nation building, they were popular shopping destinations in the 1970s and 1980s. In recent decades, they have slowly lost their appeal to developer-managed malls as the latter provides better management and tenant mix. As a result, strata malls have fallen into disrepair and lost their functional purpose, with many slated for redevelopment.
On the other hand, several interest groups are campaigning for the conservation of these buildings together with their everyday heritage. By revealing the spatial and socio-economic values of these strata malls, this installation will serve as a platform to discuss the future and possible adaptive reuse of these buildings.
Utilizing standard aluminum frames, one of the basic elements present in strata malls, the exhibition content will be divided into various rooms while allowing visitors to experience a microcosm of a strata mall. Through drawings, models and video interviews of existing tenants and architects working in these strata malls, the installation hopes to question the very essence of the strata-ownership model and its resultant architectural typology in relation to today's social and cultural context.

People's Park Complex © Darren Ho / Spatial Anatomy

영주, 대한민국
도시 축소에 대응하다

영주시 도시건축관리단
C21-1F

영주시는 인구 10만 6천의 작은 도시이지만, 2009년 전국에서 가장 먼저 건축·도시분야 민관거버넌스(총괄건축가제도)를 마련해 도시경관을 지속적으로 관리해 왔다. 이를 통해 조성된 대표적 사례가 장애인종합복지관과 노인복지관이 자리한 '삼각지' 영역과 실내수영장과 복싱장이 들어선 '영주 시민운동장' 영역이다. 전자는 철로로 후자는 옹벽으로 둘러싸여 쇠락해가던 지역으로, 주변공간과 입체적으로 연계된 공공건축 클러스터를 계획하여 각각의 장소로의 접근성과 인지성을 높인 프로젝트다. 공공건축을 통한 장소가치 향상에 초점을 둔 이와 같은 계획은, 개별 거점 건축물을 넘어서 도시의 가로, 공원, 광장 등의 공공공간과 건축물이 이루는 공간환경으로 확대되고, 이들 장소들의 네트워크 구축에 초점이 맞추어질 전망이다.

영주시의 최대과제는 축소되는 도시에 탄력적으로 대응하는 도시공간구조로 전환하는 것이다. 변화하는 환경에 유연하게 대응할 수 있도록 거점 장소를 선택적으로 집중하고, 도시 유휴공간을 체계적으로 개발하는 동시에 지속적인 민관파트너쉽을 통해서 사용자, 자원, 도구를 연결하고 혁신적인 대안을 생산하는 플랫폼 구축이 요구된다. 특히 부석사, 소수서원, 무섬마을 등의 관광자원, 중앙선 복선화에 따른 교통 인프라의 변화, 첨단베어링 국가산업단지 유치 등과 맞물려 생활 SOC 중심으로 다양한 라이프스타일을 지원하고 이를 공유하는 플랫폼들을 만들어가고자 한다. 이들 공유플랫폼과 공유공간들로 연계된 탄력적인 도시 네트워크 구축이야말로 제3기 도시건축관리단이 목표로하는 영주의 미래상이다.

YEONGJU, SOUTH KOREA
Resilience to Shrinking

Urban Architecture Management Team,
City of Yeongju
C21-1F

Although the city of Yeongju has a relatively small population of 106,000 people, in 2009 it was the first city in the country to develop a governance system between the private and public sectors for architecture and urbanism. Since then, this system has supported the city's continued investment in urban landscaping. The most notable of such projects include the establishment of a community rehabilitation center and senior center in the Samgakji district and an indoor swimming pool and boxing gym in the Yeongju Civic Stadium district. This project, with the implementation of urban planning in collaboration with public architecture clusters, introduced a three-dimensional design incorporating the surrounding areas to improve accessibility and the public image of these locations. The core of these efforts was to improve the value of the aforementioned spaces through public architecture, eventually expanding to improve the architectural integrity and quality of surrounding roads, parks, plazas, and other public areas throughout the city. The ultimate goal of this project is to establish a network centered on these locations.

The most important task that lies ahead of the city of Yeongju is its transition to a city space structure that enables the implementation of flexible response measures to the ever-decreasing size of the city. This project aims to introduce SOC-driven sharing platforms that support various lifestyles, thus particularly aligning with aid efforts for tourism sites, such as Buseoksa, Sosu Seowon, and Museom Village; for transportation infrastructure, such as renovating and restoring the central railroad; and for developing the national industrial complex for advanced bearings technology. Through this project, the third official Urban Architecture Management Team aims to establish an adaptive urban network that links sharing platforms with shared spaces, thus paving the road for Yeongju's future.

Aerial picture of Yeongju
Civic Stadium district,
Photo by Youngho Chun

생산도시 빈

스튜디오 블레이 스트리루위츠

C21-2F

본 작품은 도시에 존재하는 종합적인 생산의 장이 지속가능한 도시 발전에 어떤 역할을 하는지 조명하고, 생산적 도시를 만들기 위한 혁신적인 도시계획과 건축을 들여다본다. 이번 비엔나 출품작은 브뤼셀과 마드리드의 작품들과 공통점이 있다. 바로 지식, 서비스, 문화의 깨끗한 장소로서의 유럽 도시의 파괴적 자화상을 질병으로 인식하고 '치료'하고자 하는 바람이다. 이렇게 자기 파괴적인 증상이 확산되면서 20세기 유럽의 생산활동은 다른 대륙으로 빠져나갔고, 그 결과 사회구조·경제·생태·문화적으로 막대한 피해를 가져왔다.

본 작품은 이에 대한 패러다임의 변화를 추구한다. 비엔나를 생산의 도시로 바꾸어 나가기 위한 특별한' 치료법'으로 근본적인 '복구 프로젝트'를 제시한다. 어떻게 하면 우리 도시가 '가치'라는 중요한 키워드를 되찾을 수 있을까? 어떻게 생산을 지속가능한 도시 발전 목표와 융합하는 도시의 핵심 요소로 정착시킬 수 있을까?

본 작품은 말 그대로 유럽 도시의 '깨끗한' 자화상 아래 묻혀 있던 사각지대를 조명한다. 관람객이 어두운 방에 들어서면 불이 켜지고, 밝은 색의 벽지로 장식된 벽과 테이블보가 씌워진 원탁의 공간이 관람객을 환영한다. 관람자들은 공간의 숨겨진 창문을 통해 생산 도시의 미래에 관한 이야기를 발견하게 된다. 테이블보 위에서는 비엔나의 생산적인 정경이 펼쳐치고, 90초 길이의 영상'생산적 도시'는 비엔나가 꿈꾸는 생산의 미래를 위한 계획을 제시한다. 초콜릿-맥주처럼 다양한 요소가 혼합된 상품들은 사회문화적 요소로 생산을 더욱 활성화시키고자 하는 비엔나의 염원을 보여준다.

Productive City Vienna: Hidden Windows, Productive Screens

Studio Vlay Streeruwitz

C21-2F

The exhibition highlights the opportunities that the comprehensive field of production provides for a sustainable development of cities, focusing on innovative urban-planning and architecture.

Vienna's contribution shares with Brussels and Madrid the ambition to "treat" the destructive self-portrait of the European city as a clean place of knowledge, services, and culture. This symptom has resulted in production's 20th century-displacement to other continents, implicating dangerous structural damage, economically, ecologically, and culturally. The exhibition promotes a change in paradigm. It stages Vienna's specific treatment of the productive city as a fundamental "repair-project": how to repair the dominant idea of "value" in our cities? How to establish the world of production as a main ingredient to be integrated in the program of sustainable city development?

The installation literally brings into the limelight what has been a blind spot in the European city's "clean" self-portraits. A bright wallpapered wall and a light tablecloth on a round table welcome the visitors, when they enter the darkened room. Hidden windows make them discover narratives of the productive city's future. The tablecloth shows Vienna's productive landscape, the video Productive City reveals playfully Vienna's plan for the production's future in 90-seconds. Hybrid products such as chocolate-beer underline the city's desire to shortcut production with cultural and social affairs.

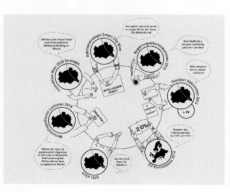

Vienna Round Table

뉴욕, 미국
임시도시: 청년, 거주, 도시
ＮＨＤＭ 건축 + 도시
C37-2F

Interim Urbanism: Youth, Dwelling, City
ＮＨＤＭ Architecture + Urbanism
C37-2F

청년층은 현대 사회의 역동적이지만 불안정한 부분을 상징한다. 대부분의 청년들은 더 이상 유년기의 안전한 공간에 속하지 않지만, 그렇다고 다른 사람들의 기대처럼 전통적인 가족 구조 속에 편입되어 있지도 않다. 언뜻 보면 마음 내키는대로 살아가며 자유롭게 이동성을 추구하는 것 같지만, 실제로는 어디에도 소속되지 못한 채 '기나긴 과도기'라는 취약한 공간에 놓여 있다. 이번 작품은 뉴욕의 청년들이 살아가는 공간에서 나타나는 '과도기'와 '소속 부재'를 비롯한 여러 특징을 짚어 보고, 이를 통해 전세계 청년층이 겪고 있는 비슷한 상황의 담론을 이끌어내고자 한다 .

본 작품은 뉴욕 청년층의 주거 공간을 들여다보고, 이들이 끊임없이 사회·정치·경제적 불확실성, 불평등에 직면하고, 한편으로는 새로 부상하는 생활 양식을 접하는 과정을 살핀다. 건축의 중요한 임무 중 하나는 지속적으로 변하는 공공 영역의 안락함을 위해 새로운 이상을 그려내는 것이다. 본 작품은 이러한 인식을 바탕으로 이에 관련한 역사적 배경과 현대의 사례를 제시하고, 청년층과 주거 그리고 도시를 위한 새로운 공간적 틀을 제안한다.

C37

2F

Youths represent a dynamic yet precarious section of today's populations. No longer belonging to safe spaces of childhood, but not yet, if ever, integrated into the expected paradigms of traditional family structures, a large portion of today's youths while seemingly spontaneous in lifestyle choices and welcoming mobility, occupy the vulnerable spaces of the in-between and the prolonged interim. The project/exhibit will examine these and other selected specificities of youth dwelling in New York City as a way to provoke conversations on parallel and similar conditions around the globe.

The project investigates the spaces that youths reside in, as they intersect with sustained sociopolitical and economic uncertainties, inequalities, and emergent lifestyles. Considering the role of architecture in defining new ideals of domesticity in an ever-transforming public realm, the exhibition presents selected historical contexts and critical contemporary instances, and propositions for new spatial frameworks for the youth, dwelling, and the city.

Collective Typologies

취리히, 스위스
분산된 주택 협동조합

스콧 로이드, 알렉시스 칼라가스, 네만자
지몬직
C37-2F

취리히는 2050년까지 비영리 도시 임대주택
비율을 1/3까지 확대하는 계획을 수립했다.
그러나 저렴한 주택 공급 계획 마련은 아직
더딘 상태이다. 또한, 스마트폰, 주문형 경제,
플랫폼 기술이 등장하면서 집과 소유권에
대한 개념이 바뀌는 한편, 도시 생활의 핵심
요소이자 집합적 실천의 매개체가 되는 디지털
기반 서비스의 역할이 커져가고 있다. 이미
취리히 임대주택 시장의 20퍼센트를 차지하는
주택 협동조합은 도시의 주택산업 지형을 점점
많은 사람들이 핵가족 구조를 탈피하려는
사회적 현실에 따라 재편하려는 최선두에
서 있다. 하지만 이런 협동조합은 아직 지역
차원에서 공동체를 형성하고 자원을 공유하며,
역동적인 환경을 구축하기 위한 디지털 기술의
가능성을 완전히 활용하고 있지는 못하다.
취리히 전역, 특히 중심부에 남아 있는 '소규모
단독 건물들'은 오늘날의 취리히 도시 구조에
스마트한 인구밀도 증가를 추진할 여지가 아직
많이 남아있다는 사실을 반증한다. 본 작품은
부동산 투기 목적의 개발회사들이 선호하는
대규모 주택 단지 대신, 기존의 소규모
복합용도 건물을 찾아내고 증축하며, 이러한
건물들을 디지털 기술 활용을 통해 네트워크로
연결함으로써 '탈중앙적 협동조합'을 구성하기
위한 새로운 전략을 모색한다. 또한 시민
중심의 스마트 미래도시의 바탕이 되는
지속가능한 공동생활의 신규모델을 제안한다.

ZURICH, SWITZERLAND
The Distributed Cooperative

Scott Lloyd (TEN), Alexis Kalagas,
Nemanja Zimonjic (TEN)
C37-2F

One-third of Zurich's rental housing
must operate on a non-profit basis by the
year 2050. But new affordable projects
are not being developed fast enough.
At the same time, the emergence of
smartphones, an on-demand economy,
and platform technologies has reshaped
understandings of home and ideas
of ownership, while digital services
increasingly mediate the collective
practices that underpin urban life.
Housing cooperatives, which already
represent twenty percent of Zurich's
rental market, are at the forefront of
attempts to rethink housing typologies
in line with evolving social realities,
as more people live outside nuclear
family structures. But they are yet to
fully embrace the possibilities of digital
technologies to build community,
share resources, and generate vibrant
environments at the neighborhood level.
Throughout Zurich, centrally located 'odd
lots' represent an unrealized opportunity
for smart densification amid the existing
urban fabric. Bypassing the need for the
kinds of large consolidated sites favored
by speculative developers, the exhibition
explores a new strategy to identify,
build out, and digitally network a series
of small-scale mixed-use architectural
interventions into a 'distributed
cooperative', proposing an updated
model of sustainable shared living for the
citizen-centered 'smart city' of the future.

Collective Typologies

C37 2F

집합적 유형

Design proposal for a shared living
space within a distributed cooperative.

베오그라드의 과거회귀-유토피아적 대안

집합건축스튜디오 (로드리고 세사르만, 시드니 시날리, 스트래턴 코프먼, 보리앙 두, 개비 허퍼넌, 벤 호일, 에이탄 레비, 캐서린 리에, 애나 미야츠키, 유탄 순, 마리사 워들, 새라 와그너)
C36-1F

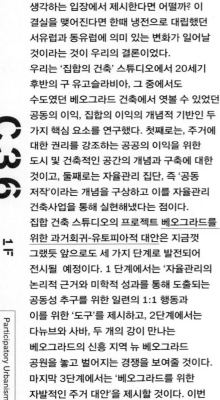

2019년 봄, MIT '집합의 건축' 스튜디오는 하나의 가정을 해 보았다. 베오그라드가 사회주의 도시였던 시절의 건축과 정치적 역사에 관한 이야기를 새롭게 쓰고, 도시와 건축의 대안을 그 시대의 유산을 소중하게 생각하는 입장에서 제시한다면 어떨까? 이 결실을 맺어진다면 한때 냉전으로 대립했던 서유럽과 동유럽에 의미 있는 변화가 일어날 것이라는 것이 우리의 결론이었다.

우리는 '집합의 건축' 스튜디오에서 20세기 후반의 구 유고슬라비아, 그 중에서도 수도였던 베오그라드 건축에서 엿볼 수 있었던 공동의 이익, 집합의 이익의 개념적 기반인 두 가지 핵심 요소를 연구했다. 첫째로는, 주거에 대한 권리를 강조하는 공공의 이익을 위한 도시 및 건축적인 공간의 개념과 구축에 대한 것이고, 둘째로는 자율관리 집단, 즉 '공동 저작'이라는 개념을 구상하고 이를 자율관리 건축사업을 통해 실현해냈다는 점이다.

집합 건축 스튜디오의 프로젝트 베오그라드를 위한 과거회귀-유토피아적 대안은 지금껏 그랬듯 앞으로도 세 가지 단계로 발전되어 전시될 예정이다. 1 단계에서는 '자율관리의 논리적 근거와 미학적 성과를 통해 도출되는 공동성 추구를 위한 일련의 1:1 행동과 이를 위한 '도구'를 제시하고, 2단계에서는 다뉴브와 사바, 두 개의 강이 만나는 베오그라드의 신흥 지역 뉴 베오그라드 공원을 놓고 벌어지는 경쟁을 보여줄 것이다. 마지막 3단계에서는 '베오그라드를 위한 자발적인 주거 대안'을 제시할 것이다. 이번 프로젝트에서는 단계 별 자료와 함께 본 스튜디오의 집단 작업에 사용된 '결정을 돕는 주사위', '집합적 기록', '가족 앨범' 등과 같은 주요 '도구'들이 전시된다.

C36
1F

BELGRADE, SERBIA

Retro-Utopian Alternatives for Belgrade, Serbia

Collective Architecture Studio (Rodrigo Cesarman, Sydney Cinalli, Stratton Coffman, Boliang Du, Gabby Heffernan, Ben Hoyle, Eytan Levi, Catherine Lie, Ana Miljački, Yutan Sun, Marisa Waddle, Sarah Wagner)
C36-1F

In the Spring of 2019, MIT's Collective Architecture Studio hypothesized that by engaging in retelling the pertinent aspects of socialist architectural and political history of Belgrade (Serbia) and by offering urban and architectural alternatives from the position that valued this heritage, the fruits of the studio's labor would have a critical function on both sides of the former Cold War divide. The Collective Architecture Studio explored two key registers on which the concept of the common, collective good played out in Yugoslavian, and specifically Belgrade's architecture, in the second half of the 20th century: first, the production and conception of urban and architectural space for the common good (with an emphasis on the historical "right to housing") and, second, the conception of self-managed, group, or "common authorship" that was implemented and performed through self-managed architectural enterprises.
The Collective Architecture Studio's Retro-Utopian Alternatives for Belgrade were developed and will be presented in three acts. Act One: a set of 1:1 exercises and devices for commoning, drawing from the logic of and aesthetic products of self-management. Act Two: a competition for a park in New Belgrade at the confluence of the Danube and Sava rivers. Act Three: Unsolicited Housing Alternatives for Belgrade. The exhibition will present the material generated in response to these prompts, as well as some of the devices ("decision dice," "collective archive," "family album," etc.) that facilitated the collective work of the studio.

스톡홀름, 스웨덴
건축 프로젝트: 셉스브론

제임스 테일러-포스터(큐레이터), AT–HH,
에센시알(카르멘 이스키에르도), 엘리자베스
B. 하츠, 헤르만손 힐러 룬드베리, OKK+,
크루핀스키/크루핀스카, 토르 린드스트란드,
닐손 람
C36-1F

예측은 건축 행위에 있어 중대한 부분이며,
건축가의 가장 중요한 역할은 다른 사고를
하기 위함에 있다. 건축 프로젝트: 셉스브론은
스톡홀름에서 활동하는 건축가 5인과 건축
스튜디오 3곳이 셉스브론(Skeppsbron)의
새로운 가능성을 보여주기 위해 제시한
예술적인 방법론과 예지력 있는 아이디어들을
조명한다. 셉스브론은 스웨덴 수도
스톡홀름의 심장부에 있는 대형 부두로서
슬루센(Slussen)에서 왕궁까지 이어진
구시가지 감라스탄(Gamla Stan)을 끼고
있다.

스톡홀름에서 가장 오래된 부두 셉스브론과
셉스브로카옌(Skeppsbrokajen)은 스웨덴의
'쇼윈도우'로서 독특한 역사를 자랑한다.
이곳은 수입상품과 수출상품이 집결되는
거점이자 장터이고 수로이며 스톡홀름과 유럽
다른 곳을 잇는 은유적 교량이기도 하다. 한때
셉스브론은 사람들 사이에서 스톡홀름의 가장
중요한 교환 장소로 꼽혔으며 아이디어와
상품이 거래되는 현장이었다. 아직까지도
스톡홀름의 모습을 상징적으로 보여주고
있지만 이제 셉스브론은 주차장과 유휴공간에
점유되기에 이르렀다.

대형 그림과 모형으로 구체화된 여덟 가지
구상은 도면, 모형, 텍스트, 참고자료 형태로
건축·상징·공적 경관으로서 셉스브론의 미래
가능성에 대해 고찰한다. 여덟 가지 구상은
스톡홀름을 중층적이고 다면적이며 뜻밖의
제안을 수용할 수 있는 환경으로 고찰한
건축가의 역량을 한껏 드러내며, 제시되는
환경을 통해 이 도시의 잠재력을 새로운
시각으로 바라보게 된다.

STOCKHOLM, SWEDEN
Architecture Projects: Skeppsbron

James Taylor-Foster (Curator), AT–HH,
Esencial (Carmen Izquierdo), Elizabeth B.
Hatz, Hermansson Hiller Lundberg, OKK+,
Krupinski/Krupinska, Tor Lindstrand,
Nilsson Rahm
C36-1F

Speculation is crucial in architectural
practice and a central role of the architect
is to think differently. Architecture
Projects: Skeppsbron highlights the
artistic methods and visionary ideas of
eight Stockholm-based architects and
studios to propose new possibilities for
Skeppsbron — a large waterfront site
at the heart of the Swedish capital that
stretches across Gamla Stan (the old
town) from Slussen to the Royal Palace.
As Stockholm's oldest quay, Skeppsbron
and Skeppsbrokajen have a unique
history as the "shop window" of Sweden:
a hub for the import and export of goods,
a marketplace, a thoroughfare, and a
metaphorical bridge between the capital
and the rest of Europe. Between people,
ideas and commerce, this was once
Stockholm's foremost site of exchange.
Although it still stands as a defining
image of the city, it has come to be
dominated by car parks and underused
public space.

Through drawings, models, words
and references, these eight visions —
rendered by way of large-scale drawings
and models — consider possible futures
for Skeppsbron as an architectural,
symbolic, and public landscape. They
celebrate the architect's ability to address
the city as a layered, multifaceted
environment capable of absorbing
unexpected proposals and, when read
together, can help us to see the potential
of the city in a new light.

Market Inheritance

C36 1F

도시의 유산, 시장

Architecture Projects: Skeppsbron at ArkDes (Stockholm), 2019 © Johan Dehlin

도시전 인트로
집합적 결과물
라파엘 루나, 임동우
C35-1F

이 전시는 도시전 전체의 인트로로서 비엔날레의 메인 주제인 집합도시를 재해석하여 '집합적 결과물'로서의 도시에 대한 아이디어를 어떠한 과정을 통해 풀어냈는지 보여준다. 도시는 계획적인 혹은 비계획적인 여러 레이어들이 겹쳐진 공간으로 이해되곤하며, 우리는 이것을 집합적 결과물로서 해석한다. 때문에, 세밀한 마스터플랜에 의해 계획된 도시일지라도, 우리가 실제 살고 경험하는 도시는 도면에 그려진 도시의 모습과는 상이하다. 이는 이번 도시전에서도 마찬가지이다. 전세계 80여개의 도시들을 초대하면서, 이번 도시전에서는 하나의 결과를 의도하기 보다는 결과를 모르는 방식으로 큐레이팅하였다. 이를 통해 하나의 도시는 엄격한 잣대의 가이드나 해석의 제한 없이 다른 도시와의 자유로운 대화와 교류를 만들어 낼 수 있다. 때문에, 우리는 이번 도시전이 완결된 각 도시의 전시의 집합이기 보다는, 여러 도시가 서로간에 만들어 내는 대화에 의해서 만들어지는 집합적 결과물로서 인식되기를 바란다. 이러한 맥락에서, 도시전 인트로전시는 현대의 도시에 대한 이슈들에 대해 논하기 위한 여러 방식들에 대한 이해를 돕기 위하여 도시를 읽는 방법론에 대한 내용을 포함하고 있다.

As an introduction to the Cities Exhibition, this space exhibits the curatorial efforts and decisions that have been made to develop the idea of collective consequences as an interpretation of the main theme, collective city. Cities are understood as an accumulation of multiple layers that result from planned as well as unplanned intentions, which we define as collective consequences. This is why even in a super master planned city, the city we actually live in is much different from what has been drawn on paper. This is the case for the exhibition as well. Having more than eighty cities from around the globe, the Cities Exhibition intends to have an open ended curation, instead of a fully controlled one, so as to create a dialogue and chemistry between individual exhibits without a strict guidance or limitations on their interpretation. Therefore, our intention is for the whole exhibition to be conceived as consequences that are being made by the dialogues between cities rather than as complete singular exhibition of a city. For this, the introductory exhibition provides methods for reading a city as a platform to understand different ways for discussing the contemporary topics that run our cities.

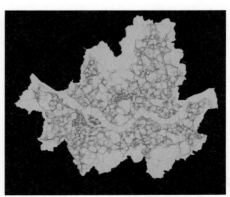

Networks of Convenience Stores in Seoul, copyright PRAUD

출판 전시

김유빈
C35-1F

출판전시는 2019서울도시건축비엔날레
도시전의 특별전으로서, 전 세계 다수의
도시에서 수집한 도시건축 저널, 잡지, 출판물
등을 통하여 도시의 현황을 살펴볼수 있는
집합적 무대이자 플랫폼이다. 동시에 도시전
참가자들의 출판물도 함께 기획함으로써,
도시전의 응축된 장이 되기도 한다. 아울러
출판전시와 연계하여 피에르 아리시오
리자르디의 "2019 서울 건축의 현황"
이라는 작업은 올해 비엔날레 개최지 서울에
기반한 국내 건축가들의 인터뷰영상과
다큐멘터리의 방식으로 소개된다. 출판전시는
현대도시건축의 다양한 면모를 탐구하며,
출판매체를 통해 투사되는 심도 깊은 연구와
비평을 제시한다.
출판전시 공간은 새로운 마이크로-플랫폼
개념의 공간을 제안하며, 그곳에서 작가,
건축가, 예술가, 학생들이 도시전 출판물들을
통하여 서로의 생각을 나누고 혜안을 얻길
기대한다. 아울러 모든 방문객이 비엔날레
기간 뿐만아니라 전시 후에도 도시건축에 대한
경험이 확장되고 인식을 높이는 계기가 되기를
또한 기대한다.

Publication Exhibition

YouBeen Kim
C35-1F

As one of the pavilions of the Cities
Exhibition at the Seoul Biennale 2019,
the publication pavilion is a collective
showcase and a platform for the most
recent collection of urban architecture
journals, magazines, and publications
from multiple cities around the world.
The Publication Exhibition also features
titles from various participants of
the Cities Exhibition thus presenting
a condensed version of the Cities
Exhibition. In addition, the Publication
Exhibition features a special collection
from Pier Alessio Rizzardi's "The
Condition of Seoul Architecture" through
documentary video, interviews and
publications highlighting the city of Seoul
where the 2019 Biennale is held. The
overall publication exhibition explores
the variety of contemporary urban
architectural issues and celebrates its
projections through printed matter that
contains a broad range of research and
criticism.
The publication pavilion provides a new
micro-platform to share thoughts and
insights from authors, architects, artists,
and students about the Cities Exhibition
through selected/featured publications.
Also, as an outro exploration of the Cities
Exhibition, the publication exhibition
extends visitors' experience and ideas to
generate continuous urban architectural
discourse about multiple cities during the
period of the 2019 Seoul Biennale and
beyond.

C35

서울의 건축 현황

피에르 알레시오 리차르디
C35-1F

서울은 뭄바이와 라고스 다음으로 세계에서 가장 인구밀도가 높은 도시에 속한다. 1965년 350만 명이었던 서울 인구는 2019년 2,500만 명을 넘어섰다. 대한민국 국토 면적의 12%에 불과한 이곳에 전체 인구의 50%가 살고 있는 것이다. 서울 도심의 땅값은 한국전쟁 이후로 680배가 뛰어올라 현재 1제곱미터 당 8만 달러라는 어마어마한 금액에 달한다.

이러한 극단적인 상황에서 활동하고 디자인하며 구축해야하는 건축가들의 현실은 서울의 건축 현황의 이론적 기반이 된다.

본 연구는 서울에서 가장 혁신적인 건축가 18인의 프로젝트를 보여주고 서울의 건축 환경에 대한 각자의 입장을 분석하는 식으로 건축가들의 관점을 소개한다. 동시에 전문가, 교수, 큐레이터 등의 노스탤지어, 미니멀리즘, 빈 공간에 대한 설명을 통해 도시의 문제를 구체적으로 짚어보고, 2008년 세계 금융위기 이후에 고객들이 용적률을 극대화하기 위해 비용이 덜 드는 제도사 대신에 독창적인 건축가를 선택하면서 생겨난 기회와 '용적률 게임'의 제약을 피하기 위해 채택된 전통적인 파사드(외벽) 개념에 대해 알아본다.

이처럼 개인의 이야기와 건축학, 미학, 철학, 정치학, 사회학을 비롯한 다학제 간 연구가 결합된 결과로서 획기적이고도 다층적인 고찰이 탄생했다.

The Condition of Seoul Architecture

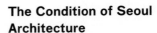

Pier Alessio Rizzardi
C35-1F

Seoul runs just behind Mumbai and Lagos in the list of the densest cities in the world. From a population of 3.5 million in 1965, the city has exceeded 25 million in 2019, hosting 50% of the population in an area of merely 12% of South Korea's land mass. The central areas have witnessed the cost of land rising by 680 times since the Korean War, reaching a prohibitive $80,000 per square meter.

This extreme situation in which architects have to operate, design, and build, creates the theoretical basis for The Condition of Seoul Architecture.

The research insists on the point of view of eighteen most innovative South Korean architects, showcasing their projects and analyzing their position on the built environment of Seoul, while experts, professors and curators frame urban issues through descriptions of nostalgia, minimalism and empty spaces, and opportunities arising since the 2008 crisis, when clients moved from low-cost drafters to ingenious architects to extend the floor area to the fore, and the traditional concept of façade adopted to bend the rules of the F.A.R Game.

The result is a groundbreaking, cumulative reckoning teased out through layers of personal accounts and cross-disciplinary research, drawing on architecture, aesthetics, philosophy, politics, and society.

청진, 조선민주주의인민공화국
청진, 이동성과 도시주의

홍민
C33-1F

청진은 '철의 도시'였다. 청진제강소와 김책제철소는 철을 뿜어내는 청진의 심장이었다. 정치적 소외에도 불구하고 공업화의 상징이었던 청진. 경제난이 덮치며 1990년대 퇴락한 도시가 되었다. 그러나 청진은 놀라운 생명력으로 북한 최대의 시장도시로 재탄생했다. 2019년 청진, 해체된 청진제강소 위에는 놀라운 도시건설 프로젝트가 모습을 드러내고 있다. 이제 '시장'과 '건설'은 청진의 새로운 심장이다. 개방과 유행을 이끄는 청진의 다이내믹을 본다.

CHUNGJIN, NORTH KOREA
Chungjin, Mobility and Urbanism

Min Hong
C33-1F

Chungjin used to be a steel city. Chungjin Steel Mill and Kimchaek Steel Mill were the two major factories in the region. The city was known as the industrial powerhouse of the nation even though it failed to draw much attention from the political arena. In the 1990s, the city suffered greatly from the economic recession, but it managed to overcome its difficulties and became the largest business city in North Korea. In 2019, an amazing urban project was initiated at the site where Chungjin Steel Mill factory was once located. Now, business activities and architecture are the two lifelines that sustain the region. We look forward to seeing dynamic openness and pioneership in Chungjin.

코펜하겐, 덴마크
작은 행성의 거대한 건축
씬 그린 라인 프로덕션
C34-1F

덴마크 전시는 덴마크 다큐멘터리 영화
작은 지구의 위대한 건축의 상영을 통해
이루어질 예정이다. 이 영화는 기후변화, 공해,
거대도시, 이산화탄소 배출 등 네 개의 다른
주제를 바탕으로 한다. 덴마크 건축가들은
미래 세대를 위한 지속 가능한 건축 개발의
선두자다. 우리는 본 1시간짜리 다큐멘터리를
통해 세계 곳곳을 여행하며 세계적인
도전과제에 대응하는 동시에 우리 일상을
개선하는 프로젝트와 건축물을 연구한다.
BIG의 비야케 잉겔스는 해수면이 상승할 경우
뉴욕 맨해튼을 보호할 수 있는 종합 계획을
제안한다. 덴마크 건축가들로 구성된 SLA는
식물로 대기를 정화하고 차량 소음을 줄이며
아파트의 열기를 식힌다는 목표로 파리에서
가장 공해가 심한 지역의 순환 도로 위에 녹색
건물을 세울 계획이다. 부에노스아이레스시는
겔 건축 사무소와 협력하여 빈민가였던 지역을
재개발하는 중이며 그곳의 거주자들을 다른
도시 지역에 물리적, 사회적으로 통합시키고
있다.
미래에는 이산화탄소 배출을 줄일 수 있는
건축물이 가능해질까? 코펜하겐에서는
혁신적인 건축가 안데르스 렌다게르가 대규모
주택 단지 건설에 재활용 소재를 활용하여
이산화탄소의 영향을 80% 완화했다. 이 모든
사례는 어떻게 하면 미래에 보다 지속 가능한
방식으로 생활하고 건물을 구축해 나갈지 대한
영감이 된다.

COPENHAGEN, DENMARK
Great Architecture on Small Planet
Thin Green Line Productions ApS
C34-1F

Danish participation will be through a
screening of Danish documentary film
called Great Architecture on a Small
Planet with four different themes: climate
change, pollution, megacities and CO2
emission. Danish architects are on
the forefront developing sustainable
architecture for future generations. In
this, one-hour documentary we travel
around the world exploring projects
and buildings that address the global
challenges but at the same time improve
our everyday life.
Bjarke Ingels from BIG has a masterplan
on how to protect Manhattan when the
sea rises. In Paris, Danish architects SLA
will place a green building on top of the
most polluted highway with hopes that
the plants will clean the air, reduce traffic
noise and cool down the apartments. The
City of Buenos Aires together with Gehl
Architects have transformed a former
slum area and integrated the residents
physically and socially with the rest of
the city.
Can we build structures that reduce CO2
emission in the future? In Copenhagen,
recycled materials are used by innovative
architect Anders Lendager in a large-
scale housing project reducing the CO2
footprint by 80%. All inspiring examples
of how we can live and build more
sustainably in the future.

Great Architecture on a Great Planet

개성, 조선민주주의인민공화국
고동치는 도시

황두진, 홍수영, 윤현철, 우경선
C33-2F

도시는 다양한 리듬으로 고동친다. 시민들의
심장 박동, 매일매일의 출퇴근 방식, 계절에
따라 오르내리는 기온, 한 해 혹은 그보다 더
긴 기간에 걸쳐 일어나는 인구변화 등이 모두
리듬의 일부이다. 펄서(pulsar)가 주기적으로
박동하며 전파를 발산하듯, 모든 도시는
각자의 독특한 리듬으로 고동친다.
우리는 2019 서울도시건축비엔날레를 맞아
개성공단 사례를 통해 '고동치는 도시'를
소개하고자 한다. 유서 깊은 도시 개성의
남동쪽에 자리 잡은 이 공단은 2016년 2월
갑작스럽지만 놀랍지는 않았던 폐쇄 조치가
내려지기 전까지 약 55,000명의 북측
노동자와 수백 명의 남측 관리자가 일하던
곳이었다.
개성공단에는 주거시설이 없었기 때문에, 북측
노동자들은 모두 외부에서 출퇴근했다. 공단은
말하자면 인스턴트 도시가 되었다. 사는
사람은 한 명도 없는 도시. 펄서가 주기적으로
전파를 발산하듯 매일 사람들이 왔다가 떠나는
도시.
우리는 출품작 고동치는 도시를 통해 시민들의
더 나은 생활과 도시 인프라에 대한 보다
효율적인 투자를 위해서는 도시 내부 주거
공간에 대한 정책이 보다 강조된 도시 계획이
필요하다는 사실을 역설하고자 한다.

GAESEONG, NORTH KOREA
Urban Pulsar

Doojin Hwang, Suyoung Hong, Hyunchul
Youn, Kyeongsun Woo
C33-2F

Cities pulsate in combinations of different
rhythms: heartbeats of the citizens, daily
commute patterns, seasonal fluctuations
of temperature, annual or longer-term
changes in population, etc. Each city is
like a pulsar, a star with a unique pattern
of pulsed appearance of emissions.
The Gaeseong Industrial District (GID)
is a subject of our case study to explore
the theme of an urban pulsar for the
2019 Seoul Biennale of Architecture and
Urbanism. Located to the southeast of
the ancient city of Gaeseong, the GID had
approx. 55,000 North Korean workers and
a few hundred South Korean managing
staff by the time of its sudden but
expected closure on February 2016.
All North Korean workers at the GID
commuted to work from outside, since
no residential facilities existed. The
GID became an instant city, a city
without residents, gaining and losing its
population every day, like a pulsar with its
repeating emission pattern.
We hope to take this Biennale as an
opportunity to show why cities should be
planned to have more emphasis on the
inner-city residential programs for higher
quality of life of its citizens and also for
more efficient investment initiative on
urban infrastructure.

C33 2F

쿠웨이트, 쿠웨이트
쿠웨이트 민간 외교관 Q8
시빌 아키텍처 (하메드 부캄신, 알리 이스마일 카리미)
C33-1F

KUWAIT CITY, KUWAIT
Foreign Architecture / Domestic Policy
Civil Architecture (Hamed Bukhamseen + Ali Ismail Karimi)
C33-1F

인프라스트럭처

C33

1F

Infrastructure

쿠웨이트는 유럽 전역에 약 5천 개의 주유소를 소유 및 운영하고 있다. 'Q8'이라는 — 한 눈에도 어느 나라 회사인지 바로 알 수 있는 — 회사가 운영하는 이 주유소는 쿠웨이트의 석유 수출시장 확보를 위한 투자 목적으로 지어졌다. 주유소는 걸프 전쟁 이전에 일어난 1990년 이라크의 쿠웨이트 점령에 대응하기 위한 재원 조달과 로비에 핵심적인 역할을 하면서, 쿠웨이트에게 점차 더 중요한 자산이 되었다. Q8은 이제 단순한 주유소 기업이 아니라 쿠웨이트의 정치·경제적 자율성을 확보하기 위한 핵심 요소가 되었다.

1973년 당시 백만 명도 안 되는 인구로 세계 석유 수출량의 10%를 차지했던 쿠웨이트라는 나라를 이해하려면 쿠웨이트의 국가 경제 계획, 그리고 외향적인 국가 정체성에 관한 이야기를 빼 놓을 수 없다. 본래 Q8은 유럽 자동차 운전자들에게 직접 쿠웨이트산 석유를 판매하여 수요를 창출하고 시장을 확보하려는 목적으로 설립되었다. 소비자는 휘발유를 (역주: 차량 주유 이외의 목적으로는) 직접 취급하지 않기 때문에, 특정 지역에서 생산된 휘발유를 선호하는 일은 없다.

그러나 특정 주유소 브랜드에 대한 선호는 분명하다. 소비자들이 Q8을 주유소 브랜드로 받아들이게 되면서, Q8은 기업이 가지는 정치적 중립성과 쿠웨이트라는 국가 차원의 경제 계획 사이에서 완벽한 균형을 유지하는 데 성공했고, 그 덕분에 오일 쇼크 이후 시대에 세계 여러 나라의 적대감과 의구심을 자극하지 않으면서 쿠웨이트산 석유를 판매할 수 있게 되었다. 이번 작품에서는 유럽 전역에서 쿠웨이트 민간 외교관 역할을 하는 Q8 주유소를 조명하고 앞으로 이 주유소가 전세계에 쿠웨이트라는 국가 브랜드를 알리는 데 어떤 역할을 할지 짚어본다.

Kuwait owns and manages approximately five thousand gas stations across Europe. Under the not-so subtle company name 'Q8', these stations began as an investment to secure a market for the country's oil export, but their role continued to grow as they became a key player in funding and lobbying against the occupation of Kuwait prior to the Gulf War. Q8 is not simply a series of serving stations across the European landscape, but a road map for the nation's political and economic autonomy.

For a country of less than a million people in 1973, which was exporting 10% of the world's oil, the narrative of the state is inextricably tied to the projections of wealth and identity outwards. The creation of Q8 was a way to build, demand, and secure a market for Kuwaiti oil by selling directly to the European driver. Since gasoline is a product one never deals with directly, the customer does not choose one source of gasoline over another, but the brand of gas station. The adoption of Q8 as a brand provided the perfect balance of corporate neutrality and nationalistic projection, a way of selling Kuwait without risking hostility or suspicion in light of a post-OPEC crisis world.

This exhibition tells the story of the stations that act as embassies across Europe and their role in branding Kuwait abroad in the years to come.

공유 도시화, 협력적 도시 실현

파비용 드 아스날, 컬렉티브 어프로치.스!
C33-1F / R

집단적 지성을 한데 모아 시민의 필요에 부합한 도시를 구상하는 일은 오늘날 떠오르는 협력적 도시 실현을 위한 하나의 도전과제다. 도시 구성에 시민의 영향이 반드시 필요한 요소로 자리 잡은 이때에 공유 도시화(Co-Urbanism) 전시는 시민과 전문가의 공동생활(cohabitation)을 개선한다는 목표로 프랑스와 그 외 나라에서 추진된 21가지 다양한 접근법을 통해 다양한 경로를 모색할 뿐만 아니라 공유 도시화 실현의 조건을 찾는다. 공공장소에 대한 구상과 관리, 사회 활성화를 위한 단기간 지원, 외국 대학과 계획되지 않은 공간, 공동 계획 도구, 계획 로드맵, 열린 구상으로 이루어진 본 프로젝트는 재정적 규제의 제약으로 사라져버린 원칙을 긍정적인 방향으로 되살려낸다. 이 프로젝트를 고안하고 실행한 팀은 도시와 인간이 만나는 지점을 고려하여 다원적인 방법을 개발하고 있다. 큐레이터 집단인 컬렉티브 어프로치.스! (Collective Approche.s!)에 따르면 이러한 접근법에 따르는 어려움은 각각의 제안으로 고전적인 도시 계획 전략을 보강할 수 있는 방안을 찾아내는 것과 그러한 제안들을 실행에 옮기는 것이다. 개별 프로젝트는 특수한 상황을 감안하여 고안되며 전문가들 간의 만남을 통해 이루어진다. 또한 적응, 방법론적 혼성화, 공동 도구의 전환에서 영감을 얻는다. 우리의 도시화 전략은 특수하고도 일반적인 상황을 고려하여 모든 영역에서 친근한 관계를 만들어내는 시도를 한다. 실험적 도시화를 통해 관습과 계획뿐 아니라 사회적 관계와 행위자 간의 관계 측면에서 가능성을 시험하고자 한다. '이미 그곳에 존재하는 것(déjà là)'에 미묘한 형태적 변화를 주어 장소의 잠재력을 증대하는 것이다. 이처럼 시간과 공간을 넘나드는 이들의 개입 방식은 파리를 하나의 생태계로 인식함으로써 시민의 주도권에 초점을 둔다.

PARIS, FRANCE
Co-Urbanism,
21 Collaborative Practices of the City

Pavillon de l'Arsenal, Collective Approche.s!
C33-1F / R

At a time when the role of citizens in the composition of the city has become essential, the exhibition Co-Urbanism explores various paths as well as seeks out conditions for collective urbanization to improve the cohabitation of citizens and experts through twenty-one different approaches from France and beyond. The projects, which entail the co-conception and management of public spaces, short-term equipment support for social revitalization, foreign universities and non-programmed spaces, co-programming tools, program roadmap, and open initiatives, revive the principles that have been erased by financial or regulatory constraints. Devised by a diverse team of architects, urbanists, multidisciplinary research institutes, public institutions, designers, and cultural actors, they seek to innovate pluralistic approaches at the intersection between the city and the human.
According to the curatorial collective, Collective Approche.s!, the challenge of this approach is to find ways to enrich the classical urban planning strategies with such proposals and to put them into action. General solutions do not apply in these cases as individual projects are designed with particular circumstances in mind and realized through the collaboration of experts. They also draw inspiration from adaptation, methodological hybridization, and the transformation of collaborative tools. This urbanization strategy seeks to increase the potential of what is already there (déjà là) by subtly changing its form. This way of intervening across time and space recognizes Paris as an ecosystem and puts the focus on citizen initiative.

A child jumping on the art installation "Selfishness" of Tore Rinkveld, in the frame of the event "Mobile Homes," located on the banks of the Seine in Paris. © Artevia

베를린, 독일 + 뭄바이, 인도

베를린 + 뭄바이_도시 인터미디어: 도시, 기록, 이야기

예브 블라우, 로버트 제라드 피에트루스코, 스콧 마치 스미스
C34-1F

베를린+뭄바이_도시 인터미디어: 도시, 기록, 이야기에서는 현대 도시 환경의 복합성과 역학관계를 이해하기 위해 새로운 '인터미디어' 방식을 연구한다. 베를린과 뭄바이에 초점을 맞춘 이번 전시는 도시 연구에 대한 새로운 접근법을 추구하는 장기적 프로젝트의 일환이다. 도시의 '사각지대'를 집중 조명하는 이번 작품에서는 도시 연구와 관련된 기존의 이야기와 주류적 사고방식으로는 포착하지 못하는 장소와 주제를 탐구하고, 사각지대를 비추기 위해 새로운 미디어를 풍부하게 활용한다. 특히 미디어와 디지털 미디어, 즉 기록 문서부터 디지털 데이터, 사진, 지도, 건축 도면, 그래픽 자료, 텍스트, 애니메이션, 영상 자료에 이르는 다양한 기술과 미디어를 사람들 간의 대화와 소통에 활용하여 서로 다른 미디어의 다양한 언어가 매우 심도 깊게 상호작용한다. 미디어 간의 결합 과정에서 새로운 '인터미디어' 언어가 탄생하고 도시에 관한 지식을 습득하고 창출하는 새로운 방식이 생겨나게 되는 것이다. 이러한 새로운 방법론을 통해 다양한 비판적 시각으로 복잡하고 역동적인 도시 현상을 들여다볼 수 있으며, 시간과 공간 속에서 펼쳐지는 복잡다단한 이야기를 시각적으로 그려낼 수 있게 된다.
베를린과 뭄바이, 두 도시의 이야기는 하나의 공통 주제로 묶인다. 이야기들은 계획성과 무계획성, 공식적 행위와 비공식적 행위 사이의 상호의존성을 다루고, 사람들의 이주와 유동성이 남기는 도시의 흔적을 살펴보며, 도시에서 벌어지는 포용과 배제의 양상을 말하고, 자연과 기술, 도시 생태와 도시 인프라 간의 상호관계를 고찰한다. 이번 전시에서는 실제 도시 현장의 생생한 이야기를 우리에게 익숙한 도시 생활과 환경에서 벗어난 '인터미디어' 방식으로 제시함으로써 베를린과 뭄바이 두 도시가 당면하고 있는 다양한 주제들을 소개한다. 관람자들은 본 전시의 이야기들을 자유롭게 해석하여 설명하고, 상호작용하여 각자의 방식으로 의미를 부여할 수 있을 것이다.

BERLIN, GERMANY + MUMBAI, INDIA
BERLIN + MUMBAI_URBAN INTERMEDIA: CITY, ARCHIVE, NARRATIVE

Eve Blau, Robert Gerard Pietrusko, Scott March Smith
C34-1F

Berlin + Mumbai Urban Intermedia: City, Archive, Narrative explores new 'intermedia' methods for understanding the complexity and dynamics of contemporary urban environments. Focusing on Berlin and Mumbai the exhibition is part of an ongoing project of new approaches to the study of cities. Focusing on 'blind spots', we explore sites and topics that are outside the conventional narratives and dominant conceptual frameworks of urban research, using new media-rich methods to bring those blind spots into focus. In particular, we use a range of technologies to bring physical and digital media — archival documents, digital data sets, photography, cartography, architectural drawings, graphics, text, animation, and film and video — into dialogue and interaction with each other. Through these methods, the distinct languages of different media interact at the deepest structural levels. They hybridize, and in the process, generate new 'intermedia' languages — and new ways of acquiring and producing knowledge about cities. These methods allow us to examine complex and dynamic urban phenomena through multiple critical lenses, and to visualize them in densely layered visual narratives that unfold in place and time. The narratives in the two cities are bound together by common themes: they examine the interdependence of the planned + the unplanned and of formal and informal practices; the urban imprint of migration + mobility and the modalities of inclusion and exclusion; and the interrelation of nature + technology and of urban ecology and infrastructure. The exhibition explores these themes in Berlin and Mumbai through site-based intermedia narratives that challenge our understanding of urban environments and processes. The stories they tell are radically open to interpretation, elaboration, and the interactive construction of meaning by viewers of the exhibition.

TYOLDNPAR
헤테로토피아의 도시

이바네스 킴, 마리아나 이바네스, 사이먼 킴,
앤드류 호믹, 애덤 슈로스, 새라 데이비스,
앙헬리키 치파, 티안 위양, 규현 킴
C34-2F

헤테로토피아의 도시에서 소개하는 주제는
다음과 같다.

1. 종합계획: 도시와 건축에 대한 단일한
   비전은 결코 시대의 변화를 반영할 수
   없다.
2. 시민성: 포스트모던 사회에는 사회적
   집단의 수만큼 다양한 커뮤니케이션
   양식이 있다. 진실은 더 이상
   지배계층만이 독점하는 명분이 될 수
   없다.
3. 자연: 기술이 인간의 생물학적 영역과
   융합되는 환경에서는 인간과 비인간,
   야만과 문명, 공적인 것과 사적인 것,
   내부와 외부의 정체성에 대한 구분이
   사라질 것이다.

우리는 이에 대한 해결책으로 암호 도시를
제안한다. 익숙한 장소들로 구성된 이
암호 도시는 새로운 집합적 공통분모를
형성하고, 인간과 자연을 융합하며, 다양한
요소가 혼재된 환경을 만드는 공간이다.
'헤테로토피아'라고 일컬을 수 있는 이러한
도시는 현실에서 대도시를 구성하는 요소들을
오늘날 인간 본성에 따른 여러 모호함과
가능성으로 압축하고 대안적인 행동을
제시한다.

TYOLDNPAR
The Heterotopial

Ibañez Kim, Mariana Ibañez, Simon Kim,
Andrew Homick, Adam Schroth, Sarah
Davis, Angeliki Tzifa, Tian Ouyang,
Kyuhun Kim
C34-2F

The Heterotopial City addresses the
following concerns:

1. Masterplanning: Single visions of
   urbanism and architecture could not
   be more out of touch.
2. Citizenship: In a postmodern
   society, there are as many forms
   of communication as there ㄹㅏㅕ
   are social groups, and truth can
   no longer be the alibi of one ruling
   class of citizen.
3. Natures: In a posthuman
   environment, there are no longer
   separate identities of human-
   nonhuman, wild and civilized,
   public-private, inside-outside.

To address this, a crypto-city made
of familiar places is the location of
new collective commons, synthetic
natures, and hybrid environments. As
a heterotopia, this city compresses
elements of real metropolises into
incongruities and opportunities that
reveal our current human nature, and
suggests alternate actions.

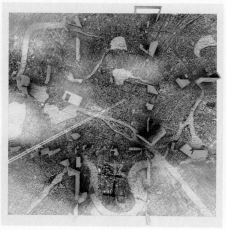

Heterotopial-plan

홍콩, 중국

도시의 여러 가지 얼굴: 홍콩에 관한 네 편의 짧은 이야기

오션 CN [톰 베레베스(팀 리더), 앤드루 하스, 황타오, 천치잰, 처우 앤드루 와이왓, 웡 첸방]
C34-1F

도시의 여러 가지 얼굴: 홍콩에 관한 네 편의 짧은 이야기는 홍콩을 다양한 도시적 패러다임으로 해석하며, 여러 차원에서 바라본 홍콩의 모습과 일반적인 도시의 현실을 이해하는 방법을 제시한다. 이번 전시작은 홍콩이라는 도시를 가까이에서 들여다보는 데 그치지 않고, 도시의 패러다임이라는 큰 개념적 틀에서 해석하고 설명한다. 또한 전시 작품의 콘텐츠와 배치, 그리고 다양한 목소리의 반영과 표현 방식을 통해 '불협화음'이라는 홍콩의 특성을 의도적으로 충실하게 구현한다. 홍콩에 관한 네 편의 짧은 이야기는 홍콩 및 다른 도시들을 이해하기 위한 다양한 개념·기술적 방식, 특성, 사고의 틀을 제시한다.

– 제한형 – 적응형 – 맞춤형 –

홍콩에 관한 네 편의 짧은 이야기는 방대한 시각·공간·문서 자료를 통해 홍콩의 다양한 도시 질서 특징을 소개한다. 정지 사진과 영상, 홍콩의 세 지역을 담은 항공 사진 기반 모형, 도표와 지도, 그리고 세 개의 3D 프린트 모형과 제작 공간에 대한 자료 속에서 홍콩의 도시적 특징을 포착하고 도시를 스마트하게 만드는 핵심 요소를 보여준다

– 스마트 –

홍콩은 세계에서 인구 밀도가 가장 높은 도시이며, 면적의 70%를 차지하는 자연 지형과 나머지 공간에 빽빽이 들어찬 도시 인프라가 뚜렷한 대조를 이루는 곳이다. 극심한 빈부격차, 도시 계획과 현실 사이의 괴리가 완연히 드러나는 도시이기도 하다.

HONG KONG, CHINA

UrbanISMS: Four Short Stories About Hong Kong

OCEAN CN [Tom Verebes (Team Leader), Andrew Haas, Huang Tao, Chen Qijian, Chow Andrew Waitat, Wong Chienbang]
C34-1F

This exhibition presents the city of Hong Kong, read through a set of urban paradigms, and mixed with different registers which include ways of comprehending actual cities in general. The specific focus on Hong Kong is understood and articulated through big conceptual brushstrokes — paradigms of the city. Attributes of the cacophonous urban experience of Hong Kong are explicitly intended to be conveyed in the content and layout of the exhibition, through a mix of voices and their varied means of representation. These Urban-ISMS are presented with different conceptual and technological methods, qualities and filters.
– parametric– adaptive – customized – UrbanISMS express the varied qualities of Hong Kong's urban order, through a saturation of visual, spatial and textual information. Hong Kong's urban attributes were mined to qualify these taxonomies through photographic stills and videos; photogammetry models, diagrams and maps, pertaining to three specific territories in Hong Kong; and three 3D print models and a description of their generative solution spaces. Ultimately, these three territories and the three models sketch out attributes of what constitutes the smartness of the city.
– smart –
As one of the densest cities, Hong Kong exudes a radical juxtaposition between its soft natural topographic landscape, covering 70% of the city, and the hard infrastructure and mass of the city. Evidence also arises of extreme economic inequity and the conflicting forces between planning and emergence.

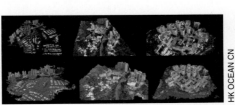

HK OCEAN CN
photogammetry heat maps & models

타이산, 중국
중국 타이산이 추구하는 유연한 집합적 도시화

일레인 욜람 퀑, 제이슨 호
C34-2F

중국 타이산이 추구하는 유연한 집합적 도시화는 타이산의 장마당(market-village, 墟)과 치러우(building unit, 骑楼, 역주: 상가 건물 1층이 안쪽으로 움푹 들어가 보행자 통로가 되는 건물)를 묘사하여 이 도시의 이야기를 관람자들에게 소개한다. 타이산의 집합적 형태(collective form)는 오래 전부터 내려온 중국의 씨족제도(clanship)에서 시작된 공간적 명료함(legibility)을 간직하고 있으며, 이는 현재까지 도시의 변화를 수월하게 진행되도록 하는 요소였다. 타이산의 집합적 형태를 들여다봄으로써 우리는 이 도시가 겪은 역사적 우여곡절을 살펴볼 수 있으며, 거시적 차원(the macro)과 미시적 차원(the micro), 유형과 무형, 과거와 미래를 아우르는 기억의 조각들을 하나의 큰 과정으로 이해할 수 있다.

오늘날 타이산은 커다란 변화를 맞이할 광둥성의 여러 주변 지역 가운데 하나이다. 홍콩, 마카오, 광둥성을 아우르는 'Greater Bay Area(GBA)' 계획이 이러한 변화의 원동력이다. 세 구역으로 나눠 진행되는 GBA 계획의 목표는 사회·경제적 발전의 혜택을 광둥성 전역이 누릴 수 있도록 하는 것이다. 이를 통해 한결 복잡하고 다양한 도시 유형이 탄생하고 있다. 도시가 이렇게 탈바꿈하는 과정에서 발생하는 환경, 집합적 정체성, 일상의 변화가 오늘날의 집합적 형태를 사회·경제적으로 재구성하는데 어떤 영향을 주게 될까? 어떻게 하면 변화를 수용하며 기존의 집합성을 유지하여 시대에 맞게 조정해갈 수 있을까?

이번 작품은 치러우에서 살아가는 타이산 사람들의 생활상을 들여다보면서, 치러우의 건물 형태가 도시의 모습과 경제, 그리고 농촌 사회 구조의 변화에 맞추어 나가는 유연성(malleability)을 갖췄음을 보여준다. 치러우의 특징인 사물 사이의 공간, 그리고 각 부분이 연결되는 구조를 통해 이 유연성이라는 뉘앙스가 잘 드러난다. 유연성이 있는 곳에는 사람, 사물, 공간이 일관성 있게 빽빽히 짜여 있지 않다. 오히려 꾸준한 변화 속에서 서로 불안정하게 연결되어 있으며, 끊임없는 확장과 수축의 과정 속에서 자신의 모습을 바꾸어 나간다. 바로 시간의 흐름에 따라 나타나는 변화를 수용하는 도시화의 모습이다.

TAISHAN, CHINA
Incremental Urbanism: Evolving Collective Forms in Taishan, China

Elaine Yolam Kwong, Jason Ho
C34-2F

Incremental Urbanism tells the story of Taishan through the narratives surrounding the region's market-villages (墟) and building unit — qilou (骑楼). The collective form retains a spatial legibility that arise from traditional Chinese clanship, providing a framework that facilitates urban change beginning from the late Qing Dynasty to the present time. The mapping of the collective form exhibits the wear and tear of Taishan's historical narrative and bridges fragmented memories spanning across the macro to the micro, tangible and intangible, back and forth, in a dialogue that informs the whole process.

Today, Taishan represents one of many peripheral territories that await great transformation in response to the Greater-Bay-Area Plan (HK-Macau-Guangdong). Divided into three zones, the territorial policies aim to distribute the socio-economic benefits across Guangdong province. As a result, we begin to see an urban pattern that is more complex and varied than the common belief of thriving urban versus declining rural. Within this transformation, how do the shifting contexts, collective identities, and every-day life shape the socio-economic restructuring of the existing collective forms? How can we maintain and calibrate the existing framework within which change can operate? Observing the ecology surrounding the qilou (骑楼), the analysis displays the building unit's malleability to transforming landscapes, economies and rural social structures. The nuanced reading centers on the spaces between things and the syntax of part-to-part connections, where spaces, people and objects are read not as cohesive structure but as an unstable series of association in continual transformation, constantly reorganizing itself through processes of expansion and retraction; an urbanism that is capable of hosting change over time.

Methods and Speculations

C34 2F

생활환경 및 도시건축성

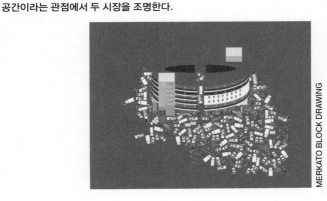

아디스아바바, 에티오피아 +
다르에스살람, 탄자니아
두 개의 시장
AC39-WO (이매뉴얼 아드마수, 젠 우드)
C S 1

두 개의 시장은 아프리카 사하라 사막 이남
지역에 자리잡은 두 개의 시장인 탄자니아
다르 에스 살람의 카리아쿠와 에티오피아
아디스아바바의 메르카토를 조명한다.
오늘날 아프리카 대륙의 건축 담론은 20세기
중반 유럽의 개입으로 인해 빚어진 변화를
찬양하거나, 숨가쁘게 변해가는 아프리카 도시
환경에서 나타나는 형식없는 현상을 독특하게
바라보는데 그치는 두 개의 시장에서는 이러한
관점에서 벗어나, 건축물과 도시를 구성하는
요소들을 그 자체의 변덕스러운 방식으로
분석하고 이 요소들이 사람, 공간, 사물,
사회적 의식과 동등한 중요성을 가진다는
시각을 제시하고자 한다. 카리아쿠, 메르카토
시장은 지역 경제활동의 장일뿐 아니라
국가 건설, 식민지 지배, 냉전 시대 정치,
신자유주의가 진행되는 현장이기도 했다.
평범한 사람들이 일상적인 환경 위에 덧붙여
가는 매력적인 이야기, 억압적인 지배를
거부하고 이에 맞서며 발전시켜 나간 이야기는
사람들의 흥미를 끄는 매력이 있다.
우리는 나이지리아의 시인, 하리 가루바가
"세계에 끊임없이 새로운 매력을
불어넣기"라고 정의한 바 있는 애니미즘적
물질주의를 바탕으로 이 두 시장에 대해 서구
건축 담론의 한계를 넘어 보다 폭넓게 이해할
수 있다. 이 새로운 사고방식은 우리로 하여금
기존의 학문적 체계를 재평가하는 동시에,
외부 엘리트 집단의 지배를 타파하고 관계를
통해 형성된 주관성 을 모색하도록 촉구한다.
두 개의 시장는 다윈의 진화론이 제시하는
선형적 발전 경로를 거부하고, 과거 역사를
재창조하는 동시에 미래를 장악 하는 반사적
공간이라는 관점에서 두 시장을 조명한다.

ADDIS ABABA, ETHIOPIA + DAR ES SALAAM, TANZANI
Two Markets
AC39-WO (Emanuel Admassu, Jen Wood)
C S 1

Two Markets proposes an examination of two urban marketplaces in sub-Saharan Africa: Kariakoo in Dar es Salaam, Tanzania, and Merkato in Addis Ababa, Ethiopia. Contemporary architectural discourse on the African continent tends to either celebrate mid-century European interventions or exoticize notions of 'informality' within rapidly transforming urban contexts. Two Markets, instead, proposes to analyze buildings and urban formations in their own mutable terms, giving equal importance to people, spaces, objects, and rituals. In addition to being sites of local economic trade, both markets have served as sites of nation building, colonial control, Cold War politics, and neoliberalism. We are interested in the magical narratives that have been superimposed on everyday environments by ordinary people, developed against and outside of oppressive regimes.
A framework of Animist Materialism, defined by Harry Garuba as "a continual re-enchantment of the world," expands our understanding of these marketplaces beyond what is possible via Western architectural discourse. It forces a reckoning with our disciplinary structures and tools while exploding the singular vantage point of the foreign elite into a collection of relationally-ascribed subjectivities. Eschewing a linear Darwinian narrative, Two Markets examines these sites as reflexive spaces that simultaneously recreate their histories and pre-possess their futures.

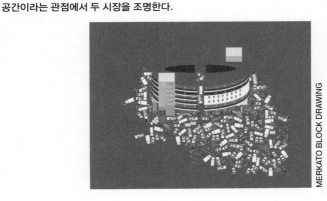

MERKATO BLOCK DRAWING

방콕, 태국
방콕의 도시화: 미래형 스마트도시를 향하여
논 알카라프라세트쿨, 쇼우헝 션
CS2

태국의 수도 방콕은 매년 수천만 명이 찾는 관광명소이다. 이곳은 대도시일 뿐 아니라 친절한 사람들과 저렴한 물가, 다양한 문화유산으로도 잘 알려져 있으며 맛있고 경이로운 놀라움들이 넘쳐나는 곳이기도 하다. 그러나 아쉬운 부분이 있다. 방문객뿐 아니라 현지인들도 방콕의 교통 혼잡과 무더위에 관해서라면 혀를 내두르곤 한다. 이동수단이 구비되어 있지 않다면 달궈진 팬 위를 걷는 것과 별반 다를 바 없을 정도다.
우리는 신기술을 사회 곳곳에 접목하여 사람을 집합시키거나 해산하는 "스마트 도시화"를 제안한다. 이를 위해 우리는 전문 인력을 구축하여 본 프로젝트가 과학적으로 증명될 뿐 아니라 공간과 장소에 관한 한 사회의 요구에 대응력을 갖도록 하였다. 데이터에 기반을 둠으로써 도시계획과 설계에서 이미 활용된 적이 있는 거대 개념을 우회하고자 하였으며, 문제의 원인과 보행자들이 겪는 고충을 도시성, 도시화, 도시주의 차원에서 접근해보았다. 본 프로젝트는 공간에 관해 물리적인 현실을 배제할 수 없는 이유를 일반적인 근거를 들어 설명한다.

BANGKOK, THAILAND
Bangkok's Urban Presence: Toward the Future of Smart Urbanity
Non Arkaraprasertkul, Shouheng Shen
CS2

Visited by tens of millions of visitors every year, Bangkok, Thailand's capital, has remained one of the best travel destinations in the world. As an urban metropolis, in addition to its being friendly, affordable, and culturally diverse, Bangkok is also delicious, exhilarating and, needless to say, full of surprises. However, there is one thing of which both the locals and visitors alike are not fond: it is not a very convenient city. From its infamous traffic congestion to unbearable temperature, Bangkok could feel like a boiling pan especially when mobility is hardly there.
We propose "Smart Urbanity" interventions through new technology vis-à-vis social factors bringing people together and pushing them apart. Combining the expertise of the key investigators, this project presents both a scientific data-driven and socially sensitive understanding of space and place. Relying on data to bypass meta-concepts overused in urban planning and design today, we get to the root cause and dilemma of pedestrians on the streets as agents of urbanity, urbanism, and urbanization. This project seeks to provide a generalization of how and why we can't and shouldn't ignore the physical reality of a certain contact in creating a sense of place.

Bangkok Presence by Shouheng Chen

바르셀로나, 스페인

바르셀로나, 도심 속 도시

AMB (바르셀로나 광역행정부)
CS3

바르셀로나는 오늘날의 역동적이고도
복합적인 경제사회를 반영한다. 다양한
소도시가 정착하면서 그 틈새에 자연공간이
조성되는 형태로 꾸며진 이 지역은 지형적
특색을 감안하고 민족의 정체성을 대물림하며
환경의 지속가능성을 고려하는 동시에
경제성과 경쟁력, 사회통합을 이뤄내야 하는
과제에 직면해 있다. 이로 인해 도시계획에
관한 새로운 접근법과 수단이 요구되는
실정이다.

이처럼 다각화된 지역에서는 혁신과 변혁을
꾀하는 동시에 문화유산을 보존하고 대자연을
수호하는 방식으로 도시를 계획할 필요가
있다. 아울러 사회기반시설을 통합하고
기후변화와 기술개발 트렌드 등 미래사회의
요구에도 명민하게 대응해야 한다.

이와 같은 맥락에서, 그린 인프라는 주된
버팀목이라고 할 수 있다. 즉, 환경적 가치를
지닌 52% 이상의 도심 공간과 잠재적 녹색
지대를 연결하는 것이다. 또한 도로, 광장,
도심공원, 지붕과 건물외벽 등이 인근의
자연공원과 연계되어 도시의 75%가 그린
인프라를 갖추는 통합된 그린 공동체를
형성한다. 본 전시는 AMB의 역량과 실행력을
바탕으로 도시 속 자연과 열린 공간의 역할을
조명하고, 도로 및 거리가 인적 친밀감을
북돋우는 방식과 도시의 기능을 활성화하는
방안에 대해 살펴본다.

BARCELONA, SPAIN
Metropolis Barcelona, A City of Cities

AMB (Barcelona Metropolitan Area)
CS3

Metropolis Barcelona is a very dynamic
and increasingly complex economic and
social reality. This territory, structured
by important natural spaces between
which different cities have been settled,
demands new approaches, tools and
instruments of urban planning, based
on the geography, to preserve the
identity of the peoples and to ensure
environmental sustainability, economic
competitiveness and social cohesion. In
this highly consolidated territory, urban
planning needs to focus on the renovation
of existing fabrics, the transformation
of obsolete areas, the preservation of
heritage and landscape elements, the
urban integration of infrastructures and
the adaptation to an immediate future
marked by such profound dynamics as
those associated with climate change,
the technological revolution or the
increasingly complex social demands.
In this context, green infrastructure is
considered as the main supportive system
of the metropolis: a set of spaces of great
environmental wealth that represents
more than 52% of the metropolitan
territory plus all those urban spaces with
the potential to become green spaces.
Streets, squares, urban parks and even
roofs or facades can connect to the great
natural parks of the surroundings and
thus form an interconnected network
of green spaces, adding up to 75% of
the territory. The exhibition shows the
reflection of AMB on the strategic and
vertebral role that natural and open
spaces can have in the territory and how
connecting and re-connecting roads and
streets can bring people's activities closer
and improve the functioning of our city of
cities.

Metropolis Barcelona © AMB

런던, 영국

런던은 지금, 그리고 앞으로...

피터 비숍, 이저벨 앨런
CS4

런던은 변화를 수용하기 위해 끊임없이 적응해
왔다. 인류가 4차 산업혁명의 문턱에 접어든
현재, 런던은 생산의 장소에서 소통의 장소로
변모하고 있다. 이러한 변화는 물리적 영역,
디지털 영역, 생물학적 영역 사이의 경계를
허무는 기술 융합을 통해 극명하게 드러난다.
4차 산업혁명이 불러올 변화는 규모, 범위,
복잡성 면에서 이전과는 완전히 다른 경험으로
인류에게 다가올 것이다. 중요한 것은 새로운
산업혁명 시대의 급격한 기술발전이 시민의
삶의 질을 저해하지 않고, 오히려 더욱
향상시키도록 해야 한다는 점이다.
이번 전시에서는 힐드리 스튜디오, 웨스턴
윌리엄슨 + 파트너스, 제5 스튜디오, dRMM,
포스터 + 파트너스, PiM 스튜디오, 에잇폴드,
이그릿 웨스트, 어셈블리지 앤 IF_DO 등 영국
건축의 혁신을 주도하는 10개 단체의 작품을
통해 런던이 21세기 도시생활 패러다임의
선두주자가 될 수 있는 방안을 제시한다.
런던을 인류 행복의 도시로 만들고자 하는
이들의 노력은 '10대(大) 선언문'으로
이어진다.
런던대학교 바틀릿 건축대학 학생들도
참여하여 새로운 세대의 런던 시민이
만들어가는, 시민을 위한 도시의 청사진을
제시한다.

LONDON, UK

London is...

Peter Bishop, Isabel Allen
CS4

London has constantly re-adapted to
embrace change. We are on the cusp of
a Fourth Industrial Revolution as London
evolves from a place of production
to a place of human interaction. This
transformation is characterized by a
fusion of technologies that blur the
lines between the physical, digital and
biological spheres. In its scale, scope and
complexity, this new era will be unlike
anything humankind has experienced
before. We need to ensure that rapid
technological progress enriches — rather
than erodes — the quality of civic life.
We have invited ten of the UK's most
innovative practices — Hildrey Studio;
WestonWilliamson + Partners, 5th Studio;
dRMM; Foster + Partners; PiM Studio;
8FOLD; Egret West; Assemblage and
IF_DO — to put forward propositions to
help London become a paradigm of urban
living in the 21st century. The result is
a ten-point manifesto to make modern
London a place where humanity can
thrive.
Selected students from the Bartlett
school of architecture have produced
proposals for a city shaped by, and for, a
new generation of Londoners.

Collective Typologies

CS4

협력적 유형

Egret West - London is a city to grow
inEgret West - London is a city to grow in

멕시코 시티, 멕시코
빅 이퀄라이저

ORU (오피시나 데 레실리엔시아 어바나),
에드위나 포르토카레로
CS5

빅 이퀄라이저는 우리가 지진이라는 예기치 못한 사태로 공포를 겪을 때 마음 속 깊은 곳에서 일어나는 인식의 전환을 연구한다. 굳건하던 대지가 요동치고, 아늑하던 집은 탈출을 가로막는 장애물이 되어 버린다. 우리를 편안하게 해 주는 가구가 자리를 차지해 갈수록, 공간이 주는 여백의 미와 비상시에 대피로가 될 수 있는 여유 공간은 사라져 버린다. 우리가 아끼던 소유물도 아무 가치가 없어지고, 오히려 우리의 발목을 잡을 뿐이다. 지진이 닥치면 (역주: 도시의 사회-공간적 분리는 사라지고) 우리는 낯선 이들과 함께 지내며 마음의 위안을 찾게 된다. 관람자들은 빅 이퀄라이저를 통해 '안전'이라는 개념을 재해석하게 된다. 또한, 지진의 엄습으로 우리의 신체와 도시라는 인공 환경 사이의 관계가 완전히 뒤집힐 때, 그리고 마음은 공포에 사로잡히고 빠른 결단이 그 어느 때보다 중요해질 때, 우리가 공간을 어떻게 활용하게 되는지 — 소파 옆에 웅크리거나, 탁자 밑에 몸을 숨기거나, 지진에도 잘 무너지지 않는 문틀 밑에 서서 지진이 지나가기를 기다리거나 — 생각하게 된다. '더 좋은 방법이 있어'라며 으스댈 여유 따위는 사라지는 것이다. 관람자들이 인식이라는 관점에서 우리의 원초적 생존 본능을 연구하는 이 작품을 통해 인류가 만들어 내는 건축물 속에서 스스로가 소외당하는 현실에 대해 다시 한 번 생각해 보는 기회가 되기를 바란다. 건축물 속에서 인간은 자연에게, 사회는 환경에게, 사람들은 서로에게 낯선 존재가 되는 것은 아닐까?

MEXICO CITY, MEXICO
The Big Equalizer
ORU (Oficina de Resiliencia Urbana),
Edwina Portocarrero
CS5

The Big Equalizer explores the deep perceptual shift we undergo when subjected to the terror ensued by the unpredictability of an earthquake. Where the ground is solid and home is shelter, the ground trembles and home are an obstacle course. The furniture we rely on for comfort is traded for the safety that negative spaces around it may afford, our possessions no longer carry value but slow us down and we share with strangers' solace and comfort.
The Big Equalizer invites you to reinterpret what we understand as safety, and to inhabit the space the way we do when the relationship between our body and the built environment is upended, when our mind is struck with terror and decision making is crucial. Crouch by the couch, take shelter under the table or stand under the doorframe — we don't pretend to know what is better.
By examining our basic instinct for survival from a perceptual standpoint, we hope to bring awareness to the constructs we create that alienate us — the human as alien to the natural, the social as alien to the ecological and humans as aliens to each other.

Mexico City. Volunteers feed survivors of the 1985 earthquake

거대한 계획
메이드 포 차이나
마이클 솔킨 스튜디오
CO1

도시설계는 유토피아적, 남성적, 권위주의적인
것으로 여겨지며 한때 배척되어온 적이
있다. 그러나 우리는 이에 동의하지 않으며
이에 대한 담론을 중요시한다. 도시는
개발과 발전수준에 대해 저마다의 출발점과
판단기준을 갖고 있기 마련이다. 폭발적인
성장과 더불어 전 세계 70%의 인구가 도시
거주민이 될 것이라 한다. 이는 곧 자연재해의
가속화를 의미하기도 한다. 중국은 지난
30년간 4억 명에 달하는 인구를 도시
중산층으로 합류시켰다. 이는 대형도시가
초대형 도시로 변모하고 있음과 기존의
도시정책이 더는 실효성이 없음을 시사한다.
공원에 있는 아파트, 곳곳으로 뻗어가는
주택단지, 사슬처럼 연결된 고가도로는 중국형
도시화에 대해 심도 있는 논의가 오가던 당시
차용했던 유럽모델에 불과하며 그마저도
실패로 끝났을 뿐 아니라 도시에는 적막감을,
인간 사이에는 격리감을 부추겼을 뿐이다.
본 프로젝트는 이상과 실현 가능성을 두루
살피나 좋은 도시여야 한다는 단일 전제에
근간을 두고 있다. 여기서 말하는 좋은
도시란 이웃공동체와 인본주의를 수호하고,
자연공간과 인위적인 공간이 균형을 이루며,
지역사회가 자율성을 보장받고 환경 인프라를
갖추는 것이다.

BIG PLANS
Made for China
Michael Sorkin Studio
CO1

The design of cities in their all-
at-once has become anathema,
deplored as "utopian" (a concept
itself, sadly, in bad odor), masculinist,
authoritarian, etc., etc. We disagree
and think this discourse is vital, not
only imaginatively but on the ground:
every city has a point of departure, a set
of governing ideas that limn the terms
— however constricting or loose — for
its development and maturation. In
light of explosive growth — by mid-
century 70% of the world's population
will be urban — and of accelerating
environmental disaster, it is only rational
that new cities be a major part of the
solution. The attraction of China — which
has elevated 400 million people to the
urban middle class in thirty years — for
us is their increasing realization that
the exponential expansion of megacities into hyper-cities as well as their
current planning default — residential
"towers in the park" stretching to every
horizon, bleak at grade, connected
by clotted highways — is not simply
alienating and dreary but drawn from a
bankrupt European model, at a time when
there is intense discussion of what
exactly the qualities of an urbanism
with Chinese characteristics are, the
projects we have chosen for the Biennale
reflect a long trajectory from the quasi-
fantastical to the fully realizable. All of
them, however, spring from the same
fundamental predicates of the good city:
the neighborhood, the primacy of people
on foot, a free mix of uses, a recalculation
of the ratio of green, blue, and built space,
a high level of local metabolic autonomy,
and the most radically possible
environmental infrastructure.

CO1

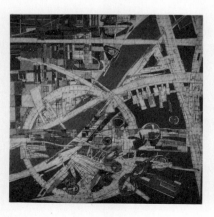

블랙 록 시티, 미국
소리의 도시: 먼지의 도시

닉 소어스
CO2

블랙 록 시티는 매년 미국 네바다 사막에
건설되었다 해체되는 도시이다. 여러 개의
동심원을 그 중심에서 바퀴살처럼 뻗어
나가는 통로가 갈라 놓은 구조로 펼쳐지는
이 도시에서 수만 명의 사람들이 일주일
동안 펼쳐지는 버닝 맨 축제를 즐기기 위해
입장권을 구매하고, 이들이 머무를 캠프가
질서정연하게 가득 들어선다. 하지만 블랙록
시티를 촬영한 위성 사진 속의 가지런한
모습과는 달리, 지상에서는 구획과 건축
규정 없이 지어진 임시 캠프와 구조물들은
시각적으로 혼돈스럽다. 따라서, 블랙 록
시티에서 길을 찾는 것은 섬세하고, 호기심을
자극하며, 때로는 염려가 되는 경험이다.
주변의 난잡한 소음을 뚫고, 심장을 뛰게 하는
비트에 따라 몸을 귀에 맡긴 채 새로운 예술적
목적지와 경험을 찾아 나서게 되기 때문이다.
어쩌면, 길을 잃어버린다는 것은 버닝 맨
축제에서 만끽할 수 있는 최고의 경험이
아닐까?
본 전시에서는 낯선 도시 환경에서 겪게 되는
'길찾기'라는 원초적인 문제를 들여다본다.
다섯 개의 아크릴 사운드 패널을 통해
만들어지는 사운드스케이프는 우리의 도시
경험에 영향을 주는 주요 음향 요소들을
구현한다. 다섯 가지 핵심 요소는 다음과 같다.

1. 소리는 공간 속에서 사람들에게 방향을
   제시한다.
2. 소리는 시간을 알리며 측정의 기준이
   된다.
3. 소리를 통해 사람들은 멀리 떨어져
   있어도 느끼고 소통할 수 있다.
4. 혼자이고 싶을 때는 소리의 벽을
   활용할 수 있다.
5. 소리는 특별한 사건이 없을 때에도
   사람들이 끊임없이 깨어 있도록 한다.

CO2

BLACK ROCK CITY, USA
City of Sound: City of Dust

Nick Sowers
CO2

Black Rock City is constructed and
deconstructed every year in the desert of
Nevada. The city is laid out in a radial grid
which is then filled in by organized camps
and tens of thousands of individuals
who purchase a ticket for the week-long
Burning Man festival. The satellite-view
order of the grid, however, is countered at
the ground level by the visual chaos of ad
hoc camps and structures built without
zoning or building codes. Navigation
becomes a delicate, curious, and at times
anxious act of filtering the background
noise, following the pulsating beats, and
leading with one's ears to find new art
destinations and experiences. Getting
lost may very well be the objective of one's
experience at Burning Man.
This installation investigates the basic
problem of way-finding within an
unfamiliar urban milieu. Five acrylic
sound panels play a soundscape which
embodies a core component of sound
mediating our experience of the city.
Those five core components are as
follows:

1. Sound orients people within a
   place.
2. Sound marks and measures time.
3. Sound produces touch at a
   distance
4. Sound provides isolation for the
   self.
5. Sound encourages exploration
   between events.

Burning Man 2007 aerial view

보스턴, 미국
보스턴의 하층 도시
랜딩 스튜디오 (댄 아담스, 마리 로 아담스)
CO3

보스턴의 고속도로는 마치 도시를 지나는 움푹 파인 참호처럼 주변 지역보다 낮은 위치에 건설되었지만, 최근에는 그 위에 천장을 씌워 고속도로를 터널로 만들고, 그 터널 위로 새로운 도시가 건설되고 있다. 고속도로 고가다리 아래에는 새로운 유형의 공원이 생겨났다. 고속도로망과 도시, 둘 중 어느 쪽에도 속하지 않은 이 공간에서 도시 관리와 고속교통망, 공공 사업이 융합된 도시가 싹트고 있다. 고속도로 교각의 접합부분에 소금이 산처럼 쌓이고, 콘크리트를 운반하는 컨베이어 벨트가 고속도로 경사로 사이로 지나가며, 스케이트보드 애호가들은 도시의 교각에서 숫자 8을 그리며 보드를 타는 혼종의 도시이다.

고속도로 공간은 '원활한 차량 소통'이라는 유일한 목적의 신호 규제체계에 통제 받아왔다. 그러나 건설시장 호황과 주택시장 위기가 동시에 닥치고 토지의 가치를 최대한 활용하기 위한 노력이 진행되고 있는 현실에서, 본 전시작 보스턴의 하층 도시는 석유 – 자동차 – 인간의 시대라는 혁신적 변화의 순간을 포착하고 새로운 신호체계를 통해 참신하고 다양성이 보장되는 규제체계를 고속도로 고가다리 아래 공간에 적용하고자 한다. 이 공간에 다양성을 불어넣고, 확대하며, 그 다양성이 더욱 뚜렷하게 부각되는 신호 관리체계를 도입함으로써 공간 생태계가 도시 이동망과 융합되고, 인프라 시설 정비 활동이 시민의 여가생활과 맞물리며, 공공 사업 공간이 어린이들의 놀이터가 되고, 애완견, 강물을 유영하는 청어, 도시의 캠핑 애호가들이 만나 새로운 집합 공간을 형성하게 될 것이다.

BOSTON, USA
Boston Understories
Landing Studio (Dan Adams, Marie Law Adams)
CO3

Boston Understories looks at spaces and activities under highway viaducts through the lens of the regulatory sign. Standard markers like the ubiquitous 'no trespassing' have been used to limit or narrowly prescribe public access and fail to reflect the actuality of these sub-infrastructural spaces, where the market forces development into infrastructural margins, ecological systems converge with mobility networks, and public works intermingle with public recreation. Boston Understories reflects on these phenomena and introduces a new, more plural taxonomy of signs to encourage, amplify, and make legible actual and imagined collective domains of urban viaduct spaces.

Infrastructure

CO3

인프라스트럭처

알도의 구상: 사회적 인프라

줄리아 잼로직, 코린 켐프스터
CO4

Aldo: a Social Infrastructure

Julia Jamrozik, Coryn Kempster
CO4

CO4

19세기 호황을 누렸던 뉴욕 주 버펄로. 그러나 20세기 중반 제조업과 산업 전반이 급격한 침체를 겪으며 사람들이 떠나가자, 도시에는 텅 빈 구역이 넘쳐나고 널찍한 거리에는 차량만이 오가게 되었다. 최근 투자를 비롯한 여러 움직임들이 일부 나타났지만 도시는 아직도 빈곤과 차별 속에서 신음하고 있다. 우리는 놀이 공간이 다양한 경제·정치·인종적 배경을 가진 사람들이 서로의 경험을 나누는 흔치 않은 무대 역할을 한다는 점과, 네덜란드 건축가 알도 반 아이크가 2차 대전 이후 암스테르담에서 추구했던 '빈 공간을 채워 나가는 도시화', 그리고 그 과정에서 '기회를 적극적으로 활용하는' 모습에 주목하여 공공 영역에서 사람들이 즐겁게 함께 할 수 있는 사회적 인프라 네트워크를 제안한다. 우리가 살아가는 이 심각한 분열의 시대에, 사람들은 이러한 아날로그적 실마리를 통해 잠시라도 공허한 메아리만이 오가는 SNS를 떠나 이웃과 소통할 기회를 가지게 된다. 놀이를 도시 변화를 위한 집합·정치적 양식이라는 시각에서 접근하고 놀이 공간을 도시의 중요한 부분으로 수용하는 것이 우리 시대의 시급한 과제이다.

본 출품작 알도의 구상: 사회적 인프라는 반 아이크 알도의 유명한 '정글짐'에서 착안한 세 점의 설치물을 새롭게 해석하여 구성한 집합적 공간이다. 관람자들이 거리낌 없이 들어와서 올라타고, 기대며, 혹은 아예 걸터앉아 상호 소통할 수 있다.

Buffalo, New York, a nineteenth-century booming center, experienced a sharp decline in manufacturing, industry, and population in the middle of the last century and finds itself with a surplus of empty lots and oversized car-centric streets. Despite a recent spur of investment and activity, many areas remain unaffected by the positive changes, leaving the city still blighted by poverty and segregation.

Based on observations that spaces of play are rare moments where people from different economic, political and racial backgrounds share experiences, and influenced by the opportunistic infill-urbanism of Aldo van Eyck in post WWII Amsterdam, we propose a network of social infrastructures for playful encounters in the public realm. In a time of grave disunity, these analogue prompts provide a chance to, however briefly, leave the echo chamber of social media and interact with a neighbor. There is an urgency to this approach that considers play as a collective and political form of occupying the city, and playspaces as an essential urban amenity.

Our installation Aldo appropriates van Eyck's signature climbing-frame and reconfigures three of them to construct a collective space, which visitors are encouraged to occupy and interact with by climbing, leaning, and lounging in it and on it.

프랑크푸르트, 독일
하이퍼시티

피터 트루머
CO5

도시는 인류가 가진 가장 거대한 유물이며 이 유물은 다양한 모습의 도시들로 이뤄진다. 도시의 모습은 사회·경제·정치적 역학관계에 따라 변한다. 각각의 도시는 특별한 형태를 갖으며, 형태는 도시 건축의 역사를 통해 잘 드러난다. 그러므로 모든 도시는 하이퍼건축에 바탕을 둔 하이퍼도시라고 할 수 있다.
조각 작품 하이퍼시티는 프랑크푸르트라는 초도시의 다양한 건축 형태를 소개한다. 각 조각의 크기는 건축물의 최대 규모를 반영하였으며 실물 대비 1/100 비율로 제작되었다. 이 작품에서 표현하는 초도시건축 양식은 다음과 같다.

1) 정치 도시:
   '벽' 건축 양식으로 표현되는 중세 도시의 모습을 담고 있다

2) 자본주의 도시:
   '바둑판 모양의 구획' 건축 양식으로 표현되는 19세기 도시의 모습을 담고 있다.

3) 사회민주주의 도시:
   '평평한 판자 모양' 건축 양식으로 표현되는 1920년대 도시의 모습을 담고 있다.

4) 복지국가 도시:
   '거대한 틀' 건축 양식으로 표현되는 2차 세계 대전 이후 도시의 모습을 담고 있다.

5) 신자유주의 도시:
   '대규모로 늘어선 탑' 건축 양식으로 표현되는 1980년대 도시의 모습을 담고 있다.

6) 금융 자본주의 도시:
   '산처럼 솟아오른 거대한 건물' 건축 양식으로 표현되는 세계화 도시의 모습을 담고 있다.

7) 비(非)인간 도시:
   '컨테이너' 건축 양식으로 표현되는 초(超)사물 도시의 모습을 담고 있다.

FRANKFURT, GERMANY
HYPERCITY

Peter Trummer
CO5

The city is the largest human artefact. This artefact is formed by multiple cities. The content of these cities derive from their socio-economic politics. These cities emerge in a particular form and these forms are articulated by the history of their city's architecture. Every city is therefore a HyperCity made out of HyperArchitecture.
The HyperCity Sculpture represents the architecture of the hypercity of Frankfurt. The size of the sculpture represents the largest real estate project currently built as a single structure. The model is scaled in 1/100 and it consists of the following HyperCityArchitecture.

1) THE POLITICAL CITY:
   The Political City is the city of the Middle Ages. Its architectural form is the WALL.

2) THE CAPITALIST CITY:
   The Capitalist City is the City of the 19th century. Its architectural form is the GRID.

3) THE SOCIAL-DEMOCRATIC CITY:
   The Social- Democratic City is the city of the 1920's. Its architectural form is the SLAB.

4) THE WELFARE-STATE CITY:
   The Welfare-State City is the city after WW2. Its architectural form is the MEGAFORM.

5) THE NEO-LIBERAL CITY:
   The Neo-liberal City is the city of the 1980's. Its architectural form is the TOWER in the Park.

6) THE FINANCIAL-CAPITALIST CITY:
   The Financial-Capitalist City is the city of globalisation. Its architectural form is the build MOUNTAIN.

7) THE NON-HUMAN CITY:
   The Non-Human City is the City of hyper-objects. Its architectural form is the CONTAINER.

Layers and Collective Memory

CO5

집합적인 도시기억과 레이어

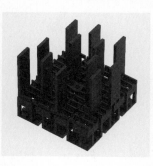

하이파, 이스라엘
정원 도시의 계단

라피 세갈 A+U

CO6

전시 작품 <u>정원 도시의 계단</u>은 하이파를 연결하는 1,000개의 계단을 도시의 현재와 미래를 공유하는 공간이라는 관점에서 들여다보고, 계단이 도시의 유적, 시장, 영화관, 주택 지역을 가로지른다는 사실에 특별한 의미를 부여한다. 계단에서 보이는 지중해의 풍경과 푸른 나무가 있는 통로는 도시의 공공·민간 공간을 서로 이어준다. 계단은 서로 다른 사람들이 만나는 장소가 되기도 하며 지나가도 좋지만 잠시 머물러도 좋은 곳이 되기도 한다. 복잡한 시내의 소음에서 벗어나 '정원 도시'의 녹색 정경을 만끽할 수 있는 하이파의 계단이야 말로 진정한 집합성의 공간이다.

계단에는 '한 장소와 다른 장소를 잇는 통로' 이상의 시각적, 물리적 실체가 있다. 계단은 사람들이 머무르고, 이웃과 만나고, 오밀조밀한 열린 공간에서 즐거운 시간을 보낼 수 있는 도시의 역동성이 있는 곳이다. 이번 작품에서는 자유롭게 설치된 '계단 구조물'과 더불어 관람자 시선에 따라 움직이는 렌티큘러 사진을 함께 배치했다. 또한 경치, 위치, 규모가 모두 다른 계단들의 네 가지 공통 요소를 연구하고 그 안에서의 집합적 순간을 포착한다. 사진 작가 길리 메린(Gili Merin)이 시각 자료 제작에 도움을 주었다. 관람자들은 각각 다른 장소에 설치된 네 점의 작품을 통해 계단과 접하고 그곳에서 보이는 경치를 즐길 수 있다. 서로 떨어져 있지만 하나의 작품을 이루는 이 설치물은 관람자들에게 도시의 이야기를 들려주고, 하이파의 멋진 정경을 선사하며, 서울의 계단도 즐거운 공간이 될 수 있음을 보여준다.

HAIFA, ISRAEL
Sit(e)lines of a Garden City

Rafi Segal A+U

CO6

<u>Sit(e)lines of a Garden City</u> explores Haifa's system of urban stairs as present and future shared spaces, monumentalizing the city's 1000 stairs which traverse monuments, markets, cinemas, and residential homes. Suturing public and private with vistas that sweep towards the Mediterranean and green corridors, the stairs create pockets and corridors of heterogeneous mingling, uniquely offering both passage and pause. Away from the noise of the street and surrounded by 'garden city' vegetation, the stairs are a true collective space.

Rather than merely a segue between one site and another, the stairs, visually and bodily, become active places to be in, meet neighbors and enjoy open space at an intimate scale within the city. The installation of Sit(e) lines in the Seoul Biennale integrates lenticular photography of views into the free standing 'stair structures', capturing moments of collectiveness in four typical stair elements that share different views, locations, and sizes. The views are composed by photographer Gili Merin, commissioned for this project. These four pieces, separated in various locations, offer encounters with the stair space and view. Separately and together they tell a story about the city, inviting visitors into the landscape of Haifa, extending this space of conviviality into Seoul.

146

홍콩, 중국
교집합 도시, 홍콩
제럴딘 보리오
C07

제럴딘 보리오는 서울도시건축비엔날레의
"집합도시"라는 주제 아래 홍콩을 교집합이
발생하는 공간 차원에서 살펴보았다.
밀도 높은 도심지역의 뒷골목은 정의가
불분명한 완충지대다. 열린 공간이자
사적인 공간이고 엄격한 규율이 존재하나
불법행위가 난무하기도 한다. 짜임새를 갖춘
곳이 아니기에 도시의 뒷골목은 그 특색과
역할이 모호하고 도시행정가들의 주목을
크게 받지 못하는 경향이 있다. 하지만 이와
같은 사각지대는 수많은 잠재력을 갖고
도시거주민에게는 시기적절한 찰나의 공간이
되기도 한다.
본 프로젝트는 선정된 특정 지역을 안팎으로
조명한다. 관객은 개별성과 집합성 사이를
오가며 교집합을 형성해가는 도시를 경험할 수
있다.

HONG KONG, CHINA
Inverted Hong Kong
Géraldine Borio
C07

Representing Hong Kong at the Seoul
Biennale of Architecture and Urbanism
2019, Géraldine Borio will examine the
theme of 'The Collective' through the
lens of her research on the city's liminal
spaces.
The interstitial network of back lanes
within the dense urban fabric are
ambiguous buffer zones that oscillates
between the classical binary opposition
of public/private, legal/illegal, inside/
outside. Not planned as such, they are
not assigned to particular functions
or programs and are often ignored
by the city's officials. However, these
in-between territories are open to
multiple interpretations and offer to the
city's inhabitant a momentary space to
appropriate.
Inverted Hong Kong is a site-specific
installation that speculates on the shifting
boundaries between inside and outside.
The audience will be engaged and
challenged in a representation of a city as
system of thresholds where the domestic
and the collective overlap.

Appropriations

C07

도시공간의 홍콩

메데인, 콜롬비아
숨 쉴 수 있는 도시 메데인을 위해

카밀로 레스트레포
CO8

인프라스트럭처

메데인은 강이 흐르는 좁은 계곡에 자리하고
있다. 협곡을 지나는 물줄기는 모두 이 강으로
흐른다. 이 협곡은 평소에는 사람들의 눈에
띄지 않다가 우기가 되어 물이 불어나고
홍수와 재난이 닥칠 때가 되어서야 모습을
드러낸다.

메데인은 공공 공간 부족, 오염, 불평등, 가난
등 여러 문제로 신음하고 있다. 만약 메데인의
협곡에 지역 공동체를 위한 집합 공간을
조성하고, 그곳을 강변 공원으로, 또 급격한
기후 변화에 대응할 수 있는 완충 공간으로
활용하면 어떨까? 만약 도시 공간을 활용해
지역 고유의 생물종을 키워내어 대기의 질을
개선할 수 있다면 어떨까?

메데인 협곡 사진들이 반투명 천에 인쇄되어
외벽면에 커튼처럼 설치되어, 중첩된 새로운
풍경을 조성된다. 이번 작품의 핵심 요소는
전시물이 실제 바람과 빛을 만나게 되는
것이다. 드리운 커튼 사이로 부는 바람에
작품이 '움직이면서' 사람들이 잊고 있던
생명의 근원, 공기가 얼마나 중요한지 깨닫게
될 때 이 작품은 비로소 완성된다.

이번 작품은 관람자들에게 도시 공간과 그
속에서 살아가는 식물들이 어우러져 조성하는
정취를 보여주고, 이러한 공간을 활용하여
더 나은 미래 도시를 만들어가야 한다고
역설한다.

MEDELLIN, COLOMBIA
The Air Between Us_Breath

Camilo Restrepo
CO8

Geographically, Medellin is located in a
narrow valley with a river in the middle
that receives all the watercourses of
ravines. Frequently, the citizens are
unaware of the ravines, making them only
present in the rainy season when they
cause floods and catastrophes.
Medellin is struggling with many
problems such as lack of public space,
pollution, inequality, and poverty. What
would happen if Medellin uses the ravines
as collective space for local communities,
as linear parks, and at the same time as
climate equalizers? And the land is used
for growing endemic species and plants
to improve the air quality?
Our proposal overlaps photos
of Medellin's ravines to create a
new landscape through pieces of
semitransparent fabrics, hanging from
the walls facades, as curtains. The
external props complete the scene: wind
and light. The wind between the fabrics
will generate movement, an essential
component to complete the experience
of this installation. Remembering Us the
quality of an invisible vital element: Air.
The big picture of the installation conveys
the topography of the city, the vegetation,
and the mood of these types of spaces if
they are thought of as an opportunity for a
better city in the future.

Infrastructure

선전, 중국
다시 채워 나가는 도시: 선전 중심가의 재구성

얼버너스 아키텍처 앤드 디자인
C09

1980년대와 90년대의 야심 찬 중국 젊은이들에게 40년 사이에 30만 인구의 외딴 마을에서 1,700만 인구의 초거대 도시로 성장한 선전은 꿈의 도시였다. "시간은 돈, 효율은 생명"이라는 경제 개혁 모토가 중국 곳곳에 울려 퍼졌다. 선전의 중심인 4.4 킬로미터 길이의 '푸텐 중심지구'(Futian Central District)는 전적으로 중국 중앙정부의 하향식(top-down) 계획과 설계에 따라 형성된 곳으로서, 그야말로 '선전의 속도'(Shenzhen Speed)로 진행되는 도시화의 모습뿐 아니라 '선전의 기적'을 극명하게 보여주는 지역이다.

1980년 처음 지정된 '푸텐 중심지구'는 현대 도시의 이성적이고 기능주의적인 원칙에 입각하여 건설되었으며, 세계 자본주의의 우상(icon)과 사회주의 이념, 그리고 중국의 전통적인 상징주의(symbolism)가 뒤섞인 공간이 되었다. 넓은 녹지가 있는 거대한 중심축(Grand central axis) 지역, 차량으로 들어찬 넓은 도로, 서로 고립된 도시 구역과 눈부신 고층빌딩이 한데 모여 현대적이고, 언제나 젊으며 미래를 바라보는, 그러나 한편으로는 사회·심리적으로 버림받은(deserted) 도심의 이미지를 형성한다.

오늘날 도시가 처한 상황을 면밀하게 조사하고 진단함으로써, 우리는 지금의 도시 구조를 대체할 새로운 공간적 내러티브(narrative)를 그려내고자 한다. 새로운 도시 구상을 통해, 모바일 장비를 활용하는 집합적 프로젝트에 도시의 다층적인 이야기와 사회적 유대를 강화하는 요소가 더해진다. 마지막으로 본 전시작은 열린 플랫폼을 구축하여 대중, 사회, 정치의 힘을 활용하고, 집합적인 방식으로 푸텐 중심지구의 대안적이며 장기적인 방향성을 확립하고 관련 관리 방안을 모색하고자 한다.

SHENZHEN, CHINA
Title: Fu[+] Infill Plan: Re-making the Center of Shenzhen

URBANUS Architecture and Design
C09

Growing from a remote town of 300,000 residents to a megalopolis of 17 million in forty years, Shenzhen was China's dream city for the young and ambitious in the 1980s and 90s. "Time is money, efficiency is life" was a widespread propaganda of China's economic reform. As Shenzhen's city center, Futian Central District covers 4.4 kilometers that not only urbanizes in "Shenzhen speed" but also epitomizes the "Shenzhen miracle" as a city center built through entirely top-down planning and design.

Futian Central District was first announced in 1980, fully adopting the rational and functionalist principles of the Modernist city and became a spatial manifestation of a hybrid mixture of global capitalist icons, socialist ideology, and traditional Chinese symbolism. A grand central axis with vast green spaces, vehicle driven wide-boulevards, isolated city blocks, and glittering skyscrapers all together produce an image of a modern, forever young, and future looking city while manifesting a socially and psychologically deserted Center. Through careful inspections and diagnoses of the urban conditions, we hope to choreograph a new spatial narrative to superimpose onto the existing urban structure. These new urban nodes will add multiple layers of urban scenarios and social condensers to hold mobile programs and collective projects. Finally the Fu[+] Infill Plan tries to build an open platform to engage the public, social, and political forces to collectively establish an alternative long-term guidance, founding, and management for Futian Central District.

C09

글로벌
스튜디오
GLOBAL
STUDIO

**글로벌
스튜디오:
집합도시
출현의 기회와
형태적 실험**

세계 40여개 대학교의 연구진들이 참여하는 글로벌
스튜디오는 현재 우리의 도시를 '집합'의 관점에서
창의적인 시각으로 연구하고 분석한 미래지향적인
담론들과 실험적인 대안들을 소개하고, 이것을 통해
우리의 도시가 만들어지는 과정에서 건축이 할 수
있는 역할이 얼마나 다양하고 창의적일 수 있는가를
소개한다.

함께 만들고 함께 누리는 '집합'의 도시를
형성해나가는 단서는 여러 사람들의 '집단'이 다져내는
관계에 따라 형성되는 건축의 패턴에서 찾을 수 있다.
즉, 거대 담론의 도시 계획이나 상업자본이 지배하는
도시형성의 논리에 어쩔 수 없이 종속되는 건축이
아니라, 생활 속에서 서로 반응하는 사람들의 행태가
집단적인 패턴으로 표출되는 공동주거, 시장, 혹은
공공 인프라스트럭처 등에서 발견되는 즉흥적이고
창의적인 형식을 연구의 대상으로 삼는다.

연구자들의 참신한 관점과 창의적인 분석작업을
통해 서울도시건축비엔날레 글로벌 스튜디오는
'집합도시'라는 비엔날레의 주제의 외연을
확장시키고, 우리의 도시를 새로운 시각으로
바라보며, 그 안에서 미래 도시와 건축의 새로운
가능성을 확인하는 창의적 발전소 역할을 담당한다.

큐레이터.˙.˙.˙최상기
협력큐레이터.˙.˙.˙이희원
보조큐레이터.˙.˙.˙최영민
전시 디자인.˙.˙.˙(주)건축사사무소오드투에이

INTRO

GLOBAL STUDIO: EMERGING FORMS OF COLLECTIVE CITY

The Global Studio section of the Seoul Biennale of Architecture and Urbanism showcases research projects from forty schools around the world under the theme of "Emerging Forms of the Collective City." The works displayed here show how architecture functions as a collective tool in the process of forming a city.

Contrary to common belief, where the role of architecture is observed to be somewhat passive in our city, which has become more commercial and corporate-driven, how can a city become more relevant to the everyday life of its inhabitants? Instead of erroneously seeking the answer from top-down urban planning or from the real estate markets, the participating studios suggest new emerging forms that can be found in the collective patterns extracted from the spontaneous and responsive architecture that interacts with the inhabitants: communal housing, markets, urban infrastructure, and public buildings.

Global Studio exhibits various collective forms that emerge as byproducts of such social interaction among the collective mass. We expect this collection of compelling architectural proposals to shed light on the important role of architecture in formulating our future cities.

INTRO

Curator.·.·Sanki Choe
Associate Curator.·.·Heewon Lee
Assistant Curator.·.·Youngmin Choi
Exhibition design.·.·ODETO.A

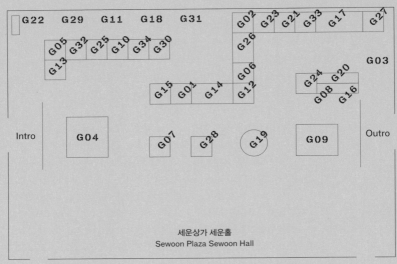

세운홀

G01 베이루트 아메리칸대학교
G02 아키텍처럴 어쏘시에이션 건축대학교
G03 어썸션대학교 + 서울시립대학교 + 호치민 건축대학교
G04 바틀렛 건축대학교 + 한양대학교 에리카
G05 캘리포니아예술대학교
G06 컬럼비아대학교
G07 홍콩 중문대학교
G08 동아대학교
G09 이화여자대학교 + 라드바우드대학교
G10 한양대학교
G11 하버드대학교
G12 홍익대학교
G13 출랄롱코른대학교 INDA 프로그램
G14 게이오대학교
G15 국민대학교
G16 쿠웨이트대학교
G17 싱가포르국립대학교
G18 영국왕립예술대학교
G19 서울대학교
G20 싱가포르과학기술대학교
G21 성균관대학교 + 계명대학교 + 영남대학교 + 카를스루에 공과대학교 + 카를스루에 응용과학대학교 + 슈투트가르트 응용과학대학교

Sewoon Hall

G01 AMERICAN UNIVERSITY OF BEIRUT
G02 ARCHITECTURAL ASSOCIATION SCHOOL OF ARCHITECTURE
G03 ASSUMPTION UNIVERSITY + UNIVERSITY OF SEOUL + UNIVERSITY OF ARCHITECTURE HO CHI MINH CITY
G04 BARTLETT SCHOOL OF ARCHITECTURE + HANYANG UNIVERSITY ERICA
G05 CALIFORNIA COLLEGE OF THE ARTS
G06 COLUMBIA UNIVERSITY
G07 CHINESE UNIVERSITY OF HONG KONG
G08 DONG-A UNIVERSITY
G09 EWHA WOMANS UNIVERSITY + RADBOUD UNIVERSITY
G10 HANYANG UNIVERSITY
G11 HARVARD UNIVERSITY
G12 HONGIK UNIVERSITY
G13 INDA – CHULALONGKORN UNIVERSITY
G14 KEIO UNIVERSITY
G15 KOOKMIN UNIVERSITY
G16 KUWAIT UNIVERSITY
G17 NATIONAL UNIVERSITY OF SINGAPORE
G18 ROYAL COLLEGE OF ART
G19 SEOUL NATIONAL UNIVERSITY
G20 SINGAPORE UNIVERSITY OF TECHNOLOGY AND DESIGN

MAP

베이루트 아메리칸대학교

버티컬 스튜디오: 여가 활동과 기반 시설의 생태학에서 소형 대안주택까지

스튜디오리더.∴.∴.카를라 아라모니, 니콜라스 파야드, 라나 사마라, 크리스토스 마르코폴로스

학생들이 만든 이 작품은 버티컬 스튜디오에서 두 가지 주요 테마로 진행됐으며 서로 다른 규모의 디자인을 보여준다. 첫 번째로 여가 활동의 생태학은 여가 활동과 생태학을 결합한 형태로서 케서완(Keserwan)에 있는 주크 모베(Zouk Mosbeh) 해안의 경관과 기반 시설에 이루어진 대규모 디자인을 보여준다. 두 번째 작품인 보이지 않는 자투리땅은 베이루트의 '건축 불가능'한 자투리땅에 대한 소규모 디자인 개입에 초점을 맞춤으로써 새로운 형태의 주택과 주거를 제시한다. 여가 활동과 생태학이라는 거시적인 요소부터 거주라는 미시적인 요소를 아우르는 이 두 가지 규모의 설계안은 다양한 규모와 도시 환경을 넘나드는 이번 전시회의 주제다. 첫 번째 작품은 해변 건축, 해안 가장자리의 상태, 자연의 해양 생태계와 경관 생태, 수자원 기반 시설, 새로운 형태의 여가 활동 등 다양한 하위 주제를 탐색한다. 학생들은 탐색을 통해 해안 지역의 새로운 미래를 구현한 대규모 하이브리드 프로젝트를 제안한다. 두 번째 작품에서는 디자인을 통해 건축이 불가능할 법하고 '보이지 않는 것'으로 간주되는 자투리땅에 대한 건축법의 제한 규정을 비판한다. 이처럼 문제가 있는 부지의 활용 방안으로 그에 맞는 실험적인 주택 유형의 제안을 통해 베이루트의 자투리땅에 대안적이며 포용적인 거주지를 제시한다.

G01

AMERICAN UNIVERSITY OF BEIRUT

Vertical Studio: From Ecologies of Leisure and Infrastructure to Small Alternative Housing

Studio Leaders.∴.∴.Carla Aramouny, Nicolas Fayad, Rana Samara, Christos Marcopoulos

The work of the students is divided under two main thematics that were undertaken in the Vertical Studio and represent two main scales of design interventions. The first Ecologies of Leisure deals with large landscape and infrastructural interventions on the coast of Zouk Mosbeh in Keserwan, integrating programs of leisure and ecology. The second Invisible Plots focuses on small-scale interventions in residual 'unbuildable' plots in Beirut, proposing new modes of housing and dwelling in the city. The two scales, from macro leisure and ecology to micro dwellings form the basis of this student exhibition presented across scales and urban contexts. The work in the first section explores various sub-themes from beach architecture, coastal edge conditions, natural marine and landscape ecology, water infrastructure, and new forms of leisure. The students propose large-scale hybrid projects that attempt to speculate on a new future reality for the coastal region. In the second section, the students critique through design the building code's restriction on small plots, which are deemed 'invisible' and as unbuildable parts of the city. Accordingly, they propose experimental housing typologies adapted to these problematic sites, which can house alternative and inclusive habitats in Beirut.

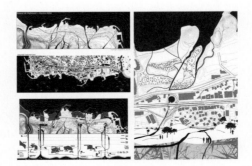

아키텍처럴 어쏘시에이션
건축대학교

투사된 도시

스튜디오리더.∴.∵샘 자코비

현대 사회에서 공동주택만큼 정치적 주체성이
뚜렷하게 드러나는 것은 없다. 공동주택은
국민을 해방하기도 하고 억압하기도 하는
일종의 무기 역할을 해왔다. 공동주택 설계는
동질성, 사회 평등, 원활한 계층 이동을
증진하는 데 없어서는 안 될 요소가 되었다.
공동주택의 목표는 근면한 사회와 생산성
강한 국민을 만드는 것이었다. 이러한 목표는
건축과 계획을 통해 공동주택이 통치성을
공간화하여 정부의 기술과 불가분의 관계가
있는 사회와 공간 형태를 창조할 수 있다는
생각의 바탕이 되었다. 그러나 지금 우리는
가족을 대신할 공동 활동, 사회 인프라, 돌봄
네트워크에 대해 되돌아보는 과정을 통해
새로운 다세대 주택 모델을 만들어내야 한다.
사회·경제적 변화와 인구 변화를 감안하면
전 세계 정치인, 개발업자, 사회 운동가,
노동 운동가 사이에서 공동체가 주도하거나
관리하는 공동주택 단지를 조성해야 한다는
주장이 쏟아져 나오는 것도 당연한 일이다.
공동주택의 개념이 통치에서 관리로 변화하고
있다는 뜻이다. 공동주택 단지와 집합주택
단지는 주택의 집단적 주체성을 다양한
형태로 구현했다. 우리 도시를 설계하고
핵가족 이후에 새로운 형태로 변화하고 있는
유권자들의 주택을 설계하는 과정에서 이러한
집단적 주체성이 점점 더 중요한 고려사항으로
떠오르고 있다.

ARCHITECTURAL ASSOCIATION SCHOOL OF ARCHITECTURE
Projective Cities

Studio Leader.∴.∵Sam Jacoby

Arguably nothing has shaped political
subjectivities in modern societies
more than housing. Housing has been
weaponised: it has liberated as well as
oppressed its subjects. Housing design
had become instrumental to fostering
social homogeneity, social equity, and
social mobility. Housing was to create
an industrious society and productive
subjects. This built on the understanding
that housing, through its architecture
and planning, is capable of spatialising
governmentality, of creating social and
spatial forms that correlate and become
a technology of government. We have
to rethink shared activities, social
infrastructures, and social networks of
care needed in place of the family to
draw and design new multi-generational
housing models. Given these socio-
economic and demographic changes,
it is unsurprising to see a global rise
of rhetoric by politicians, developers,
and social and labour activists alike
that make claims for community-led or
community-controlled housing projects.
It reflects a shift from government to
governance. Communal and collective
housing projects have shaped different
collective subjectivities in housing, which
are increasingly gaining new importance
in the way we have to design our cities
and housing for changing constituencies
beyond the nuclear family.

G02

어썸션대학교 + 서울시립대학교 + 호치민 건축대학교
집합도시의 이동성
스튜디오리더.∵.∵윤정원, 한부디홍,
응우엔홍광, 헝레디투, 프리마 비리야바다나

이 전시는 베트남 호치민시에서 세 학교가
함께 진행한 워크숍의 집단경험을 전달하기
위해 기획되었다. 워크숍 기간 동안 학생들은
연구자로서의 예리한 감각, 분석적인 관찰력
및 참신한 시각으로 무장한 채 호치민씨티의
일상 속으로 뛰어들었다. 연구의 결과물은
연속성 있는 집합도시를 생성해나가는 사이공
시민들의 창의적인 전략들을 담은 단서를
제공한다. 그런 면에서 이 연구는 자연스럽게
이 도시의 전문가들 즉, 현지인들이 공공의
공간을 향유하는 습관과 행동패턴 및
불문의 규율을 배움의 원천으로 삼는다.
워크숍 기간동안 연구자들을 사로잡은
키워드가 있다면, 그것은 역동성, 유동성,
선형, 완충공간, 지표면 등의 개념들이다.
얼핏보면 호치민씨티는 열대 아시아의 어느
도시와 차이 없이 고밀도 소규모 난상개발과
양극화된 건물 유형으로 특징 지어질 것이다.
그러나 불과 몇 블럭을 걸어보는 것 만으로도
이 도시의 차별적인 매력은 그 건물군에
있는것이 아니라, 도시의 지표면에 새겨진
사람들의 유동적이고 역동적인 집단 흔적에
의해 활성화되어지고 있다는 점을 알 수
있고, 이것이 바로 '집합도시'를 만들어내는
근원적인 에너지임을 알 수 있다. 이 전시는
이와 같은 배움들을 시각적으로 표현한
결과물이다.

ASSUMPTION UNIVERSITY + UNIVERSITY OF SEOUL + UNIVERSITY OF ARCHITECTURE HO CHI MINH CITY
Mobilizing Collectivity
Studio Leaders.∵.∵Jungwon Yoon, Hanh Vu Thi Hong, Nguyen Hong Quang, Huong Le Thi Thu, Prima Viriyavadhanna

This exhibition displays a compilation of collective efforts by students from three schools during a design workshop held in Ho Chi Minh City. The students ventured out into the streets of this fascinating city armed with a keen sensibility for analytical observation and fresh perspectives to collect evidence on how residents demonstrate creative strategies to form a continuous and collective city. Urban research here resorts to learning from the experts: the locals. And from the unwritten, implicit and temporal rules that govern the appropriation of public and private spaces for common use. A few keywords that captivated the minds of the researchers were mobility, fluidity, linearity, buffer, and ground. At first glance, Ho Chi Minh City may appear similar to any tropical Asian city characterized by its density, aggregation, and polarized spectrum of building types. However, it does not take more than a few walks around the corner to realize that the true secret that makes this city mesmerizing lies not only in its physical built environment, but in the linear patterns of fluidity and mobility engraved on its ground level by the people of this continuous collective city.

G03

바틀렛 건축대학교 + 한양대학교 에리카

공유도시

스튜디오리더.∴.∴.사빈 스토프, 패트릭 웨버, 김소영

오늘날 세계 여러 도시에서 공유 문화가 진화하고 있을 뿐 아니라 우리가 일상에서 생활하고 일하며 통근하고 학습하며 여행하고 식품을 생산하는 방식에 큰 영향을 끼치고 있다. 공유 패러다임은 공유 공간, 인간의 상호작용과 만남, 재화와 서비스 교환 등의 형태로 이미 도시 속에 존재하고 있었지만 오늘날에는 우리 생활에 반드시 필요한 요소로 확장되는 추세다. 개인의 이익과 소유를 특징으로 하는 기존의 소비 형태는 인터넷 사회 연결망, 빠른 기술 발전, 공유 경제 덕분에 협업, 협력, 공동체를 기반으로 하는 새로운 소비 형태로 변화하고 있다. 학생들은 서울과 수도권의 공유 문화와 협력적 소비에 대해 연구했으며 건축학적 개입과 도시 개입을 통해 그처럼 교환과 사회화가 강화된 새로운 생활양식에 접근한다. 학생들은 서울과 런던의 워크숍에서 집단 작업을 몸소 체험했으며 도면과 모형으로 서로의 아이디어를 절충해야 했다. 학생들의 통찰력 있는 연구와 도발적인 디자인 제안은 도시의 구축환경이 어떻게 사회 결속력(social cohesion)을 증진할 수 있는지를 보여주며 그 방법은 새로운 공간을 창조하고 코리빙(co-living, 주거지 공유), 코워킹(co-working, 사무 공간 공유)의 유형 분류 체계를 구축할 뿐 아니라 공유와 협업에 도움이 되는 기반 시설을 새로이 구축한다.

BARTLETT SCHOOL OF ARCHITECTURE + HANYANG UNIVERSITY ERICA

Sharing Cities

Studio Leaders.∴.∴.Sabine Storp, Patrick Weber, So Young Kim

Nowadays in cities around the world, a sharing culture has evolved and has a profound impact on our everyday life, in terms of the ways we live, work, commute, learn, travel, produce food and more. The sharing paradigm, existing already in cities in shared space, human interaction and encounter, and the exchange of goods and services, has expanded to become a necessary part of our lives. Because of increasing online social networks, rapid technological advancements, and the rise of a sharing economy, the old form of consumption characterized by private interests and individual ownership has changed to a new form that is built on collaboration, cooperation, and community. The students have been exploring this topic of sharing culture and collaborative consumption in cities like Seoul and its surroundings, and their architectural and urban interventions address new emerging lifestyles, which enhance exchange and socialization. From workshops in Seoul and London, they experienced firsthand how to work collectively and negotiate their ideas in drawings and models. Their insightful research and provocative design proposals reflect how the built environment in cities can improve social cohesion by creating new spaces and building typologies for co-living, co-working, or new infrastructures to facilitate sharing and collaboration.

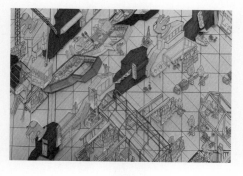

캘리포니아 예술대학교
영역 도시: 거대도시의 가장자리
스튜디오 리더.∴∵니라지 바티야

전경 뒤에는 항상 배경이 있다. 캘리포니아의 배경은 센트럴 밸리(Central Valley)다. 캘리포니아 해안은 멋진 경관, 진보적인 환경 운동, 자유로운 문화, 밀도를 더욱 돋보이게 한다. 반면에 캘리포니아 내륙은 자원 수확과 채굴, 보수적인 가치관, 노후한 사회 기반시설을 특징으로 한다. 같은 캘리포니아이면서도 지형, 자산, 인종, 기후, 오염도 면에서 뚜렷이 다른 이 두 지역은 갈수록 극과 극으로 갈리고 있는 자원 소비와 그러한 자원을 뽑아내기 위한 토지 및 공동체 착취 실태를 상징적으로 보여주는 곳이기도 하다. 이처럼 캘리포니아 해안과 내륙은 뚜렷하게 다른 모습으로 남아있지만 고속 철도가 건설되면 이 두 세계 사이의 공간적인 벽이 허물어지고 독자적인 신세계인 광역 도시(territorial city)가 만들어질 것이다. 기반시설로 연결된 지역들은 연쇄적 도시화를 통해 대도시권으로 거듭나게 된다. 이처럼 대단위 도시화로 만들어지는 지역권이 거대 지역(megaregion)이다. 본 스튜디오는 캘리포니아에 건설될 고속철도의 설계, 영향, 기회를 지역과 건축학적인 측면에서 조사하고자 한다. 이러한 기반시설은 대규모 공동 건설 사업이지만 그 과정에서 (인간 또는 인간이 아닌) 새로운 행위자가 기존의 하향식(top-down) 시스템 내에서 권한을 발휘할 기회를 얻을 것으로 보인다.

CALIFORNIA COLLEGE OF THE ARTS
The Territorial City: Edge of the Megalopolis
Studio Leader.∴∵Neeraj Bhatia

Every foreground has a background. In California, our background is the Central Valley. Coastal California adorns an image of scenic landscape, progressive environmental movements, liberal culture, and density, while inland California is characterized by resource harvesting and extraction, conservative values, and a deficiency in social infrastructure. Separated by topography, wealth, race, climate, and pollution, these two Californias are emblematic of the increasing divide between the realities of resource consumption and the exploitation of land and communities to extract these resources. While these two Californias have remained distinct, the construction of high-speed rail infrastructure will produce a spatial collapse between these two worlds, and create a new world of its own—a territorial city. This creates a new form of territorial urbanization that has been termed the megaregion, which is a continuous map of urbanization connected by infrastructure. This studio will examine the design, impact, and opportunities of the high-speed rail in California both at a territorial and architectural scales. While these infrastructures are large-scaled collective constructions, it provides a venue to empower new agents—human and non-human—into this once top-down system.

G05

컬럼비아대학교

스페큘리티브 시티: 위기와 혼란의 건축적 투사

스튜디오 리더∴∵데이비드 유진 문

지난 12년 동안 잇따른 경제 혼란으로 거품 붕괴 시대가 출현했다. 금융 시장 폭락으로 사회·정치적 갈등이 더해지고 기후 변화가 현실로 다가오고 있다. 이러한 위기와 혼란 속에서 전형적인 건축 생산 방식이 뒤집히는가 하면 하향식 위계구조와 상향식 위계구조, 공적 영역과 사적 영역의 구분, 소유권, 건축 계획 등의 개념에 대한 이의가 제기되고 있다. 스페큘리티브 시티 프로젝트의 연구 주제는 위기와 혼란 이후의 건축 담론과 변화된 관행을 새로운 가능성의 기회로 간주하고 그 변화가 사회·정치·경제적 맥락에서 건축과 도시화에 미치는 영향을 연구하는 것이다. 본 프로젝트는 위기에 처한 기존 건축 관행뿐만 아니라 큰 충격을 겪었으면서도 생산적이었던 과거 시대의 역사적 중요성을 알아본다. 구체적으로는 제2차 세계대전 종전 직후와 70년대 후반에서 80년대 초반까지를 비롯하여 현대의 환경을 예측케 하고 형성한 주요 순간들을 살펴본다. 본 프로젝트는 선별된 시나리오와 관련 지형을 검토함으로써 거품 붕괴, 무절제, 신자유주의, 공백, 비정형 건축, 재사용, 회복 가능성 등으로 대표되는 신경제 환경에서 건축이 나아갈 길을 연구한다. 이를 통해 위기 속에서 집합 영역을 재통합할 수 있는 기회를 찾아내고자 한다.

COLUMBIA UNIVERSITY
Speculative City: Crisis, Turmoil, and Projections in Architecture

Studio Leader∴∵David Eugin Moon

Most recently, the disruptive events of the last twelve years have contributed to the latest post-bubble era, with a dramatic financial collapse compounded by socio-political conflicts, and the realization of predictions in climate change. These crises and turmoil upended traditional notions of the typical processes of architectural production and question top-down and bottom-up hierarchies, public and private realms, ownership, and the architectural program.
The inquiries in Speculative City outlines changes in the discourse and practice of architecture following crises or turmoil as an opportunity for new potentials, exploring the social, political, and economic contexts and their influences on architecture and urbanism. The projects investigate the current practices of the architecture of crisis as well as the historical significance of the previous post-traumatic yet productive periods from post-World War II, the late 70s and early 80s, and other seminal moments that shaped the modern environment. Examining a selection of scenarios and their geographies, the projects explore architecture in the context of the new economic landscape of post-bubble conditions, excess, neo-liberalism, vacancy, informal architecture, re-use, and resilience, highlighting hidden potentials within the crises to reassemble meaningful collective realms.

홍콩 중문대학교
실제 마을의 상태: 다시 생각해보는 중국 농촌
스튜디오 리더.˙.˙피터 페레토

본 디자인 스튜디오는 지속적으로 확대되는 도시 영역에서 농촌 문화유산 역할을 살펴본다. 우리가 집합도시라는 주제의 맥락에서 농촌 문화유산의 역할을 고찰하려는 까닭은 이를 통해 집합이라는 개념과 관련된 중요하고도 시급히 해결해야 할 문제들을 파악할 수 있기 때문이다. 역설적이게도 오늘날의 도시가 안고 있는 여러 문제들은 우리가 농촌을 어떻게 대하느냐에 따라 해결되기도 하고 악화되기도 한다.

기존의 국가가 주도하는 정책 바탕의 농촌 보존 패러다임은 관광을 농촌 문제의 해결책으로 도입하는 데 초점을 맞췄다. 우리는 버려진 마을에 다시 생기를 불어넣음으로써 기존 패러다임에 맞서고자 한다. 이러한 종류의 마을은 이미 수없이 많으며 하나같이 암울한 현실과 마주하고 있다. 마을이 완전히 없어지거나 '디즈니화(Disneyfication)' 등의 건축 보존 방식을 통해 고유한 특징을 잃고 박제된 건축 박물관으로 전락하는 것이다.

지난 30여 년간 소외되고 오용되고 생기를 박탈당한 중국 농촌의 상태를 본격적으로 재조명해본다. 통계에 의하면 오늘날 중국의 농촌은 하루 300개 마을이 사라지는 위험에 처해있다. 우리의 연구는 정량적인 단계를 초월하여 농촌 마을의 독특한 집합성 표현의 기회를 재현할 수 있을 것으로 기대한다.

CHINESE UNIVERSITY OF HONG KONG
CONDITION / REAL VILLAGES: Rethinking China's Countryside
Studio Leader.˙.˙Peter Ferreto

This design studio looks at the role of rural cultural heritage in an ever-expanding urban territory. We propose this in the context of the "Collective City" theme because we believe it raises important issues about the notion of the collective. In a paradoxical way, many of the problems that face the contemporary city can be solved or exacerbated by how we address the rural.

We set out to challenge conventional preservation paradigms, instigated by a national rural heritage policy focused on introducing tourism as a rural solution. In short, we aim to make these abandoned villages operational again. An abundant number of villages of this kind exist, all of which are confronted with a dark reality: i.e. face certain extinction or lose their soul by giving way to the practice of "Disneyfication," a form of architectural conservation that mutates villages into an architectural taxidermy museum.

It brings to the table, the condition of the Chinese village, which for more than thirty years has been neglected, abused and depleted of life. Today, Chinese villages are disappearing at alarming rates, 300/day by some statistics. Without entering into quantitative research, our proposal wishes to demonstrate that rural villages offer unique opportunities for collective readings.

동아대학교
우암동 소막마을 재생
스튜디오 리더 ∴ 차윤석

이 스튜디오는 건축과 도시를 논함에 있어 가장 중요한 요소는 무엇일까라는 의문에서 시작했다. 우리가 찾은 대답은 다음과 같았다: 현대 도시와 건축의 가장 큰 특징은 다양성이며, 이 다양성들의 집합이 집합도시(Collective City)라는 결과물로 나타나며, 이를 제대로 구현하기 위해서는 집합도시의 가장 중요한 요소는 다양한 개인과 그들의 집단이라는 점을 이해해야 한다. 사람을 이해하고 그들의 삶의 자취를 좇는 것이야 말로 민주적 다양성이 보장되는 집합도시를 만드는 첫걸음일 것이라고 믿기 때문이다.

본 스튜디오의 목적은 고령화사회에 대비하고 노년층에게 요구되는 여러 사회적 물리적 대안들을 제안함으로써 또 다른 의미의 지속가능성에 대해 고찰하고자 하는 것이다. 구체적인 제안으로는 소막마을의 역사적 자원인 소막사의 역사적 가치를 복원하기 위한 건축적 제안과 더불어 마을에서 단순하게 노인들이 살기 안락한 환경을 만드는 것이 아니라 노인들이 자급자족할 수 있는 환경을 만들어 노인들 스스로의 힘으로 살 수 있도록 하고 경제적인 가치 또한 창출할 수 있는 프로그램을 제시하고자 한다. 또한 이를 실현하기 위한 방법론과 실행단계 계획을 통하여 집합도시가 지향해야할 방향을 제시하고자 한다.

DONG-A UNIVERSITY
Country for Old Men
Studio Leader ∴ Youn Suk Cha

The most important feature of modern cities and architecture is diversity, and the collection of diversity appears as a result of the "Collective City." To realize this, it is essential to understand that the most important element of the "Collective City" is the diverse individuals and their social groups. We believe that understanding people and their lives will be the first step in creating a "Collective City" that ensures democratic diversity.

The purpose of this studio is to examine sustainability in another sense by preparing for an aging society and proposing several social and physical alternatives for older people, as well as architectural proposals to restore the historical value of Somaksa, a historical resource of Somak Village. We are going to present programs that will not simply create a comfortable environment for the elderly in the village, but also create an environment for the elderly to live on their own and create economic value. It is also intended to suggest the direction that the "Collective City" should be oriented towards through method and process. We hope our plan will be a good alternative in the face of our urban problem.

G08

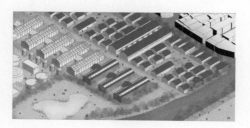

이화여자대학교 +
라드바우드대학교

PILaR(+): 을지로의 미래를 위한 후회 없는 시나리오

스튜디오 리더.∴.∴클라스 크레세, 에르빈 판데르 크라벤

유엔이나 세계은행 같은 국제개발기구는 오늘날 여러 도시에 닥친 금융 위기와 지속 가능성 위기를 해결하는 개발정책에 점차 큰 관심을 보이고 있다. 유엔 해비타트(UN-Habitat)와 세계 토지 제도 연합(Global Land Tools Network, GLTN)이 특히 장려하는 정책은 PILaR(Participatory, Inclusive Land Readjustment, 참여적, 포용적 토지 구획 정리) 프로그램이다. 해당 프로그램은 한국과 다른 나라의 토지 구획 정리 경험에서 출발한다. 두 대학의 도시계획 및 디자인 전공생들이 참여한 이 프로젝트는 PILaR 프로그램에 도시의 지속 가능성을 위한 방안을 접목하여 을지로 부지에 적합한 일련의 시나리오를 개발하는 것을 목표로 한다. 본 실험에서 학생들은 토지 계획 관련 규제와 제도 환경 등 비물리적인 요소까지 감안하여 부지를 평가한다. 그 이후 다양한 이해 관계자들의 입장을 확인하여 이들이 원하는 여러 미래 시나리오를 설계한다. 해당 시나리오들은 신뢰가능하고 후회를 남기지 않을 차세대 설계 구상의 바탕이 된다. 이 같은 방법을 통해 을지로의 미래에 새로운 가능성을 제시하는 다양한 설계 시나리오를 창출하고 있다. 프로젝트는 이해 관계자들의 공감대를 찾아내어 현재의 교착 상태를 돌파하는 것을 목표로 한다.

G09

EWHA WOMAN'S UNIVERSITY + RADBOUD UNIVERSITY

PILaR(+): No-Regrets Scenarios for the Future of Euljiro

Studio Leaders.∴.∴Klaas Kresse, Erwin van der Krabben

International development organizations, such as the United Nations or the World Bank, have become increasingly interested in development policies that managed to deal with the financial and sustainability challenges cities are facing today. One tool that is specifically promoted by UN-Habitat and the Global Land Tools Network (GLTN) is the PILaR (Participatory, Inclusive Land Readjustment) program, which takes Korean and other countries' experiences with land readjustment as a point of departure. The PILaR(+) that planning students from Radboud and design students from Ewha are taking on includes sustainability measures into the PILaR program and develops a series of scenarios for the site in Euljiro.
In this experiment students assess the site, including the non-physical givens such as the planning regulations and institutional setting. Then, they identify different stakeholders' positions in order to develop designs for stakeholders' desired future scenarios. These stakeholder scenarios in the next step are the basis for the design of a new generation of scenarios we call believable, no-regrets future scenarios. With this method the project creates a range of design scenarios that offer alternative opportunities for the future of Euljiro. Our goal is to identify the common ground among stakeholders and break the stalemate.

인프라 건축의 하이브리드 스튜디오

스튜디오 리더.∵.라파엘 루나

21세기 도시 건축 이론의 현실에 대한 비판적 대응으로 등장한 '집합' 개념은 우리가 현대 '도시'라고 부르는 곳을 생성하는 중층 구조에 대한 해석을 골자로 한다. 중층 구조는 정치든 사회든 경제든 기술이든 각기 다른 영역에서 활동하는 수많은 행위자에 의해 구축된다. 여기서 진화된 집합 개념은 데이터와 기반 시설 등의 중층 구조를 토대로 도시를 새롭게 이해하는 도시 이론 담화를 발전시켰다. 빅데이터와 스마트 시티 개념이 현대 도시의 관리와 효율성뿐 아니라 신진대사 기능을 이해하기 위한 합리적인 차세대 방안으로 떠오르고 있다. 그러나 스마트 인프라와 도시 공간 사이에서 집단 공유지(collective commons)를 만들어 내는 연결고리는 여전히 탐색의 여지가 많다. 인구 이동, 교통, 편의 시설, 부동산 가격의 흐름을 추적해 보면 데이터를 통해 도시 축소와 도시성(urbanity) 확대라는 모순된 현상에서 비롯되는 도심 공동화처럼, 서울 등 다른 도시들이 머지않아 겪게 될 여러 도시 문제가 분명해지는 만큼 건축이 도시 담론에 참여해야 한다는 결론에 도달하게 된다.

HANYANG UNIVERSITY
Infra-Architectural Hybrids Studio

Studio Leader.∵.Rafael Luna

As a critical response to the current state of urban architectural theory in the 21st century, the idea of "the collective" unfolds as an interpretation of an amalgamate of layers that generate what we describe as the contemporary "city." These layers imposed by the many actors that operate within different dimensions, be it political, social, economic, and technological, have evolved an urban theoretical dialogue from a new understanding of cities via the basis of data and infrastructure. Although the notion of Big Data and Smart Cities has been evolving as the next logical step in understanding the management and efficiencies of the contemporary city and its metabolism, there is still much to be explored as a link between smart infrastructure and the urban space that generates the collective commons. The tracking of flows in population displacement, transportation, amenities, and real estate prices, has made evident the need for architecture to engage the urban dialogue as data also manifests urban problems such as the paradoxical condition between shrinking cities and an expanding urbanity that produces urban voids, which is imminent in cities like Seoul.

G10

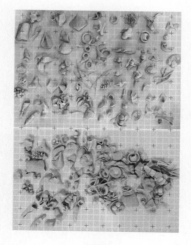

하버드대학교
거리의 미래

스튜디오 리더 ∴ 안드레스 세브츠크

본 스튜디오는 새로운 이동성 기술이 로스앤젤레스의 기존 건축 환경에 미치는 영향을 조사했다. 우리는 조사를 통해 사회적으로 포용성 있고 지속가능한 환경을 극대화하는 다중 방식의 해결책을 모색하고자 한다. 본 스튜디오는 산타모니카에서 시내로 이어지는 엑스포선(Expo Line)의 주변 지역을 집중 조사했다. 그뿐만 아니라 자율 주행 자동차, 공유 자동차, 전기 자동차, 개인 이동 수단, 자동 택배 배송 시스템 등장에 따라 시내 이동 경험이 어떻게 변화할지를 탐색했다. 본 스튜디오는 엑스포선 주변의 다양한 건축 환경과 사회·경제적 환경에서 기존의 이동 수단을 조사했다. 또한 새로운 이동 수단이 도입되면 공간에 대한 접근성, 기존 이동 패턴, 거리의 기능과 공공장소의 기능에 어떠한 차질이 빚어질지 검토했다. 본 스튜디오는 새로운 지하철 노선 주변의 6개 지역에 초점을 맞춰 새로운 이동성 기술이 대중교통을 보완할지, 아니면 대중교통의 경쟁 상대가 될지를 탐색했다. 그 이외에도 새로운 이동 수단이 다양한 건축 환경과 사회·경제적 환경에 미치게 될 영향을 조사했다. 대상은 2028년 LA 올림픽의 중심지 역할을 할 USC (서던 캘리포니아 대학교) 지역과 피코, 버몬트, 크렌쇼, 라시에네가, 세풀베다, 팜스 주변의 환승 가능한 역세권 지역이었다.

HARVARD UNIVERSITY
Future of Streets
Studio Leader ∴ Andres Sevtsuk

This studio investigated the impact of new mobility technologies on the built environment of Los Angeles, seeking solutions that maximize multi-modal, socially inclusive, and environmentally sustainable outcomes for the city. The studio focused on sites surrounding the Expo Line, leading from downtown to Santa Monica and explored how the experience of urban travel is likely to change with the advent of automated, shared and electric vehicles, personal mobility devices and automated package delivery systems.
The studio examined current modal choices in different built and socio-economic environments around the Expo Line and explored how the introduction of new mobility options is likely to disrupt spatial accessibility, existing trip patterns, as well as the functions of streets and public spaces. By focusing on half a dozen neighborhoods adjacent to a new Metro line, the studio also explored how new mobility technologies could complement, rather than compete with public transit and explored the impact of new mobility options in varying built environments and socio-economic settings. Sites included the USC area, which will serve as the epicenter of L.A.'s 2028 Olympics, as well as transit catchment areas around Pico, Vermont, Crenshaw, La Cienega, Sepulveda, and Palms.

G11

홍익대학교
도시 생활형 공장
스튜디오 리더.∴.∴ 김주원, 임동우

서울이 탈 도시화 시대에 접어드려는 오늘날, 건축가가 탈 도시화 현상에 어떻게 대처해야 하는지 생각해 볼 필요가 있다. 그 사이에 서울에서는 인구가 줄어들고 있고 있을 뿐만 아니라 이미 일부 산업이 사라졌다.

그러나 정밀 가공, 인공 지능, 초연결 소셜 미디어 등의 다양한 분야에서 새로운 산업혁명이 일어남에 따라 생산 산업을 도시로 되찾아올 수 있는 기회가 생겨나고 있다. 새로운 산업은 '굴뚝 산업'으로 불리는 전통적인 제조업보다는 개인이나 작은 집단이 생산을 담당하는 신 가내수공업(neo cottage industry)에 가까울 것이다. 또한, 생산의 개념조차 제조업 분야에 국한되지 않고 콘텐츠 생산으로까지 확대될 것이다.

본 스튜디오는 생산 기능을 겸비한 최첨단 도시 주택 유형을 제안함으로써 눈앞으로 다가온 탈 도시화 시대에 대비하고자 한다. 이 도시 주택 유형은 도시와 지역을 재생하는 촉매 역할을 할 것이다. 본 제안은 서울의 도시 재생 사업 수단 가운데 하나로서 1인 가구의 증가와 인구 감소 추세를 두루 해결할 수 있는 방안이다.

HONGIK UNIVERSITY
Factory for Urban Living
Studio Leader.∴.∴ Juwon Kim, Dongwoo Yim

As Seoul is entering into post-urbanization era, it is crucial to think about how architects should address the phenomenon. The city not only is losing its population but also already has lost some of its industries.

However, along with the new industrial revolution in many fields, such as micro manufacturing, artificial intelligence, or hyper-connected social media, there are chances to bring in production industries back to the city. The new industries may not be of the same type as so called chimney-industries, but will more likely be neo-cottage industries where individuals or a small group of people can produce. And even the concept of production may not be limited only to manufacturing industries but will be expanded and include contents production.

The studio will address this upcoming era by proposing a new urban housing typology that includes a production function. It will be used as a catalyst to reform our city and neighborhoods. It is one of the tools of urban regeneration projects in Seoul that tackles both the growth of single households and the trend of losing population.

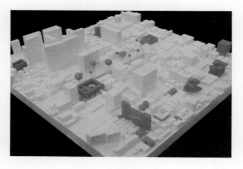

출랄롱코른대학교 INDA
취약 상태
스튜디오리더.∴.∵알리샤 라차로니

본 연구의 출발점은 매 학기마다 유일하고
특별한 순간들(의례)이나 다양한 일과
중에서의 대중 행동을 조사하는 것이었다.
여기서 우리는 대중 행동이 어떻게 다양한
형태의 집단성을 촉진하는지에 초점을 맞췄다.
공공 공간은 의식이나 특정한 맥락의 일상
행위에서만 생겨나는 것이 아니다. 언어, 장식,
유형물, 도상학과 같은 집단 정체성의 표상이
의미 있게 사용될 때도 공공 공간은 탄생한다.
본 스튜디오는 취약 계층 보호라는 목표에
따라 어떻게 설계 방식을 변경하면 언어, 문화,
원리, 감정의 취약성을 보완할 수 있을지를
핵심 질문으로 제기한다. 학생들은 방콕에서
젠트리피케이션, 부동산 가격 압박, 진보적
동질성, 시각적 정체성 감소 등과 같이 '지연된
폭력' 현상의 타격을 받는 도시의 취약 집단에
접근했으며, 기존 현실과의 긴밀한 소통으로
이들의 복잡한 사회·경제·형태·자연적
구조를 이해하여 새로운 공간을 제안하는
단서를 제안한다.

INDA – CHULALONGKORN UNIVERSITY
Vulnerable States
Studio Leader.∴.∵Alicia Lazzaroni

The public acts studied as starting points
have constituted, in different semesters,
either very singular and special moments
(rituals), or various ranges of everyday
life routines, with particular attention to
how these acts can help support different
forms of collectivity.
In line with the reflection on the
importance of cultural visual references,
public spaces are not generated only
by the rituals and everyday routines
rooted in a particular context, but
also by the meaningful use of certain
features of collective identity, like
language, ornamentation, materiality and
iconography. One of the core questions
raised in the studio asks how the
protection of vulnerable realities would
require from us a change in our design
habits, in order to include vulnerability as
a language, as a cultural principle or as a
type of sensibility. Students approached
fragile urban communities in Bangkok,
affected by "slow-violent" phenomena
like gentrification, real estate pressure,
progressive homogenization, or decrease
of visual identities. Understanding the
complexity in their social, economic,
morphological, and natural structures has
been the key to propose new spaces in
close dialogic relationship with existing
realities.

게이오대학교

Global Studio

스튜디오 리더.∴.호르헤 알마산

도시는 설계될 수 있는가? 모더니즘의 이상이었던 도시 종합 계획이 실패한 후 집단 프로젝트로서의 도시 계획은 점차 강력해지는 신자유주의 지배 체제에서 시장의 힘에 밀려 차츰 중단되어왔다. 오늘날 그에 따른 부정적인 영향이 나타나고 있는 만큼 우리는 건축과 어바니즘의 타당성에 주목할 필요가 있다. 건축가와 어바니즘 전문가들은 다시 한 번 통합적 능력을 발휘하여 집단 프로젝트를 구상하고 그에 맞는 건축 환경을 구상해 내야 한다.

구시대적 전체주의 기법에 본 프로젝트의 토대를 두어서는 안 된다. 오히려 새로운 질서, 효율성, 미학을 찾아 발전시켜야 한다. 이를테면 현대 도쿄의 도시 구조를 사례 연구 대상으로 삼을 수 있다. 도쿄는 다공성, 포용성, 적응성을 갖춘 도시로서 종합 계획의 결과물도, 기업 도시화의 결과물도 아니기 때문이다. 도쿄는 최근에 부상하는 도시의 패턴을 보여준다. 소규모 행위자에 의한 다양한 변경과 차용이 역동적인 과정 속에서 상호작용 하면서 복합적 결과를 만들어내고 전체적으로 통합적 도시 형태를 창조하는 것이다.

그 가운데서 우리는 작쿄 거리, 요코초(골목길), 철로 아래 주택, 대도시 주변의 마을, 도로 주변 등의 다섯 가지 형태를 살펴본다. 우리는 도쿄의 문화적 고유성을 다룬 여느 담론과는 달리 문화 혼재의 타당성에 초점을 맞추고자 한다.

KEIO UNIVERSITY

Tectonics of the Spontaneous City: Transcultural Survey on Tokyo as an Alternative to Corporate Urbanism

Studio Leader.∴.Jorge Almazan

Can the city be designed? After the failure of the Modernist dreams of total urban planning, and under the dominance of an increasingly neoliberal regime, the city as a collective project has been gradually abandoned to the so-called market forces. In the face of the negative consequences of this abandonment, the relevance of architecture and urbanism needs to be claimed. Architects and urbanists need to provide again their synthetic capacity to shape the built environment according to a self-conscious collective project.
This project cannot be based on old totalitarian techniques. A new inventory of orders, efficiencies and aesthetics need to be found and developed. The urban fabric of contemporary Tokyo can be a source of case studies, since it provides examples of porous, inclusive and adaptive urban patterns that are neither the result of master planning nor the consequence of corporate urbanism. These patterns are emergent: the combined result of multiple modifications and appropriations by small agents interacting in a dynamic process, which as a whole create an integrated urban pattern.
We examine five of these patterns: zakkyo streets, yokocho, undertrack infills, metropolitan villages, and flowing streets. Unlike much of the discussions on Tokyo that emphasize its cultural uniqueness, we aim at transcultural validity.

G 14

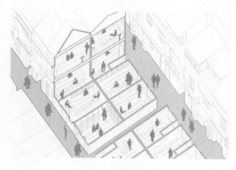

국민대학교

슈퍼마켓 : 초집단 주거

스튜디오 리더.∵.∵.최혜정, 봉일범, 김우일, 이규환

시장은 전통적으로 체제로서든 장소로서든 도시 형성과 성장에 큰 역할을 해왔다. 중심가, 마을 회관, 시장, 주택가는 건축 환경에서 특정 문화와 정체성을 발생시키는 지주이자 주체 역할을 해왔다. 그러나 최근 세계 시장들은 심각한 타격을 입고 있다. 상거래에서 가상 메커니즘과 원격 메커니즘이 등장하면서 도시 생산이 도심에서 멀어지게 되고 이로 인해 지역이 변화하고 일상용품의 생산과 소비 방식이 변화하고 있다. 상업과 사회적 교환 형태 대부분이 온라인으로 옮겨가는 동안 소매상점, 공공장소, 중심가는 점점 비어가고 활력을 잃고 있다.

소비와 생산을 온라인 상거래, 신용카드, 소셜 미디어, 스마트폰, 정보 기술이 담당하는 상황에 따라 물리적인 영역과 가정생활의 전통적인 개념은 재정립되어야 한다. 건축은 이러한 사안에 어떻게 대처할 것이며 새로운 집단 참여 환경에 맞춰 어떻게 그 역할을 강화할 것인가? 본 스튜디오는 우리가 생각하는 전통적인 가정생활 영역과 도시 형태를 '대체' 할 수 있는 건축학적 가능성을 찾는다. 그뿐만 아니라 새롭게 부상하는 교환 방식 유형과 시장 네트워크에 대응하기 위해 새로운 형태의 건축적 조건을 탐색한다.

G15

KOOKMIN UNIVERSITY
Supermarket : Seeking Super-Collective Living

Studio Leader.∵.∵.Helen Hejung Choi, Ilburm Bong, Wooil Kim, Kyu Hwan Lee

Market in tradition, whether as system or place, has played a significant role in urban formation and growth. The main streets, village centers, marketplaces and housing communities have served as anchors and agents generating specific a culture and identity for our built environment. Recent global phenomenon of the market, however, is heavily influenced by the virtual and remote mechanism of commercial transaction, forcing urban production out of the inner city, and thus changing our neighborhoods and the way we produce and consume our everyday goods. Much of the commerce and form of social exchange has shifted to online space while our retail centers, public domain, and main streets are becoming vacant and losing their vitality.

With consumption and production relying more on online commerce, credit cards, social media, mobile phones and information technology, the traditional conception of our physical territories and domesticity must be reconsidered. How can architecture confront these issues and potentiate its role towards a condition of new collective engagement? The studio seeks architectural possibilities that can 'supersede' the way we conceive our traditional boundaries of domesticity and urban forms, and explores new forms of architectural conditions in response to emerging forms of exchange and the network of the market.

쿠웨이트대학교

메카 스튜디오: 진행형으로서의 매니페스토

스튜디오 리더.∴.∴.샤이카 알 무바라키

본 스튜디오는 모든 이슬람교도에게 변함없이 큰 의미가 있는 성도이지만 논의가 제대로 되어 있지 않고 정보가 부족한 도시인 메카에 대해 보다 심도 높은 이해를 이끌어내고자 한다. 종교 의식의 성지인 알 카바 중심가와 인근 지역은 시간이 흐름에 따라 유동 인구가 급속도로 증가하고 있으며 5성 호텔과 상점가가 즐비하게 늘어서 1년 내내 순례자들을 맞이한다. 이처럼 도시 개발이 종교적 중심지의 발전에만 편중되다 보니 그곳에서 불과 몇 킬로미터 떨어진 곳에 난립해 있는 촌락은 점점 열악해져 가는 상태인데도 거의 주목을 받지 못하고 있다. 이에 본 스튜디오는 메카라는 도시를 주제로 하여 도시를 조사하고 묘사하며 평가하고자 한다. 이슬람교 성지가 되기 전에 메카를 지탱했던 내러티브를 조사하면서 알게 된 일련의 역사적 사실을 통해 시간의 흐름에 따른 도시 발전 과정을 새롭게 그려낼 수 있었다. 여행 서류, 지도, 시, 동양학 학술지, 서신 등을 검토하여 현재의 도시, 건축, 종교, 정치에 두루 작용하고 제약에 구애 받지 않는 새로운 이야기를 만들기 위해 심층적인 현장 방문을 통해 메카의 문제를 파악하면서 현대의 서사에 의문을 던져본다. '진행 중인 매니페스토'를 통해 본 스튜디오는 오늘날 메카의 지속 불가능한 측면을 지적한다. 그 가운데는 지속적인 지역 파괴, 알 하람 중심지의 끝없는 확장, 유적지와 역사적 가치가 있는 시장 파괴, 저소득층 지역을 비롯한 여러 지역에 반복적으로 홍수를 일으키는 무계획적인 수자원 관리 체계 등이 포함된다. 본 스튜디오는 임시 해결책 마련보다는 상태 평가를 목표로 하는 프로젝트를 통해 도시가 당면한 문제를 조명하고 서술하며 타파한다. 메카의 과거와 현재에 대한 진정한 이해로부터 출발하는 '진행 중인 매니페스토'를 통해 이 도시의 가능성을 소개하고자 한다.

KUWAIT UNIVERSITY
Mecca Studio: An Ongoing Manifesto

Studio Leader.∴.∴.Shaikha Al Mubaraki

The studio formulates a deeper appreciation of Mecca, the misunderstood holy city where great significance for all Muslims remains, but is poorly discussed and un-interrogated. A beacon of religious rituals, the central district of Al Kabba and its vicinity has increased drastically in footprint overtime, to accommodate the array of five star hotels and malls, that cater mainly to affluent pilgrims year-round. This focus on the development of the religious heart of the city leads to minimal examination of the unplanned, deteriorating settlements that stand only a few kilometers away from this celebrated center.

In an 'ongoing manifesto' the studio points out the unsustainable aspects of the current city, putting in focus, the continuous demolition of neighborhoods and the ever expansion of Al haram center, the demolition of old historic sites and markets, and the unplanned water systems that allows for repeated flooding, particularly in low-income neighborhoods to name a few. Through projects that aim to critique rather than design band-aid solutions, the studio highlights, narrates, and negotiates the issues at hand in hopes of outlining the potentials of the city in an ongoing manifesto, that culminates from a true understanding of Mecca's past and present.

G16

싱가포르국립대학교
열기: 적도 도시의 집합 건축
스튜디오 리더.∴.에릭 루뢰

기후 변화, 인구 증가와 함께 적도 도시, 기후, 대기상태의 관계는 점점 더 복잡한 접점에 이르고 있다. 본 스튜디오는 호치민에 존재하는 '열기'를 만들어내는 대기 매체를 조사한다. 포화 상태에 이른 비형식적인 도시화, 두꺼운 외피, 적도 도시의 '열기'를 조절하고 여과하는 지붕 집합체 등의 세 가지 특징이 본 연구의 핵심 주제다. 적도 도시가 입자 형태의 다공적이고 비형식적인 도시보다 형식적이고 조절되었으며 '현대적인' 대도시로 발전함에 따라 적도에서의 집합도시 개념은 대규모 자본, 세계화 물결, 수입된 기술 시스템의 위협을 받고 있다. 본 설계 연구는 20세기 중반의 트로피컬 모더니즘과 그로 발생한 집합성, 규모, 식생, 습도, 더위, 강우 논쟁의 경쟁 관계에 초점을 맞춘다. 우리의 주제는 이처럼 '열기' 환경을 만들어 내는 여러 매개체를 통해 시각적이고 상징적인 것을 초월하여 기후와 대기라는 구체적인 대상으로 확장된다.

대기를 도시와 건축에 동일하게 침투하는 접착 물질로 간주한다면 도시, 건축, 대기를 하나로 묶어서 생각할 필요가 있다. 이 세 가지 요소는 기후적이고 문화적인 매체로서 빌딩 집합체와 건축 외피에 두루 영향을 미칠 뿐만 아니라 건축과 도시 설계를 재고하는 방식에도 영향을 준다.

G17

NATIONAL UNIVERSITY OF SINGAPORE
HOT AIR: The Equatorial City and The Architectures of Aggregation
Studio Leader.∴.Erik L'Heureux

As the equatorial city's relationship to climate and atmosphere becomes an increasingly complex interface in relation to climate change and great population growth, the studio will research the atmospheric mediums of 'hot air' situated in Ho Chi Minh City. Three features will guide this work: saturated informal urbanisms, thick envelopes, and aggregated roofs that modulate and filter the "hot air" of the equatorial city. As the equatorial city evolves from the granular, porous, and informal, to a more formal, conditioned, and 'modern' metropolis, the idea of the collective city on the equator stands threatened in the face of larger-scale capital, global aspirations and imported technological systems. The design research will focus on the contested relationship with mid-20th century tropical modernism and the resulting contestation of aggregation, scale, vegetation, humidity, heat, and rain—the mediums that produce a "hot air" environment that will expand our repertoire beyond the optic and iconic to the climatic and atmospheric.
If atmosphere is the glue that permeates both the city and architecture alike, then it is imperative to think of the city, architecture, and atmosphere together, as a climatic and cultural medium that impacts both the aggregation of buildings and the architectural envelop simultaneously, impacting the way in which architecture and urbanism might be rethought.

영국왕립예술대학교
현대의 의례

스튜디오 리더 : 다비드 사코니, 잔프란코
봄바치, 마테오 코스탄초, 프란체스카 로마나
델랄리오

본 전시는 학생들이 추진한 엄선된 프로젝트를
보여준다. 해당 프로젝트들은 현대의 의례로
알려진 동시대 관습과 기능을 분석한 새로운
도시 원형을 제안한다. 이러한 의례가 일련의
움직임으로서 반복적이고 율동적으로
수행되면 물리적 형태를 만들어내는
패러다임이 된다. 이러한 물리적 형태는
인간의 몸과 구조화된 환경의 상호작용을
일으키고 이를 통해 건축물은 구조물과
일상적인 사회관계의 영구적인 보존물로서
의미를 갖는다. 이와 같이 의례는 공간 창조의
도구가 되며 집단 또는 단독으로 치러진다.
의례는 특정한 물리적 흔적을 남김으로써 건축
형태를 설계하는 중요한 재료가 된다.
의례가 남기는 물리적 흔적들은 다양한
규모의 도시에서 개인과 집단 영역을 두루
포용하는 공간을 창출할 수 있다. 이때
창출되는 공간은 가정 영역만큼이나 공간
복지의 장이 될 수 있고 이를 통해 주체성이
부상된다. 떠오르는 주체성은 때로 인간이
아닌 영역에 초점을 맞추며 인간과 인간 이외
존재의 공생에 이의를 제기하기도 한다.
그렇지 않을 경우에는 한층 더 규범적인 도시
원형과 동일시되기도 한다. 본 프로젝트들은
새로운 행태에 대응하여 새로운 방식의 집합적
형태화나 진보적인 건축 형태를 제안한다는
점에서 공통점이 있다.

ROYAL COLLEGE OF ART
Contemporary Rituals

Studio Leader∴∴Davide Sacconi,
Gianfranco Bombaci, Matteo Costanzo,
Francesca Romana Dell'Aglio

The exhibition shows a selection of
students' projects that propose new urban
prototypes generated from the analysis
of contemporary practices and functions
that are recognised as contemporary
rituals. These actions, when performed
repetitively and rhythmically as a
sequence of movements, become a
paradigm that generate a physical form,
which gives meaning to architecture
as a constitution and perpetuation of
quotidian social relations through the
interaction of the body with a structured
environment. As such, ritual becomes
a space-making device, which whether
performed in a group or in solitary, leaves
certain physical traces behind, becoming
a crucial ingredient in the design of
architectural forms.
They have the potential to generate
spaces at different scales of the city,
embracing both the individual and the
collective sphere. These are spaces that
can be interpreted as a spatial welfare as
much as a domestic realm; what they all
provide is an emerging subjectivity, which
sometimes focuses on a non-human
sphere, while challenging the cohabitation
between human and other-than-human,
otherwise identified with a more canonical
archetype of the city form. What these
projects have in common is that they all
propose new collective formalisations or
new progressive forms of architecture in
response to new behaviours.

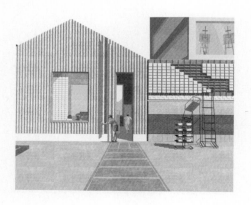

서울대학교
도시 리노베이션과 제3의 공간
스튜디오 리더.∵.∵.존 홍

1930년대부터 1980년대 사이에 급속도로
현대화된 서울 도시 구조물들이 최근 자리를
잡아가고 있다. 사상 처음으로 서울 인구가
감소하고, 도시 재생 정책에 따라 과거와
같은 구조물을 만들어냈던 경제 발전의 틀을
벗어나는 오늘날 우리가 생각해보아야 할
질문이 있다. 지금과는 다른 문화, 정치, 경제
맥락의 유산으로 남은 현대적 공간을 보존하고
리노베이션한다는 것은 어떠한 의미일까?
우리는 도시 부지를 확실히 개발할 예정인
'민관' 제휴 세력을 실질적으로 살펴봄으로써
코워킹(공유 근무) 공간, 사무실, 식음료
프로그램, 문화 시설, 호텔 등과 같이 외견상
포괄적인 재생 복합 건축물들을 새로운 형태의
집합체 생성의 촉매로 활용하는 방안을
모색하고자 한다.
특히 제3의 공간에 대한 아이디어는 도시
영역의 내·외부 사안을 두루 조율하는
데 활용될 것이다. 영역 외부의 사안을
조명하자면 해당 부지 자체는 그 역사가
1930년대로 거슬러 올라갈 정도로 유서 깊은
장소다. 그 당시는 영등포가 한강 남쪽에서
최초의 공업 지대로 발전하기 시작했던 때다.
해당 부지의 발전은 단편적인 맥락까지는
아니지만 다양한 맥락 속에서 이루어졌으며
교통 인프라, 쇼핑몰, 거주지, 역사적으로
중요한 시장, 경공업, 공방, 홍등가 등으로
상징된다. 이 지역의 역사는 대선 제분 공장을
구심점으로 하는 일제 강점기 이후의 이질적인
개발 세력으로 이루어져 있다.

SEOUL NATIONAL UNIVERSITY
Urban Renovation and the Third Space
Studio Leader.∵.∵.John Hong

The urban artifacts of Seoul's rapid
modernization from the 1930s to the
1980s are recently coming of age. As
the population of Korea shrinks for the
first time in history and the policies
of regeneration shift away from the
economic version of progress that initially
created these structures, we have to
ask: How should we define preservation
and renovation when applied to these
inherited modern spaces from a different
cultural, political, and economic context?
By recognizing the actual 'public-private'
partnership forces that will inevitably
develop our site, we will explore how
seemingly generic regeneration mixed-
use programs such as co-working spaces,
offices, food and beverage programs,
cultural facilities, and hotels can be used
as catalysts to generate new forms of
collectivity.
In particular, the idea of third space
will be used to coordinate both extra-
and inter-territorial issues. In light of
the former, the site itself has a deep
history stemming from the 1930s when
Yeongdeungpo was the first industrial
area to be developed south of the Han
River. The site's evolution is marked
by a rich, if not fragmented, context of
transportation infrastructure, shopping
malls, residences, historical markets,
light-industry, artist lofts, and red-light
districts. Its history is a tapestry of
sometimes conflicting post-colonial and
developmental forces with the Daesun
Flour Mill at its epicenter.

G19

싱가포르 기술디자인대학교
생산적 주변부
스튜디오 리더.∴.∴캘빈 촤

새로운 유형의 도시 개발단지가 주변 지역에 속속 들어서고 있다. 이러한 개발단지는 전통적인 도시-농촌 이분법에서 벗어난 공간 형태와 시스템으로 구성된다. 광저우 농업 지구에 새로운 지식 캠퍼스가 세워지는가 하면 말레이시아 소도시에 최첨단 고속 철도가 건설되는 것에서 알 수 있듯이 이처럼 거대한 개발단지는 신흥 도시 주변부에 빠른 속도로 건설되고 있다.

본 스튜디오는 지방정부의 계획에 따라 2차와 3차 외곽 순환 도로 사이에 마련된 광저우 지식 도시의 주변 지역에 주민을 2,000명씩 수용할 수 있는 주택단지들의 다양한 설립 방안을 제시하고자 한다. 단지는 평균적인 마을 규모가 될 것이다. 말레이시아 남부에는 고속 철도가 지나가는 네 개 소도시 (세렘반, 말라카, 무아르, 바투파핫)의 외곽을 생산적인 지역으로 탈바꿈시키기 위한 몇 가지 방안을 제시한다.

개별 그룹은 건축 규모와 도시 전략의 교차점을 파고들어 구체적인 건축 원리를 개발함으로써 공간 소유권, 자원 공유, 사적인 것과 공적인 것의 관계와 같은 사안에 두루 초점을 맞추었다.

종합하자면 본 스튜디오의 목표는 전통적인 도시/농촌과 소도시/대도시의 정의를 탈피하여 새롭게 떠오르는 공간 형태와 시스템에 따른 해결책을 개발하고 이러한 신흥 지역 주변부의 미래를 제시하는 것이다.

SINGAPORE UNIVERSITY OF TECHNOLOGY AND DESIGN
Productive Peripheries
Studio Leader.∴.∴Calvin Chua

Increasingly, new urban developments are located in a peripheral territory with spatial forms and systems that fall outside the traditional urban-rural dichotomy. From the creation of new knowledge campuses in agricultural zones in Guangzhou to the development of new High-Speed Rail towns in Malaysia, these mega developments are rapidly implanted onto the peripheries of emerging cities.

Focusing on the peri-urban areas of the Guangzhou Knowledge City, located between the 2nd and 3rd ring road planned by the local government, the studio developed various proposals for a series of settlements, each supporting 2000 inhabitants—the typical size of a village. While in Southern Malaysia, a set of proposals were developed to transform the peripheral spaces in four High-Speed Rail towns—Seremban, Malacca, Muar and Batu Pahat—into a productive territory.

Working at the intersection between the scale of architecture and urbanism, each group developed specific architectural elements that addressed issues of spatial ownership, resource sharing and the relationship between the private and the common.

Collectively, the goal of the studio is to develop a critical response towards emerging spatial forms and systems that falls outside the traditional definitions of urban/rural and town/city, and project a future for the peripheries of these emerging regions.

G20

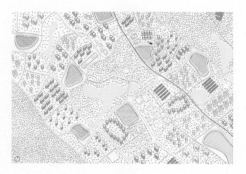

성균관대학교 + 계명대학교
+ 영남대학교 + 카를스루에
공과대학교 + 카를스루에
응용과학대학교 + 슈투트가르트
응용과학대학교

혁신 지구: 집합도시 접근법

스튜디오 리더∴∵토르스텐 슈처

본 연구와 설계 프로젝트는 대구의 예전 동촌
지구를 분석하는 한편 지속 가능한 도시
재생을 통해 동촌 지구를 집합 혁신 지구로
전환하기 위한 방안들을 논의하고자 한다.
특히 동촌을 혁신 지구로 전환하기 위해
다양한 규모의 계획을 제시했다. 본 연구에서
다루는 구체적인 주제는 집합도시, 이동성,
건축물, 재생 에너지, 지속 가능한 수자원
관리, 도시 개발, 녹색 환경, 미기후 등이다.
한층 더 집합적인 지역 사회로 변모하기
위해서는 대화와 논의가 활성화되어야 한다.
이를 위해 본 연구는 참여 방식의 조정, 공유적
접근법과 협력적인 파트너십의 공유, 만남
장소 설계 등과 같은 방안을 제시한다.
집합적 혁신 지구를 위한 계획들은 가변성,
다기능성, 개방성, 연결성, 집합성 등의 특징에
초점을 맞추며, 이러한 특징들은 창의적인
의사소통, 협업, 사회 참여를 촉진한다. 본
재개발 전략과 전망의 목표는 지역사회에
새로운 정체성을 부여하고 모든 지역사회
구성원이 공감할 만한 혁신 산업 아이디어를
구상하는 것이다.

SUNGKYUNKWAN
UNIVERSITY + YEUNGNAM
UNIVERSITY + KEIMYUNG
UNIVERSITY + KARLSRUHE
INSTITUTE OF TECHNOLOGY
+ KARLSRUHE UNIVERSITY
OF APPLIED SCIENCE
+ UNIVERSITY OF APPLIED
SCIENCE STUTTGART

Innovation District: A Collective City Approach

Studio Leader∴∵Thorsten Schuetze

The research and design projects
discuss the analysis of the existing
Dongchon district in Daegu and the
development of multiple plans for the
district's sustainable urban regeneration
to a collective innovation district. The
renovation of Dongchon to an innovation
district was planned on various scale
levels. The specific themes addressed
are, among others, collective city,
mobility, buildings, renewable energies,
sustainable water management
and urban development, green and
microclimate.
The transformation to a more collective
local society was addressed by the
adaptation of participation methods,
sharing approaches and cooperative
partnerships, and the design of
meeting places, fostering dialogue and
consultancy.
The plans for the collective innovation
district are characterized by variance,
multi-functionality, openness,
connectivity and constellation. They
facilitate creative communication,
collaboration and engagement. The
proposed redevelopment strategies
and perspectives aim to provide new
identities to the community and offer
ideas for innovative industries that every
community member can sympathize with.

G21

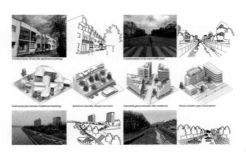

시라큐스대학교

설계 | 에너지 | 미래: 중국, 슝안 지구

스튜디오 리더∴페이 왕

도시가 인류의 가장 중요한 혁신이라는 것은 대부분의 사람이 아는 사실이다. 도시는 지식을 비롯한 유·무형 자산을 생산하고 조직하며 분배하는 것을 돕는다. 그뿐만 아니라 도시 덕분에 우리는 도시 없이는 상상조차 하지 못했을 방식으로 자산을 활용하고 증대시키며 변화시킬 수 있다. 도시는 인간 혁신의 환경 그 자체라는 점에서 그야말로 혁신이다. 2017년 4월 건설된 슝안 신구는 베이징에서 남쪽으로 100 km 떨어진 곳이다. 슝안신구의 주된 역할은 베이징-톈진-허베이 경제 삼각벨트 징진지를 연결하는 개발 허브다. 게다가 국유 기업 사옥, 정부 기관, 연구개발 시설 등 중국 자본의 '비주력' 부문이 이곳으로 옮겨질 것으로 보인다. 슝안은 경제 성장의 자극제가 될 뿐만 아니라 다수의 중국 비정부 기능을 담당함으로써, 아직 다른 도시에서는 미처 발현되지 않은 미래 도시의 문제점들에 대한 현재의 해결책을 제안하는 도시 인큐베이터 역할도 할 전망이다.

SYRACUSE UNIVERSITY

Design | Energy | Futures: Xiong'an, China

Studio Leader∴Fei Wang

Almost everyone acknowledges that the city is humankind's single most important innovation. The city not only helps us to produce, organize and distribute material and immaterial assets, including, and especially, knowledge, but it also enables us to leverage, multiply and transform those assets in ways that would be inconceivable without the city. The city is an innovation precisely because it is the very milieu of human innovation itself. Established in April 2017, Xiong'an New Area is located about 100 km southwest of Beijing. Its primary function is to serve as a development hub for the Beijing-Tianjin-Hebei (Jingjinji) economic triangle. Additionally, "non-core" functions of the Chinese capital are expected eventually to migrate here, including offices of some state-owned enterprises, government agencies, and research and development facilities. Xiong'an will spur economic growth and assume many of Beijing's non-governmental functions. It will also serve as an urban incubator that solves current problems while it simultaneously speculates about urban futures, offering solutions to problems not yet recognized in other cities.

G22

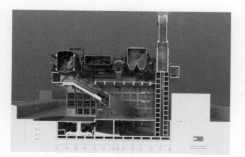

베를린 공과대학교

재형상화 지도: 이주자, 거주자, 방문자 – 새로운 집합 형태

스튜디오 리더.∴.∴도미니크 바르트만스키, 김선주, 에밀리 켈링, 마르티나 뢰브, 세브린 마르갱, 티머시 파프, 다그마르 펠거, 요르그 슈톨만

우리는 세 연구 스튜디오에서 시각적 분석과 지도제작 도구를 활용한 공간이 생성될 때의 변화를 조사한다. 대규모 공동연구센터의 일부인 본 스튜디오들은 디지털 기술과 다국적 역학관계가 일으키는 변화에 주목한다. 우리는 일상생활의 집합적 형태가 재협상되고 재배치되는 과정을 살펴보는 한편 지도를 만들어 그러한 변화를 기록하고 해석한다. 시각적 기법을 활용하면 다양한 규모, 시간, 관계적 상호작용을 통합할 수 있다. 본 세 스튜디오는 사회학, 건축학, 도시설계를 포괄하는 다학제간 공동 연구를 진행한다. 축척 변이 방식의 지도 '감춰진 곳에 살다'는 도시로 유입되는 새로운 무주택 인구를 수용하는 베를린 호스텔 산업의 공간·조직·사회적 구조를 낱낱이 보여준다. 노숙자 수용 시설이라는 사실을 드러내지 않는 것이 호스텔 산업의 존립을 위한 필수 조건이라고 주장한다. 사진 형태의 지도 '인스턴트 스마트시티'는 '스마트 도시 송도'의 주거지역에 사는 중산층의 도시 일상을 상세히 보여준다. 이 세밀한 문화기술적 사진 시리즈는 공간 재형상화에 대한 집합적 현상을 보여준다. 일련의 지도인 '미술 연구소의 접근성'은 베를린에 있는 '세계 문화의 집'의 포용성과 배타성을 연구한다. 본 연구는 공간의 물질구조, 접근성, 전용 사이의 관계를 다양한 차원으로 보여준다.

G23

TECHNISCHE UNIVERSITÄT BERLIN
Mapping Re-figurations: Migrants, Residents, Visitors – New Forms of Collectivity

Studio Leader.∴.∴Dominik Bartmanski, Seonju Kim, Emily Kelling, Martina Löw, Séverine Marguin, Timothy Pape, Dagmar Pelger, Jörg Stollmann

In three research studios, we employ visual analysis and mapping tools to investigate current changes in the production of space. Part of a larger collaborative research center, the studios emphasize changes brought about by digital technologies and transnational dynamics. Looking at renegotiation and rearrangement of collective forms of everyday life, we develop mappings to document and interpret these transformations. Our visual techniques enable us to integrate different scales, temporalities and relational interactions. The three studios are an interdisciplinary collaboration between sociology, architecture and urban design. The trans-scalar mapping 'Residing in the Hidden' reveals the spatial, organizational and social structures of Berlin's hostel industry in its capacity of accommodating homeless newcomers. It argues that the invisibility of the phenomenon is a necessary condition for its existence. The photographic mapping 'Instant Smart Cities' elaborates on everyday urban life of the middle classes in residential neighborhoods of the 'Smart City Songdo.' Overdrawn ethnographic photo series point at collective phenomena of spatial re-figuration. The mapping series 'Accessibilities of an Art Institution' investigates the inclusivity and exclusivity of 'Haus der Kulturen der Welt in Berlin.' On different scales this analysis points at the relation between the spatio-material structure and modes of accessibility and appropriation.

빈 공과대학교

경계의 기록

스튜디오 리더.∴.므라덴 야드릭

전형적인 주거지역을 창의적으로 재단장하는 것을 목표로 하는 본 프로젝트는 두 도시의 주택단지에 동일한 방식으로 개입한다. 하나는 빈 한델스키 지역의 주택단지이고 다른 하나는 서울 잠실의 주택단지다. 주요 목표는 도시사회의 사회·물리적 '경계' 현상을 극복하는 데 도움이 될 도구를 설계하는 것이다. 지붕 배치, 외관, 내부, '사이' 공간 등의 네 개의 전략적 부분에 '사회적 보형물'로 불리는 물리적 첨가물을 설치하는 해결책이 제시된다. 본 해결책의 목표는 생활의 질을 개선하고 세입자들 사이의 사회·물리적 상호작용을 돕거나 강화하는 것이다. 빈과 서울의 오늘날 상황을 분석하는 데는 지역사회의 사회적, 보편적 경계를 기록하는 것도 포함한다. 이 작업은 집합도시의 필수 요소를 이해하는데 반드시 필요하다. 경계의 기록은 도시사회의 한계와 배타성을 보여주는 사회·물리적으로 단절된 생활 단위의 문제점을 탐색하고 이 문제를 극복하기 위한 해결책을 제공한다. 관련 조직 내의 문화 간 대화를 통해서는 익숙한 현상들을 낯선 사회문화적 맥락으로 조사한다. 이를 통해 상호 학습 경험을 제공하고 지속 가능한 사회 공동체를 만들고자 한다.

TECHISCHE UNIVERSITÄT WIEN

The Book of Limes

Studio Leader.∴.Mladen Jadric

The project is a parallel intervention in two cities, a creative refurbishment of the typical "Dormitory Town" — one housing project built in Handelskai, Vienna and the second one in Jamsil, Seoul. The main goal is to design tools helping us to overcome a social and physical phenomenon of "boundaries" in urban society. Proposed solutions are named "social prostheses" — they are physical additions installed in four strategic areas: roof-landscapes, façade, interiors and spaces "in between." Their goal is to improve the quality of living and to enable and/or intensify physical and social interaction among the tenants.
An analysis of the contemporary situation in Vienna and Seoul includes documenting social and general community boundaries. This experience was crucial for understanding the need for a collective city. The Book of Limes explores the problem of social- and physical cocooning, the phenomena of limits and exclusivity in urban society, and offers solutions as to how to overcome this process. Intercultural dialog within the relevant frameworks explores these familiar phenomena within a non-familiar cultural and social context to facilitate a mutual learning experience and help create sustainable social communities.

G24

텍사스 테크대학교
중재 장치
스튜디오 리더.∵.박건, 임리사

서울은 계속해서 변화하는 원동력에 적응하여
끊임없이 진화하고 있다. 서울의 서교동과
중림동은 오늘날의 요구에 따라 언제든
변화할 수 있는 곳이며 뛰어난 적응력과
유연성을 보이는 지역이 되었다. 두 지역에는
역사적인 흔적과 현대화의 징후가 가득하다.
신규 프로젝트로 아직은 걸음마 단계인
보행자 네트워크 '서울로7017'를 통해 이
지역은 끊임없이 움직이는 도시 공간으로
거듭났다. 2019 서울 스튜디오 조사는 두
지역의 잠재성을 밝혀내고 구현한다. 우리의
프로젝트는 창의성의 실현으로서 출발점이자
결론을 탐색하여 가장 생산적인 영향을 이끌어
낸다. 그 원동력은 재생산이나 강요라기보다는
텅 빈 듯 보이는 공터에서도 이전에 보이지
않거나 상상하지 못했던 현실을 구현하는
데서 발생한다. 고갈된 듯 보이는 지역에서도
이전에 보이지 않던 현실이나 상상하지 못했던
현실을 찾아냄으로써 그러한 지역을 영향력
있는 공간으로 만들어낼 수 있다.

G25

TEXAS TECH UNIVERSITY
Mediating Armature
Studio Leader.∵.Kuhn Park, Lisa Lim

Seoul is constantly evolving and adapting
due to the ever-changing forces it is
directed by. It allows for this specific
region of Seoul, Seogye and Jungnim,
to be highly susceptible to change
followed by current needs and formed into
already an area of high adaptability and
flexibility. There are apparent evidences
and traces of historic and modernizing
manifestations of the region. By a
newly established yet tested pedestrian
network, Seoullo7017, the region turns
into perpetual urban fields of motion.
2019 SeoulSTUDIO's investigation as
a collective enabling enterprise both
reveals and realizes hidden potential. As a
creative practice, our effort precipitates its
most productive effects through a finding
that is also a founding; its agency lies in
neither reproduction nor imposition but
rather in uncovering realities previously
unseen or unimagined, even across
seemingly exhausted grounds.

케이프타운대학교
희망의 공간

스튜디오 리더.∴.파들리 아이작스, 멜린다 실버먼

본 스튜디오의 연구는 공간의 정치적 속성에서 출발한다. 지난 20년 동안 케이프타운 델프트 지역의 촌락은 속도로 변화했다. 공공건물은 정적이고 거의 변화가 없는 반면에 정부에서 제공한 주택과 공공장소는 주민들에 의해 참신하고 흥미로우며 적극적인 방식으로 변형되고 있다. 본 스튜디오는 지역 주민들이 촌락의 변형을 통해 새로운 공간성을 창조하는 과정을 비판적 시각으로 소개하고자한다. 본 스튜디오는 이 지역에서의 지속적인 주택 수요와 공공 공간 인프라가 어떻게 공간을 활성화시키는지 연구한다. 본 스튜디오는 담론 형태의 건축학적 조사에서 사회 공간 이론과 관행뿐 아니라 건축학적 설계 구상과 제작 과정까지 포함한 접근법을 취한다.

스튜디오의 목표는 건축이 어떻게 이처럼 창의적인 전략을 낳고 사람들에게 주어진 공간을 변형할 수 있는 기회를 선사하는지 연구하는 것이다. 이러한 행위들이 모여 도시의 활기를 북돋고 도시 건설을 촉진하는 사례는 수없이 많다. 본 스튜디오는 건축이 독립된 오브제로 숭배받기보다는 더 많은 사회 참여를 이끌어낼 수 있는 생산적인 보강물로서의 역할을 가진 발전장치가 될 수 있다는 점에 초점을 맞추고자 한다.

UNIVERSITY OF CAPE TOWN
Spaces of Good Hope

Studio Leader.∴.Fadly Isaacs, Melinda Silverman

The fundamental premise of the studio is the political nature of space. In Delft, over the last twenty years the settlement has been radically altered. Other than the public buildings, which have remained static and unchanging, residents have actively transformed their government-issue houses and reconfigured public space in new and interesting ways. The studio critically maps how these have been harnessed by local residents to construct a new spatiality. The studio seeks explorations as to how public spatial infrastructure, inclusive of the ongoing demand for housing, can be used to invigorate spaces in the area. This is approached as a discursive mode of architectural inquiry that brings into dialogue socio-spatial theories and practices, with architectural design imagination and making.

The purpose of this studio is to explore how architecture can engender these creative strategies, enabling people to transform given space in ways that enhance their lives. In many instances it is these actions that collectively contribute to urban vitality and help build the city. The emphasis of the studio is on architecture as a generator, and on buildings which function as armatures for further engagement, rather than as closed and highly fetishised objects.

G26

홍콩대학교
거주 영역
스튜디오 리더.∵.제럴딘 보리오

인공적인 세상과 자연 사이의 불분명한 경계에
걸쳐 있는 서울 시내의 산은 우리에게 오늘날
도시 거주자와 환경의 관계를 보여준다.
오늘날의 자연은 오락과 여가 활동을 통해
'경험'될 뿐만 아니라 문화 정체성 구축에
중요한 역할을 한다. 우리가 추진한 사례
연구는 백악산, 낙산, 남산, 인왕산 등 네 개의
서울 시내 산 에서 이루어지는 가시적인 인간
활동을 관찰하는 것을 골자로 한다. 우리는
이러한 문화적 랜드마크가 공간으로 조직되는
방식과 이들이 유발하는 행동을 건축학적
관점에서 조사했다. 그뿐만 아니라 프로그램과
기반 시설의 지속성을 부각함으로써 이러한
자연 감상에 도움을 주는 근본적인 작동
시스템을 밝혀냈다. 연구 과정은 역할극
형태로 이루어졌다. 학생들은 각각의 전문가
역할을 맡아 장소 개선 과정에 참여하였다.
미래의 건축가들은 정치인, 역사학자,
지리학자, 엔지니어, 민속학자 역할을 통해
스스로의 미래 역할에 대해 재고할 수 있었다.
학생들은 전문가적 관점에서 인류나 문화
요소가 스스로를 자연 환경에 표출할 수 있는
도구방식을 읽어내고, 이해하며, 보일 수 있는
수단을 개발했다.

THE UNIVERSITY OF HONG KONG
Inhabitable Territories
Studio Leader.∵.Geraldine Borio

Located on the ambiguous limits between
the artificial and natural world, Seoul
Urban Mountains inform us about the
contemporary relationship between the
city's inhabitants and the environment.
Today, nature is 'experienced' and
consumed through fun and leisure
activities but also holds a significant role
in the building of cultural identity.
Seoul's four inner mountains, Baegaksan,
Naksan, Namsan, and Inwangsan, were
our case studies for observing the visible
manifestation of these human activities.
From an architectural point of view, we
have interrogated the way these cultural
landmarks are spatially organized and
the behavior they produce. The study has
also revealed the underlying system at
work that supports this appreciation of
nature by highlighting programmatic and
infrastructural continuities.
The course was taking the form of a
role-play where students endorse the
role of different specialists involved in
the place-making process. Putting on
the shoes of the politician, historian,
geographer, engineer and ethnographer
provokes architects-to-be to reframe
their perspective on their own future role.
Based on the specialist's viewpoint, the
students have developed their tools for
reading, understanding, and representing
the manner in which the human or
cultural element has manifested itself in
the natural setting.

G27

매니토바대학교 + 칼턴대학교 +
토론토대학교

블록 돌연변이: 도시 블록 다섯 곳의 투사적 진화

스튜디오 리더.∴.∵전재성

본 스튜디오는 도시 블록 크기 안에서 '집합도시'라는 주제를 살펴본다. 도시 블록은 형식적일 필요는 없지만 유기적/역동적이고 공간으로서의 성격을 갖추어야 한다. 개별 스튜디오는 스튜디오의 성향과 오늘날의 도시 전략에 대한 이해를 바탕으로 서울의 공간적, 형식적, 일화적, 프로그램적, 단편적인 도시 'DNA 사슬'을 연구하고 해당 'DNA'를 사용하여 하이브리드-투사-가공 도시 블록을 규모별로 생산/구성하고자 한다.

본 프로젝트는 모더니스트들의 이상적이고 형식적인 블록(혹은 슬래브), 영역을 넘나드는 블롭, 아키그램과 팀텐의 블록과 슬래브, 현대 주택 단지의 메가 블록이나 슈퍼 블록들이 건축에 기여한 사실을 인정하는 한편 홍콩 카오룽처럼 내러티브가 담긴 기존 공간의 진화적인 돌연변이나 진화를 구상해본다. 기존 환경을 증강하고 집합시킬 뿐 아니라 기존의 형식적 혹은 비형식적 도시, 조직 혹은 세포와 외부 혹은 외래 요소를 '구현된 집합성'이라는 잠재된 가능성에 통합·이식하는 과정을 통해 투사적 상상이라는 새로운 혼종이 생성될 것이다. 스튜디오마다 도시의 서로 다른 부분을 조사하고 공간·건축학적 DNA를 수집할 것이다. 또한 이 DNA를 개조하고 교배하여 새로워진 DNA를 서울의 블록 단위에 끼워 넣거나 접목 혹은 배치하고자 한다.

UNIVERSITY OF MANITOBA + CARLETON UNIVERSITY + UNIVERSITY OF TORONTO
Block Mutations: Projective Evolutions of Five Urban Blocks

Studio Leader.∴.∵Jae-Sung Chon

The studio examines the theme 'Collective City' in the scale of an urban block, not necessarily as a formal entity but as an organic/dynamic and spatial one. The plan is to have each studio study the urban 'DNA strands' of Seoul (spatial, formal, anecdotal, programmatic, social... fragments) and produce/compose a hybrid-projective-fictitious urban block in scale using the 'DNA' based on each studio's own biases and understandings of contemporary urbanism.

While recognizing the critical contributions of the idealized and formal blocks of modernists (Cerda or block/ slabs), the trans-territorial blobs, blocks, and slabs of Archigram and Team X, and the mega or super blocks of contemporary developments, the project imagines evolutive mutations/evolutions of existing narratives (perhaps more like that of Kowloon). It will generate projective imaginations, new hybrids, through augmentation and aggregation of existing conditions, and by incorporating and grafting of external/foreign elements on to the latent potentials (embodied collectivities) within the existing urban tissues/cells (informal and formal). Each studio will investigate different parts of the city, collect spatial and architectural DNA, recompose and hybridize toward a new condition, and nest them or situate them in a block scale of Seoul.

G28

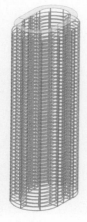

펜실베이니아대학교
몬트리올, 감각의 증강
스튜디오 리더.∵.∵사이먼 김

UNIVERSITY OF PENNSYLVANIA
Montreal, Sensate and Augmented
Studio Leader.∵.∵Simon Kim

과제나 학술 연구 대상으로서의 도시 건축은 공용 건조 환경에 한정된다. 도시 건축의 암묵적, 명시적 의미와 정서는 시간이 흐름에 따라 물질 뿐 아니라 행동으로 구현된다. 우리는 몬트리올이라는 장소의 건축과 도시 전략에 지속성, 주체성, 자율성을 포함할 것이다. 몬트리올은 역사, 신환경, 그리고 생태지능을 엑스포와 올림픽 개최에 반영하여 앞선 사고방식으로 정점을 향해 가는 도시로서의 이상적인 후기 산업화 도시의 본보기가 되었다.

본 스튜디오는 인간 중심 설계라는 고전적인 위계질서에서 벗어나 인간이 아닌 행위자에게 창작자로서의 자격과 책임의식을 부여하는 방향을 추구 한다. 구성하는 입장에서 설계를 하고 인간으로서 안락하기만 한 환경에 안주하기보다는 시간의 흐름에 따라 계속해서 혁신하는 설계 과정에 참여할 것이다. 우리의 목표는 (계절, 물, 공기, 동물 등) 인간에게 필요한 조건 이외의 것에 맞춰 변화하고 행동하는 환경을 조성하는 것이다. 우리는 기능과 용도를 하향식으로 지시하여 건축을 결정하는 일보다 인간과 인간이 아닌 행위자가 공존할 수 있는 공간의 지속적, 일시적 점유에 주안점을 둔다.

Architecture of a city as a proposition or a form of intellectual investigation is tethered to a built, shared environment. Its implicit and explicit meanings and affects are to be developed in material and also in behavior over time. To do this, we will imbue architecture and urbanism with duration, with its own agency and self-governance in the location of Montreal. Montreal's history in Expos and Olympics places it as an ideal postindustrial model with an apex towards advanced thinking of new environment and eco-intelligence.

This studio will break from the classical hierarchy of humancentric design and allow for nonhuman authorship and stewardship. Rather than design from a compositional position, and to dwell in a seamless zone of human comfort, this studio will engage in a design process with transformations over time, to produce environments that change and behave for other-than-human requirements (such as seasons, water, air, animal). Rather than determining architecture from a top-down application of function and use, we are more interested in a durational and temporal occupation of space for both the human and nonhuman.

G29

서울시립대학교

도시 하이브리드: 재생 도구로서의 사회적 콘덴서

스튜디오 리더.∴.마르크 브로사, 변효진

본 스튜디오는 상호보완적인 프로그램뿐만 아니라 관광산업의 성장에 어울리는 체험을 소개함으로써 동대문의 서울 약령시장을 사회적, 문화적, 생산적, 그리고 집합적 기반시설로 재해석한다.

시장은 생산과 상거래를 위한 기반시설일 뿐만 아니라 집합적 문화 생산과 교환의 중심지다. 사회적 관계가 응축되어 있는 곳으로서 공동체에 자율권을 부여한다. 본 설계 스튜디오는 서울 약령시장의 도시·계획·건축적 재생뿐 아니라 전략적 지속가능성이라는 중장기적 목표를 달성하기 위한 설계 해결책을 제시하고자 한다. 서울 약령시장의 재개발을 위해서는 물리적 기반시설에 대한 개선 방안 마련 뿐 아니라 그곳에서 열리는 활동과 행사에 대한 재고찰 작업이 선행되어야 한다. 기억에 남지 않는 행사들을 차례로 배열하여 정보를 소극적으로 전달하기보다는 적극적 체험을 기반으로 하는 관광 패러다임을 새로이 도입할 필요가 있다. 이러한 행사는 시간이 지나면서 보다 영속적인 도시 전략으로 발전하게 되며 지주 역할을 하는 신축 건물들로 완성된다. 이러한 신축 건물들은 임시 주거지 기능뿐 아니라 시장의 기능을 보완하고 그 영역을 확대하는 편의시설 역할까지 하면서 방문객, 시장, 노점상, 지역 주민의 접점이 된다. 이와 같은 집합 편의시설은 이 지역의 장기적이고 지속 가능한 지역 재생에 도움을 준다.

UNIVERSITY OF SEOUL
Urban Hybrids: The Social Condenser as a Tool for the Regeneration

Studio Leader.∴.Marc Brossa, Hyojin Byun

The studio reinterprets the Yakryeong Herbal Medicine Market in Dongdaemun as a social, cultural, and productive collective infrastructure through the introduction of complementary programs and experiences catered to a growing tourist industry.

Markets are not only productive commercial infrastructures, but also centers of collective cultural production and exchange, and condensers of social relationships, which empower communities. This design studio seeks to provide design solutions for the urban, programmatic, and architectural regeneration of Yakryeong Market, and for its strategic sustainability in the mid and long term.

The urban renewal of Yakryeong market does not only contemplate upgrading its physical infrastructure, but also rethinking the activities and events it hosts. A new paradigm of tourism based on active experience rather than on the passive transmission of information is introduced through phased sequences of temporal events. Over time these events develop into more permanent urban strategies, punctuated by a series of new anchor buildings, which combine temporal housing programs with complementary amenities. These collective amenities expand the scope of the market, interfacing between visitors, market vendors, and residents, and support the long-term and sustainable regeneration of the area.

G30

시드니 공과대학교
제3의 공간과 집합도시
스튜디오 리더.˙.˙.˙앤드루 벤저민, 제라드
라인무스

건축의 관계성은 우리가 본 스튜디오에서
추구해온 개념으로서 일차적으로 사람과
사건 사이의 관계를 고려하고 여기에 단순한
특수성이라는 개념을 더한다. 이러한 과정을
거쳐 각각의 건물들은 상호 관계망으로 연결될
뿐만 아니라 그 관계망의 일부가 된다.
이 맥락에서 '제3의 공간'은 정해진 논리에
따라 생성된 다른 공간의 일부이면서도 그러한
생성 논리에 의해 형성되거나 구조화되지
않는 공간을 뜻한다. 따라서 제3의 공간은
불확정성의 관계 측면에서 규정된다. 실제로
불확정성 관계가 의도적인 차단과 그에 따른
자율성의 구현물인 틈을 규정하고 생성한다.
영등포의 대선 제분 공장을 보면 교통 인프라,
쇼핑몰, 주거지, 시장, 경공업, 공방, 홍등가
등으로 둘러싸인 장소의 산업 유산 재생에
대한 몇 가지 의문이 솟아난다. 우리는 '재래식'
개발 접근법 대신, 기존 건물의 내부와 건물
사이에 '제3의 공간'을 끼워 넣고 장소에
맥락을 더하는 방안을 제안한다. 구체적으로는
소규모의 공공 예술 도서관을 만들어 인근의
예술 공동체를 지원하도록 하는 것이다.
우리는 새로운 관점으로 도서관 유형과 개발
논리를 고려한다.

UNIVERSITY OF TECHNOLOGY SYDNEY
Third Spaces and the Collective City
Studio Leader.˙.˙.˙Andrew Benjamin,
Gerard Reinmuth

Relationality in architecture is a position
we have been developing in the studio
and which considers relations among
people and events as primary, over and
above conceptions of simple particularity.
Thus, by extension, single buildings are
constituted by and form part of a network
of relations.
In this context the third space is the
creation of a space that, while internal
to other spaces with already determined
logics of creation, is not formed or
organised by those logics. The third space
is defined therefore in terms of relations
of indetermination. Such relations define
(and create) actual openings—openings
as an informed interruption and thus as a
form of autonomy.
At the Daesun Wheat Factory in
Yeongdeungpo, questions arise around
the regeneration of industrial heritage
on a site surrounded by transportation
infrastructure, shopping malls,
residences, markets, light-industry,
artist lofts, and a red-light district. We
suggest that "conventional" development
approaches might be discarded in favour
of a strategy that weaves "third spaces" in
and amongst the existing buildings and
connects the site back to its context. The
brief is for a small public arts library that
supports the nearby arts community. Both
the logics of development and the
typology of the library are rethought.

G31

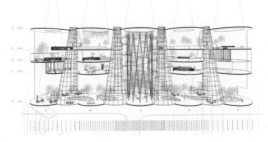

텍사스대학교알링턴
서울의 지표면

스튜디오 리더.∴.조슈아 네이슨

본 스튜디오는 도시의 '공통 기반'인 지표면에 주목한다. 여기에서 말하는 지표면은 문자 그대로 땅의 표면으로서 집합도시의 기반이 되는 물리적 요소를 뜻한다. 지표면의 본질이 도시 캔버스라는 것은 자주 간과되는 사실이다. 지표면이라는 캔버스에 도시가 그려지면 지리학·생물학·형식적으로 인간의 주거가 시작된다. 본 연구는 서울의 주요 지표면에 대한 자료를 수집한다. 서울의 지표면은 서울을 보다 나은 도시로 만들고 현대 도시 생활의 다층적이고 통합적인 본질을 파악하는 데 필수불가결한 자원이다. 그러한 점에서 도시의 지표면은 새로운 공공 공간을 탄생시킬 수 있는 활동적이고도 심층적이며 다층적인 매개체이다. 본 프로젝트는 공공 생활을 위한 공간을 활성화하고 활용하려는 목표로 도시 내부에서 얻을 수 있는 교훈을 참신하고 의미 있게 재해석하고 적용한다. 본 스튜디오는 서울의 지표면 활용을 위한 세 가지 조건을 조사했다. 기반시설로서의 지표면, 건축학적 지표면, 집합적 지표면 등이다. 우리는 도시가 지표면을 어떻게 활용하여 집합적 지표면으로서의 공공 공간을 형태와 용도 측면에서 지속적으로 개선하고 재규정하는지 살펴보았다. 우리는 그러한 지표면의 역할을 평가하고 확장하기 위해 서울을 분석적으로 연구했으며 특히 잠실의 새로운 공공 공간을 세부 연구 대상으로 삼았다. 다시 말해 우리의 목표는 서울의 집합적 도시 활동과 정체성에 이미 존재하는 풍요로운 특징들을 한층 더 강화하는 것이다.

UNIVERSITY OF TEXAS ARLINGTON
Seoul, Skin Deep

Studio Leader.∴.Joshua Nason

Seoul, Skin Deep investigates the surfaces of Seoul as its "common ground"—the physical element acting as medium for the collective city. Often overlooked, the essential nature of city surface is as canvas on which the city is painted, a geological/biological/formal beginning of human habitation. Students gathered evidence of Seoul's understanding of its surfaces as essential space: resources to better the city and identify the layered and integrated nature of contemporary urban living. City surfaces are both active and deep—a stratifiable medium for new public space. These surfaces constantly refine the redefined uses for public space as collective ground.

The semester began by cataloging space, materiality, use, and methods of making in order to explore the potentials found within the rethinking of surface as actionable medium. This coupled with research on Seoul generated process/practice overlaps that revealed the ground's inherent connectivity. Students investigated three types of surface in Seoul: Infrastructural Ground, Architectural Ground, and Collective (public) Ground. Students reinterpreted and applied lessons from within the city to activate and appropriate space for public life in new ways. Projects propose new types of strategic, activity-scaled public spaces aimed at linking citizens to their neighborhood in meaningful, experiential ways.

G32

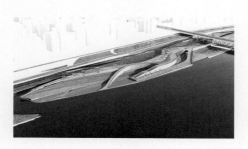

EAFIT 대학교 URBAM
두 강의 대화

스튜디오 리더∴홀리아나 퀸테로 마린

우리는 끊임없이 분쟁에 시달리는 신흥 지역에 공정한 절차를 구축하고자 한다. 신흥 지역에서는 경제 발전과 기반시설 개발, 천연 자원과 지역 공동체 보호 사이에서 갈등이 일어난다. 아트라토강 유역과 아부라강 상류 지역이 그러한 갈등을 단적으로 보여주는 사례다.

아트라토강 유역의 아프리카 후손과 토착민으로 이루어진 지역 공동체는 천연 자원 착취라는 논리에 희생되어 극심한 빈곤과 방치, 그리고 불법 무장 단체들에게 시달렸다. 이러한 상황은 푸르르고 다양한 생물들이 살고 있는 열대 자연에서 탄생한 풍요로운 문화와 다민족성과는 대조를 이룬다.

아부라강 상류 지역 거주민들은 지속되는 분쟁에 휩싸여 있다. 콜롬비아 내 무력분쟁으로 터전을 잃고 농촌지역을 떠나온 이주민들은 아부라강의 북쪽 지역 도시에서 터를 잡아 인생을 다시 일구는 기회를 발견했다. 이들은 메데인 같은 대도시의 변두리에 거주하며 역경 속에서 스스로의 힘으로 지역사회를 건설했다. 이 두 상황에 대해 이해하는 것은 행위자와 그 이해관계를 두루 살피는 것 뿐 아니라 갈등과 차이를 인정하는 과정을 통해 공동체 문제 해결에 도움을 줄 수 있다.

URBAM - UNIVERSIDAD EAFIT
Dialogue between Two Rivers

Studio Leader∴Juliana Quintero Marin

Our work aims to build fair processes in emerging areas that are in constant dispute. These areas face a tension between economic and infrastructural development and the protection of natural resources and local communities. The basin of the Atrato River and the northern area of the Aburrá River—Río Norte— are clear examples of this tension.

The basin of the Atrato River, its communities (afro descendants and indigenous inhabitants) have been affected by a logic of natural resources exploitation that have caused extreme poverty, abandonment, and presence of armed illegal actors. This condition contrasts with its cultural and multi-ethnic richness related to the lush, tropical, and biodiverse nature.

The inhabitants of the northern area of the Aburrá River have lived in constant dispute. Migrants from rural areas, displaced by the armed conflict in Colombia, found in Río Norte the opportunity to rebuild their lives and to have a place in the city. Margined in a metropolitan area like Medellín, they erected a self-built neighborhood in the middle of the adversity. Understanding these two conditions could contribute to the approach to the collective, not only as a concept that integrates actors and interests, but also as a notion that implies tensions and differences.

G33

연세대학교
더 파일론
스튜디오 리더.∴.∵이상윤

현대 도시의 특징은 보이는 정보와 보이지 않는 정보가 혼재되어 있다는 것이다. 과거의 아날로그 방식으로 기록된 정보에, 비가시적 형태의 디지털 기록들이 더해져 오늘날의 도시를 구축한다. 눈에 보이지 않는 정보와 기술은 사람들에게 편리함을 주지만, 이와 동시에 잠재적 위험요소로 작동하기도 한다. 보이지 않는 것의 존재가치를 알아야하는 상황들이 발생하며 사람들은 비가시적인 정보와 기술에 대한 두려움을 느끼고, 안정감을 찾기 위해 보이지 않는 것을 시각화하고자 노력하고 있다. 이러한 상황은 비가시적 정보를 가시화하여 시민과 공유할 수 있는 매체를 필요로 하고, 이러한 일환으로 우리는 도시의 파수꾼 역할을 감당할 수 있는 파일론(Pylon) 디자인을 통해 시민과 공유할 수 있는 매체로서의 구조물을 제안하고자 한다.

파일론은 미시적 관점의 인간 스케일에서부터 거시적 관점의 도시 스케일까지 포괄하여 작용할 수 있도록 설계되었다. 인간 스케일에서 파일론의 역할은 도시에서 필요한 비가시적 정보를 자생적으로 인식하며 시민과 정보를 공유하는 것이다. 대기환경, 소리/소음, 공중권, 와이파이 등의 무선통신망 등에 대한 비가시적 정보들은 파일론 안의 메커니즘으로 측정된다. 파일론의 기둥과 같은 형태는 거시적인 관점에서 도시의 그리드를 형상화 할 수 있다. 도로가 물리적 형태의 도시 그리드라면, 파일론은 상공의 교통체계를 정의할 수 있는 비가시적 형태의 도시 그리드로 작동하는 것이다.

YONSEI UNIVERSITY
The Pylon
Studio Leader.∴.∵Sangyun Lee

One of the main characteristics of the contemporary urban environment is that the invisible information coexists with the visible, and in many cases the two entangle each other to form highly hybridized entities. Today's cities are built with the innumerable accumulation of the invisible records on top of the visible information from the past. The invisible information and underlying technology has brought about both the convenience and potential threat to people and the society. In an effort to mitigate the uneasiness which people often experience when facing this sort of threat, they have sought the way to transform the invisible into more tangible ones. The Pylon functions as a medium by which people translate the unseeable entanglement of information into the palpable.

The Pylon was designed to operate in various scales spanning from the micro-scale of human activities to the macro-scale of urban interactions. In the micro-scale the Pylon reacts to invisible information such as atmospheric environment, noise and wifi reception, converts it into visible signs, and share them with the citizens. The Pylon forms an urban grid in the macro scale. While the road system is a physical grid on the ground level, the Pylon functions as an invisible grid that defines the traffic system in the air.

While the topmost levels of the Pylon house the essential operations, which collect and reinterpret the invisible information of the city, the bottom part is more flexible and versatile to adapt to the urban environments. The lower part functions as the structural body to support the upper parts, and at the same time is able to accommodate public programs such as toilets, community facilities, and public parking spaces.

G 34

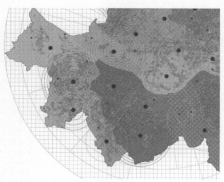

Global Studio Participant
Credit List

AMERICAN UNIVERSITY OF BEIRUT

Professor / PM
Carla Aramouny
Nicolas Fayad
Rana Samara
Christos Marcopoulos
Participating Students
Nicolas Abou Haidar
Nicol Yamin
Nawal El Katoul Al Rahbani
Karim Rifai
Sabine Doueiry
Salem Shamia
Sarah El Harouny
Lea Saad
Tala Farraj
Haya Safadi
Yara Abdallah
Rawan Koussan
Ralph Karam
Youmna Fattal Dakmak
Julia Saab
Hiba Rachidi
Yara Mortada
Rasha Haidar
Linda Yachoui
Carl Yammine
Nour Abdel Baki
Salameh Abla
Lina Akkawi
Ghida Anouti
Lama Barhoumi
Sirena Chammaa El Rifai
Zeina Chamseddine
Dina Dabliz
Hadi El Kassar
Tala El Khatib
Mahmoud Fahs
Wiaam Haddad
Lina Hammoud
Hiba Itani
Laure Jaber
Aya Meskawi
Ali Munzer
Nour Ramadan
Leah Rose Mary Rassie
Clara Saade

ARCHITECTURAL ASSOCIATION SCHOOL OF ARCHITECTURE

Professor / PM
Sam Jacoby
Participating Students
Dimitris-Ronald Chatziioakeimidis
Huace Yang
Pengyu Chen
Witthawat Prabhasawat
Wojciech Mazan
Yunshi Zhou

ASSUMPTION UNIVERSITY + UNIVERSITY OF SEOUL + UNIVERSITY OF ARCHITECTURE HO CHI MINH CITY

ASSUMPTION UNIVERSITY
Professor / PM
HƯƠNG Le Thi Thu
VIRIYAVADHANA Prima
LEELAPATTANAPUTI Veera
CHOKCHAIYAKUL Laddaphan
TANTILERTANANT Nuttee
Participating Students

POTSAWEE Pradoemkulchai
PIMONSARI Chukiatpornpongsa
BENJAMIN Milintacupt
THITAPA Chusakul
MONTHITA Krathumkaew
Attawin Hatayodo
Ornnicha Vijitkornthong
Satriya Chaipanich
Bunheng Ham
Moksha Modgill

UNIVERSITY OF SEOUL
Professor / PM
Jungwon Yoon
Participating Students
Byeonghyeon Cho
Seokwon Choi
Jongwon Jung
Juhee Han
Kyungseok Hong
Yoonjeong Kang
Mijung Park
Huiji Yang

UNIVERSITY OF ARCHITECTURE HO CHI MINH CITY
Professor / PM
Hanh Vu Thi Hong
NGUYỄN Trần Phạm Sĩ
QUANG Lê Hồng
DŨNG Đinh Xuân
PHƯƠNG Tô Thanh
AN Trần Duy An
Participating Students
VƯƠNG Phan Quốc
NHIÊN Trần Hồ Hạo
THẢO Nguyễn phương
TAYLOR Page
HÂN Nguyễn Ngọc
TÀI Lê Anh
TUẤN Trần Huỳnh
NHUNG Bùi Thị
TÀI Lâm Tuấn
HUÂN Lê Ngọc Nam
HƯƠNG Nguyễn Hoàng
NGỌC Lâm Hồng
PHÚC Nguyễn Hoàng
VỸ Nguyễn Quang
TRÂN Trần Ngọc Bảo
SIOBHAN McQueen

BARTLETT SCHOOL OF ARCHITECTURE + HANYANG UNIVERSITY ERICA

BARTLETT SCHOOL OF ARCHITECTURE
Professor / PM
Sabine Storp
Patrick Weber
Participating Students
Nicola Chan
Tom Cubitt
Naomi de Barr
Emily Martin
Daniel Medina Avilan
Giles Nartey
Robert Newcombe
Hoi Lai (Kerry) Ngan
Rebecca Outterside
Allegra Wilder
Yip Wing Sui
Shi Yin Ling
Maite Seimetz
Vitika Agarwal
Ernest Chin
Carlota Vallejo
Wei (Eugene) Tan
Tom Ushakov

Sung (Ryan) Wong
HANYANG UNIVERSITY ERICA
Professor / PM
So Young Kim
Participating Students
Hojoon Lee
Yeonju Lim
Suhyun Kim
Minju Baek
Shingu Kim
Shinyoung Kim
Donghoon Kim
Jiyoung Park
Sungjae Yoon
Woorin Lee
Sujung Cho
JeongHyun Oh
Junyoung Choi
Youngyoon Kim
Na Ha
Hyeonseon Chu
Soo Hwan Jung
Goh Zhi Yu
Yu Bing Jie
Meriem Khaldoun
Giulia Tolu

CALIFORNIA COLLEGE OF THE ARTS
Professor / PM
Neeraj Bhatia
Participating Students
Keehyun Ryu
Taamara Rath
Cassady Kenney
Jose Rodriguez Trujillo
Kshitija Chandrashekhar Nagarkar
Hannah Jane Kim
Sayer Al Sayer
Dina Elattar
Vaidehi Davda
Hardik Bhimani

CHINESE UNIVERSITY OF HONG KONG
Professor / PM
Peter Ferretto
Haoran Howard Wang (Exhibition Designer)
Ling Cai
Participating Students
— Design Team
Paula Ziwei Liu
Beryl Yuk Tsin Wong
Milly Man Yan Lam
Sungryeol Choi
— March Studio / Hakka Village
Amelie De Guerre
Jacky Chun Hin Fong
Ping Hang Bono Fung
Po Yi Bowie Ha
Wing Chuen Stanley Hung
Shakir Hussain
Pui Yeung Javee Lam
Wing Sze Gillian Ngan
Ho Fung Issac Tam
Nga Man Gloria Tam
Siu Fung Simon Tsang
Suet Ying Circle Yuen
Fengji Freddie Zhang
— March Thesis /Dong Villages
Beryl Yuk Tsin Wong
— Workshop / Dong Community — 2016
Cho Man Chan
Melanie Edith Cheng
Ming Hin Cheung
Sin Ki Fung

Ngo Kwan Koon
Kin Kwan Lam
Sai Chin Lam
Ho Yan Lee
Choi Ling Leung
Tsz Ling Leung
Weixin Li
Yim Ling Wong
Peng Yang
Man Lung Jacky Ip
— 2017
Wing Yan Chan
Chi Hi Chan
Sze Wing Chan
Hong Ting Cheung
Sheung Kwan Choi
Sze Hei Chow
Wing Yee Tam
Wai Kin To
Hoi Kei Wong
Man Yan Lam
Fai Nam Wong
— 2018
Lok Man Cho
Tianpeng Fan
Jingru Jiang
Yan Yu Kwok
Chung Hei Li
Pak Wai Ling
Tsz Kwan Liu
Ching Ching So
Wan Ting Tang
Wing Sze Tong
Yan Hei Tse
Chun Lung Wong
Chun Yin Yau
— 2019
Chun Hei Chan
King Long Chan
Yin Tat Chan
Chung Hei Ho
Keqing Jiao
Hoi Sau Koo
Pik Ying Lo
Shuhan Miao
Wai Faan Wong
Pui Shan Yiu
Hiu Man Cheng
Hoi Lun Kwan
Hoi Kan Ning
Ho Ching Wong
Ka Lok Wong
Lai Fei Choi
Koon Kau Ho
Oi Ping Tse

COLUMBIA UNIVERSITY
Professor / PM
Nayun Hwang
Participating Students
Adelaida Albir
Andrea Chiney
Don Chen
Eugene Chia-Yu Chang
Eugene Ong
Fiona Ho
Joanne Chen
Julie Christine Perrone
Kevin Hai Pham
Iara Dykstra
Masha Konopleva
Po-Han Lin
Wo Wu
Professor / PM
David Eugin Moon
Participating Students
Pooja Annamaneni
Linda Cristina Schilling Cuellar
Betsy Daniel
Thuto Durkac Somo

Hanyi Liu
Alicia French
Karina Todorova Spassova
Travis Tabak
Monica Marie Wojnowiak
Laura Huan Wu

DONG-A UNIVERSITY
Professor / PM
Youn Suk Cha
Participating Students
Hyung Moon Won
Hui Geun Oh
Min Seon Kim
Sun Hyeok Lee
Hyun Ung Shin
Ji Hoon Han
Min Seo Park
Tae Hyuk Gil
Kyung Hwan Kim
Hye Mi Kim

EWHA WOMANS
UNIVERSITY + RADBOUD
UNIVERSITY

EWHA WOMANS
UNIVERSITY
Professor / PM
Klaas Kresse
Participating Students
Jieun Jeong
Heejoo Park
Yu Seon Kim
Min Kyoung Kim
Jung Won Kim
Sang Hwa Nam
Changeun Song
Jiye Yoon
Sunho Lee
Jisu Lee
Hyunsong Lee
Kyung Ryung Choi
Tubanur Eroglu

RADBOUD UNIVERSITY
Professor / PM
Erwin Van der Krabben
Participating Students
Emil Ros
Job Wevers
Sjoerd Stolk
Willem van Wagenberg
Emma Breuker
Brent Houterman
Joost Reumkens

HANYANG UNIVERSITY
Professor / PM
Rafael Luna
Participating Students
Sung Min Kang
Hyunji Kim
Habin Park
Jun ho Baik
Ziqi Wei
Siqi Li
Zijun Zhang
Gyuhee Jo
Mengying Cong
Sangkyu Han
Natalia Jarzab
Izabela Karpiel
Hong-Siou Lin
Pol Margalef

HARVARD UNIVERSITY
Professor / PM
Andres Sevtsuk
Evan Shieh (TA)

Syed Ali (RA)
Participating Students
Weihsiang Chao
Yuebin Dong
Ruanlanming (Dora) Du
Solomon Green-Eames
Sang Yoon Lee
Sunmee Lee
Linyu Liu
Yuzhou Peng
Xin Qian
Amanda Ton
Finn Vigeland
Lanchun Zeng
Bailun Zhang

HONGIK UNIVERSITY
Professor / PM
Juwon Kim
Dongwoo Yim
Participating Students
Jaeseong Kim
Joungyun Kim
Sijung Park
Hayoon Park
Hongsun Baek
Geonyoung Lee
Eunhye Lee
Jinsoo Lee
Jiho Yoon
Sooyoung Choi

INDA – CHULALONGKORN
UNIVERSITY
Professor / PM
Alicia Lazzaroni
Participating Students
— Y3 Showcased Projects
Shompoonuth Kumpakdee
Vedant Urumkar
Lalipat Sirirat
Luksika Pratumtin
Sapanya Patrathiranond
Sitanan Teeracharoenchai
Thanjira Vimonanupong
Boontita Boonsusakul
Puntawan Suppakornwiwat
Pittaya Thamma
Panas Saengvanich
Pasinee Kerdpongvanich
Pimboon Wongmesak
Nanna Thaiboonruang
Piriyakorn Tamthong
Thunda Rerkpaisan
Kanchaporn Kieatkhajornrit
— Y2 Installation Building
Napas Simarangsun
Kamolthip Polsamak
Thanvarat Jamnongnoravut
Kodchakorn Promjaree
Athitaya Piamvilai
Tamon Sawangnate
Hattakarn Lertyongphati
Paweenda Patarathamaporn
Tida Rama
Tikumporn Panichakan
(ex-alumna, TA)
— Tutors of Showcasing
Students
Alicia Lazzaroni (coordinator)
Antonio Bernacchi
Hadin Charbel
Patrick Donbeck
Scott Drake
Sabrina Morreale
Payap Pakdeelao
Lorenzo Perri
Carmen Torres
Rebecca van Beeck
Tijn van de Wijdeven

Danny Wills

KEIO UNIVERSITY
Professor / PM
Jorge Almazan
Participating Students
Gabriel Chatel
Javier Celaya
Naoki Saito
Itaru Iwasaki
Yuichi Tatsumi
Yoshihisa Moriya
Yuki Wada
Sudeep V Zumbre

KOOKMIN UNIVERSITY
Professor / PM
Helen Hejung Choi
Ilburm Bong
Wooil Kim
Kyu Hwan Lee
Participating Students
Eunmi Choi
Hwajin Lee
Hyeseong Kim
Jieun Kim
Jinho Kim
Junhyung Lee
Kyungchul Sung
Sanghyun Koh
Seongmin Jang
Yeojin Lee
Yeonjin Kim
Yongsuk Cho
Youngbean Choi
Youngki Kim

KUWAIT UNIVERSITY
Professor / PM
Shaikha Almubaraki
Participating Students
Bedour AlMoosawi
Razan Alzayed
Maryam AlThaher
Sara AlHashash
Bedaiwi AlBedaiwi
Yasmeen Shehadeh
Deema AlOthman
Sara AlZeer
Lulwa AlBader
Wadha Al Sabeeh
Nada Abu-Daqer
Deema Al Wugayan
Danya Al-Khalaf
Zainab AlHashemi
— Special Thanks to:
Muneerah Al Rabe

NATIONAL UNIVERSITY OF SINGAPORE
Professor / PM
Erik l'Heureux
Participating Students
Cai Tong
Lee Dong Eun
Luh Astrid Mayadinta
Masita Binte Mohd Yusof
Melvin Lim Chung Wei
Mitch Goh Jie Chern
Nicholas Tai Han Vern
Oswald Hogen Salim Huang
Siti Nur Farah Binte Sheikh I
Sharlyn Hwang Ya Wen
Sun Yutong

ROYAL COLLEGE OF ART
Professor / PM
Davide Sacconi
Gianfranco Bombaci
Matteo Costanzo

Francesca Dell'Aglio
Participating Students
Nico Alexandroff
Paul Bisbrown
Kane Carrol
Cecile Diama-Samb
Alexander Mark Findley
Fraser Ingram
Thomas Jenkins
Mijung Kim
Jaehyub Ko
HoYin Lui
Ayesha Silburn
Jade Tang
Eleni Varnavides
Anastasia Whitehead
Rosa Whiteley

SEOUL NATIONAL UNIVERSITY
Professor / PM
John Hong
Jungho Kwon (TA)
Participating Students
Natalie Blom
Sihyun Jeong
Joohyung Lee
Yoon Ji Kim
Youngmyoung Kim
Jihye Park
Ha-eun Rim
Yan Xiang
Jeongwoo Song

SINGAPORE UNIVERSITY OF TECHNOLOGY AND DESIGN
Professor / PM
Calvin Chua
Participating Students
Samantha Tang
Ng Jinxi
Samantha Lim
Chan Jia Hui
Gabi Quek
Joey Tan
Melissa Mak
Neo Sze Min
Kwang Guochuan
Shobha Narendran
Tan Wei Lin
Willa Trixie Ponimin
Lisa Koswara
Mok Jun Wei
Ng Xin Ling
Neo Xin Hui

SUNGKYUNKWAN UNIVERSITY + YEUNGNAM UNIVERSITY + KEIMYUNG UNIVERSITY + KARLSRUHE INSTITUTE OF TECHNOLOGY + KARLSRUHE UNIVERSITY OF APPLIED SCIENCE + UNIVERSITY OF APPLIED SCIENCE STUTTGART

SUNGKYUNKWAN UNIVERSITY
Professor / PM
Thorsten Schuetze
Participating Students
Fabrizio Maria Amoruso
Min Hee Sonn
Sang Troung Thanh
Fitria Ressy
Yoann Sasselina
Yau Chun Hon
Bayarmaa Badral

YEUNGNAM UNIVERSITY
Professor / PM
Emilien Gohaud
Hyunhak Do
Participating Students
Dongju Lee
JooHee Moon
Juhee Park
Sungha Choe
MiRi Park
HyunAh Ahn
Yeongil Kim
Adèle Bergna

KEIMYUNG UNIVERSITY
Professor / PM
Hansoo Kim
Participating Students
Gyu nam Park
Kyung bin Park
Min jee Kim
Soo jin Sung
Joo han Lee
Hyeong gyun Kim
Do yeon Kim
Seung ryeol Lee

KARLSRUHE INSTITUTE OF TECHNOLOGY
Professor / PM
Markus Kaltenbach
Participating Students
Isabell Forschner
Sebastian Humpert
Marc Hodapp
Ina Zaloshnja
Patrick Bundschuh
Jonathan Winzek
Xiang-Ru Zhu
Julian Hübner
Julian Beck
Michael Hosch
Sarah Mutterer
Leonie Hellersberg

KARLSRUHE UNIVERSITY OF APPLIED SCIENCE
Professor / PM
Jan Riel
Participating Students
Daria Graf
Julia Paul
Heiko Fückel
Tobias Mattheis
Alina Preuß

UNIVERSITY OF APPLIED SCIENCE STUTTGART
Professor / PM
Philipp Dechow
Participating Students
Valerie Sporer
Vanessa Bleckmann
Marie Huber
Sophia Rausch
Bahar Akgün

SYRACUSE UNIVERSITY
Professor / PM
Fei Wang
Participating Students
Minglu Wei
Furui Sun
Maria Camila Andino Donoso
Radia Berrada
Anthony James Bruno
Brooke Calhoun
Chieh Wei Chiang
Jose Sanchez Cruzalegui
Karisma Dev

192

Lei Feng
Andres Feng Qian
Spencer Gafa
Rachael H Gaydos
Paul Nathan Gibson
Rongzhu Gu
Aaron Guttenplan
Ross Hanson
Hanseul Jang
Sang Ha Jung
Harshita Kataria
Nivedita Keshri
Ketaki Kini
Anna Korneeva
Chun Yen Ku
Clarrisa Lee
Chenxie Li
Haoxiang Liao
Wang Liao
Xiao Lin
Yanhan Liu
Yui Kei Lo
Yifei Luo
Ian Masters
Soravis Nawbhanich
Ryan Oeckinghaus
Christina Rubino
Demosthenes Sfakianakis
Junran Tao
Katherine Truluck
Evan Webb
Bingyu Zeng

TECHNISCHE UNIVERSITÄT BERLIN
Professor / PM
Timothy Pape
Participating Students
Flavia Biianu
Edda Brandes
Pauline Bruckner
Almar de Ruiter
Valentin Dobrun
Finya Eichhorst
Fadi Esper
Stefan File
Muhannad Ghazal
Aaron Geier
Jörn Gertenbach
Anne Gunia
Christopher Heidecke
Olga Juutistenaho
Dariya Kryshen
Anna Lesch
Andrés Reyes Kutsche
Farina Runge
Alina Schütze
Lisa Wagner
Jonas Wulf

TECHISCHE UNIVERSITÄT WIEN
Professor / PM
Mladen Jadric
Federica Rizzo
Participating Students
Hajrudin Ovcina
Andjela Misic
Andjela Kovacevic
Betül Kazanpinar
Laura Sánchez Fernández
Tamara Sztastyik
Sherif Takla
Luiza Ganeva
Preslava Slavcheva
Alexandrina Chergarska
Alexandra Mihaela Pavel
Roberta-Andreea Manea
Valentina Lucich
Redon Softa

Go-Eun Lee
Aleksandra Majkanovic
Valeria Tosa
Fabian Kompatscher
Marwa Yasin
Branislav Stojkov
Mariya Tochkova
Mila Stiliyanova Mihaylova
Simel Karabulut
Selen Cicek
Ezgisu Kuyumcu
Filippo Ossola
Sara Wessila Selicato
Emine Balci
Beyza Köroglu
Kübra Inal

TEXAS TECH UNIVERSITY
Professor / PM
Kuhn Park
Lisa Lim
Participating Students
Zachary Acosta
Vanessa Aguilar
Dannette Aguirre Tizatl
Amir Sam Halabi
Trey Kelly III
Martin Merino
Jonah Remigio
Roberto Emilio Sanchez Perez
Matthew St.Jacques
Horel Villa Hernandez
Whitney Williamson
Zeqiang Zhu

UNIVERSITY OF CAPE TOWN
Professor / PM
Fadly Isaacs
Participating Students
— Exhibition Products
Francisca Audu-Ukpoju (2019)
Marelize Lourens (2019)
Mathithia Maloba (2019)
Layla Sedick (2019)
Lena Weir (2019)
— BLUEPRINT and Slide Animation
Anees Arnold (2015, 2016)
Kayla Brown (2015, 2016)
Jade Budd (2017, 2018)
Tarryn Langenhoven (2017, 2018)
Sean Meyer (2017, 2018)
Rushdee Mohamed (2017, 2018)
Frans Pieters (2016)
Patrick Schuster (2016, 2018)
Alex Swartz (2016, 2018)
Warren Van Niekerk (2016, 2017)
— Design Research Mapping
Sihaam Ajouhaar (2016)
Klara Bezuidenhout (2015)
Subi Bosa (2017)
Luet Buys (2015, 2016)
Shafeea Chogle (2016, 2017)
Jason Chokupermall (2015)
Mishkah Collier (2017)
Simon Ferrandi (2016, 2017)
Mira Friedman (2018)
Marjan Ghobad (2018, 2018)
Natalie Goni (2017)
Justin Grove (2016)
Fierdouz Hendricks (2017, 2018)
Sebastian Hitchcock (2018, 2019)
Zaheer James (2015, 2016)
Shravan Jugurnauth (2015)
Helena Karamata (2018)
Tao Klitzner (2018, 2019)
Inam Kula (2018)

Andre Le Roux (2016, 2017)
Tawanda Madzingaidzo (2015)
Louise Malan (2015)
Mosa Molepo (2018)
Tanya Moodaley (2016)
Kirsten Moses (2016, 2018)
Ebrahim Mullajee (2015)
Zonke Mzi (2017, 2018)
Maashitoh Rawoot (2015)
Alonso Santos (2016)
Khanya Socikwa (2017, 2018)
Sideeq Samodien (2017, 2018)
Wesley Samouilhan (2015)
Baliwe Sibisi (2016)
Kenny Simbayi (2015)
Luphiwo Tanda (2018, 2019)
Stephanie Terwin (2016)
Melanie Vanz (2017)
Jonathan Wilson (2015, 2016)
Zietsman, Tiaan (2016)

UNIVERSITY OF HONG KONG
Professor / PM
Géraldine Borio
Participating Students
Bunjaya Alvin Dharma Saputra
Mabel Choi Wai Pan
Olivia Kwong Tsz Tung
Inhoo Seo
Sisay Sombo
Tjuatja Evelyn
Samantha Wang Aiyu
Tina Wang Tianduo

UNIVERSITY OF MANITOBA + CARLETON UNIVERSITY + UNIVERSITY OF TORONTO

UNIVERSITY OF MANITOBA
Professor / PM
JaeSung Chon
Participating Students
Ben Greenwood
Connery Friesen
Tia Watson
Meighan Giesbrecht
Jessica Ferrigno
Mackenzie Swope
Evan McPherson
Eric Decumutan
Matthew Peters
Carl Valdez
Marina Jansen
Matthew Rajfur
Ivan Katz
Alex Chem
Collin Lamoureux

CARLETON UNIVERSITY
Professor / PM
Ozayr Saloojee
Participating Students
Adan Achtenberg
Anissa Alami Merrouni
Haley Baric
Shih-Sung Chen
Celeste Correia
Sarania Dabee
Catherine Dela Cruz
Walaa El Sydabi
Viktor Ivanovic
Mitchell Jeffrey
Jaron Kasabian
Chesta Lalit
Jerry Mangilit
Jaykumar Patel
Kristine Prochnau
Waise Sahel
Razi Sasikaran

Brendan Schoug
Ksheel Shetty
Kristen Smith
Madelaine Snelgrove
Jasmine Sykes
Andrea Tamayo Bernal
Sma Waiganjo
Nadisha Wanniarachchige

UNIVERSITY OF TORONTO
Professor / PM
Adrian Phiffer
Participating Students
Zoal Abdul Razaq
So Ahn
Nick Callies
Kevin Chan
Annie Cottrell
Shea Gouthro
Angela Konieczkowski
Erica Nassirian
Brendan Onstad
Jin Park
Simon Rabyniuk
Elly Selby
Robbie Trakji
Alice Wong

UNIVERSITY OF PENNSYLVANIA
Professor / PM
Simon Kim
Participating Students
Mostafa Akbari
Yunyoung Lina Choi
Wenna Dai
Zihao Fang
Yuting He
Wenqi Huang
Mikyung Lee
Chuqi Liu
Bowen Qin
Yingke Sun
Si Yang Xiao
Weimeng Zhang
Chengyao Zong

UNIVERSITY OF SEOUL
Professor / PM
Marc Brossa
Hyojin Byun (tutor)
Participating Students
Phaphavee Sriapha
Yada Chatavaraha
Sapanya Patrathiranond
Akarpint Chomphooteep
Chadaporn Sompolpong
David Lasek
Jun-seob An
Sangki Jeong
Eui-seong Yun
Jae-kyung Park
Ye-ji Park
Ah-yeon Yuk
Haelim Chang
Wook Ho Jung
Woraphan Jampachaisri
Tae Hwa Kang
Siin Lee
Yamin Kosittanakiat
Gi Youn Yoon
Mónica Muro
Pattarawan Rungrattawatchai
Rita Choonhaprasert
Hathairat Kangval

UNIVERSITY OF TECHNOLOGY SYDNEY
Professor / PM
Gerald Reinmuth

Participating Students
Dinal Savinda Batapaththala
Enoch Chiu
Kevin Chuang
Yang Feng
Jed Finnane
Jassinta Fong
Eleanor Gibson
Huijie Gu
Hao Han
Jonathon Hoare
Gaurav Kapoor
Rose MacMahon
Rongtan Ni
Ivah Janine Perez
Brodie Robertson
Srujan Vichare
Asmayadinata Muljadi Putera
William Kelly
Hwa Lee

UNIVERSITY OF TEXAS AT ARLINGTON
Professor / PM
Joshua Nason
Participating Students
— Student Design Studio Team
Rameesa Ahmed
Diego Barreto
Jesus Cabrera
Luis Cortes
Arabel Cutillar
Ulises De La Cruz
Jeffrey Fudge
Ali Golnabi
Victoria Hernandez
Kaylee Lamb
Guillermo Padron De La Cruz
Gabrielle Rossato
Zainab Safri
Miguel Salmeron
Daniel Tighe
— Student Support Team
Gustavo Alejandro
Janeth Cardenas
Vanessa Chiquito
Eunhee Cho
Mariah De La Garza
Logan Engstrom
Daniel Escobar
Mariah Fairbanks
Meliisa Farrell
Tanner Harty
Ran Hong
Tristin Lott
Mark Marianos
John McGraw
Emmanuel Ogunlola
Nancy Ramirez
Rubi Sanchez
Isaiah Sigala

URBAM - UNIVERSIDAD EAFIT
Professor / PM
Alejandro Echeverri
Francesco María Orsini
Natalia Castaño Cárdenas
Juliana Quintero Marin
Participating Students
Valentina Franco
Gloria Elizabeth Toro
Jorge Iván Blandón
Gabriel Bettín Enrique
Erica Cristina Muriel
Norela Ruiz
Cesar Nicolás Mendoza
Juan Fernando Zapata
Jhony Alexander Díaz
Claudia Marcela Londoño

Juan Sebastian Bustamante
Isabel Basombrío
Carlos Andrés Delgado
Juliana Montoya
Juliana Gómez
Tatiana Ríos
Andrés García
Crsitina López
Luis Fernando Gonzalez
Gustavo Duncan
Paola Escobar
Fernanda Mesquita
Juan Esteban Vergara
Paul Pico Oteniente

YONSEI UNIVERSITY
Professor / PM
Sang Yun Lee
Participating Students
Woonghee Cho
Dongeun Hwang
Unsol Choi
Eunji Hyun
Hyeim Jun
Wonyoung Choi
Min Jae Park
Gayoung Lim
Youngho Jeong

현장프로젝트
LIVE PROJECTS

시장은 집합으로서 도시의 형태를 가진다. 도시는 교환의 장소에서 시작되었으며, 그 시장의 밀도가 높아지고, 시장 주위로 다양한 기능들이 부가되면서 현재 형태로 진화하였다. 도시의 형태는 산업혁명 이후의 철도와 도로의 발달, 분업화된 산업구조, 집약된 자본으로 변화하고 있지만, 그 안에 원초적 모습의 전통시장은 어느 도시나 아직도 여전히 그 모습을 유지하고 있다.

이 원초적인 집합도시인 전통 시장의 다양한 관점을 통해서 우리의 도시문제를 다시 한 번 들여다본다. 우리의 서울은 외적으로 글로벌하지만, 내적으로 단절되어 있고, 미시적으로는 집합적이나, 거시적으로는 파편화되어 있으며, 겉으로는 조화롭게 보이지만, 내부적으로 갈등이 산재한 도시이다. 이런 모순적인 도시문제로부터 도시적 대안을 찾아보고, 의논하며, 배우고, 알아가며, 체험하며, 즐기는 과정을 시민들과 함께하고자 한다.

또한, 자연발생적이며, 집합적 도시의 특성을 가진 전통시장이 다양한 사회·문화적 행위를 포함하며, 시민들이 더욱 좋아할 수 있고, 경제적으로도 풍요로운 장소가 되기 위해서 도시건축 디자인이 어떻게 기여를 할 수 있을 것인가 라는 질문은 현장 프로젝트를 이끌어가는 중요한 화두가 될 것이다.

큐레이터∴ 장영철
협력큐레이터∴ 유아람, 최주연, 홍주석

INTRO

A market is a meeting place, taking on the form of a city. As the density of the market increased and added on complexity and function, the markets merged with the form of the surrounding city, currently comprising a large proportion of central Seoul today. As we witness the expansion of railroads and roads, as well as the development toward capital intensive and specialized industries, the market has maintained a consistent presence and typology in many cities across the globe, serving as important socio-economic infrastructure for our dense urban cores.

The Live Projects exhibition explores the challenges currently facing our cities from this important and universally recognizable urban form. While Seoul is a global and international city, it grapples with issues of social isolation and fragmentation that is reinforced through physical separation and boundaries. The exhibition explores urban alternatives in Korea that can strategically and innovatively address these contradictory problems of the contemporary condition. Situated within the heart of the city itself it actively engages with the local citizens of Seoul, tackling the most pertinent questions of the city.

The marketplace serves as an example and case study of how questions of the collective have evolved and the socio-political, economic, and cultural impact of these market models. These spaces also offer the opportunity to study and understand how architecture and urbanism can make a significant contribution to these traditional spaces of trade and exchange and for them to continue to be enjoyed by everyone. Public engagement in these spaces and the exhibition is central to the project.

Curator.˙.˙Young Chul Jang
Associate Curator.˙.˙Aram you, Jooyeon Choi, Jooseok Hong

INTRO

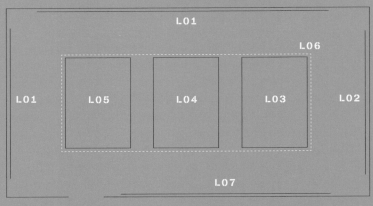

입구
Entrance
↑

서울역사박물관 기획전시실 B
Seoul Museum of History
Special Exhibition Hall B

MAP

집합도시장

L01 집합도시장 / 유아람
L02 서울 2045 / 오영욱
L03 2019 DISPLAY_02 / 토마즈 히폴리토
L04 동대문시장과 배후기지 / OOO간
L05 시장의 초상 / 노경
L06 무엇이 가만히 스치는 소리 / 오재우
L07 데이터스케이프: 서울장의 형태 / 방정인, 스튜디오 둘 셋

COLLECTIVE MARKET CITY

L01 COLLECTIVE MARKET CITY / ARAM YOU
L02 SEOUL 2045 / YOUNG WOOK OH
L03 2019 DISPLAY_02 / TOMAZ HIPÓLITO
L04 DONGDAEMUN MARKET AND THE BACKSTREETS / OOOGAN
L05 PORTRAITS FROM THE MARKET / KYUNG ROH
L06 THE SOUND OF GENTLY BRUSHING BY / JAEWOO OH
L07 DATASCAPE: FORM OF SEOUL MARKET / JEONGIN BANG, STUDIO TWOTHREE

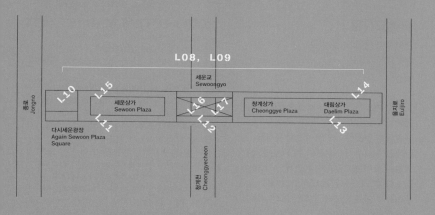

다시세운광장, 세운상가 보행데크 – 청계대림 데크, 세운교
Again Sewoon Square, Sewoon Plaza Pedestrian Deck –
Cheonggye Daelim Deck, Sewoon Bridge

서울도시장

L08   을地:공존 / 윤현상재
L09   도시상회 / 어반플레이

파빌리온 프로젝트

L10   감각 場 / UAUS: 국민대학교
L11   풍경재생 / UAUS: 연세대학교
L12   가설.가설.가설. / 서승모
L13   컵플라워 / UAUS: 서울시립대학교
L14   받히다, 바치다 / UAUS: 가천대학교
L15   내가 만드는 키오스크 / UAUS:
       중앙대학교
L16   리:커버 / UAUS: 선문대학교
L17   플로트폼 / UAUS: 한양대학교

SEOUL CITY MARKET

L08   EULJI: COEXISTENCE /
       YOUNHYUN TRADING
L09   CITY MARKETS / URBAN PLAY

PAVILION PROJECT

L10   WEAVING DOME / UAUS:
       KOOKMIN UNIVERSITY
L11   PLAY-SCAPE / UAUS: YONSEI
       UNIVERSITY
L12   PERHAPS. PERHAPS. PERHAPS
       / SEUNGMO SEO
L13   CUPLOWER / UAUS:
       UNIVERSITY OF SEOUL
L14   CRATER: CREATE WITH CRATE
       / UAUS: GACHON UNIVERSITY
L15   IKIO / UAUS: CHUNG-ANG
       UNIVERSITY
L16   RE: COVER / UAUS: SUN MOON
       UNIVERSITY
L17   FLOATFORM / UAUS: HANYANG
       UNIVERSITY

집합도시장

협력 큐레이터∴∵유아람
서울역사박물관
L01

도시에서는 수많은 것- 물건, 사람, 자본,
정보 -이 끊임없이 움직이고 서로 교환된다.
교환되는 것들은 개인의 주변에서부터
거대한 도시까지 아우른다. 우리를 둘러싼
교환이 모여들어 물리적 환경을, 지역을,
장소를, 그리고 서울을 이룬다. 그렇기에
우리는 이렇게 말할 수 있다. "도시는 교환의
장소로부터 시작되었다."

COLLECTIVE MARKET CITY

Associate Curator∴∵Aram You
Seoul Museum of History
L01

In the city, so many things—objects,
people, vehicles, capital, information,
data—are constantly moving and
exchanged. From a personal boundary
to the greater city, the environment
we experience is made up of constant
exchanges. The exchanges around us
come together to form the physical
environment, the area, the place, and
the city of Seoul. That is why we can say:
"The origin of cities lies in the place of
exchange."

COLLECTIVE MARKET CITY

모든 것이 모여들어 가득 쌓여있는 시장은 '오감'으로 서울을 느끼게 한다. 서울의 시장은 이 도시에서 일어나는 교환의 물화(物化)이며, 그 자체로 교환의 장소이다. 시장이 많은 부분을 차지하고 있는 서울이라는 도시는 교환이 물화된 곳이다. 시장에 몸을 맡기면 가지각색의 방법으로 감각을 자극한다. 각각의 시장이 갖고 있는 고유의 냄새, 습도, 온도, 색과 밝기, 그리고 대화소리가 있다. 이들은 시장 전체를 이루는 가게들에서, 가게의 매대, 매대의 상품과 작은 물건들이 모였을 때, 거기에 상인과 행인이 함께 만들어내는 미시적 화학작용의 결과다. 시장은 분명 사람이 만들었지만 통제되고 계획되지 않은, 교환 과정에서 의도치 않게 배어난 자연발생적인 총체이다. 자연발생적인 전통시장일수록 교환의 모습이 자연스레 드러난다. 우리가 보고 느끼는 시장 속에는 서로 다른 사람들이 모여 산다는 사실, 사회의 원초적인 집합성이 있다. 이 원초적이고 자연적인 집합체는 역사, 정치, 경제, 기술에 근거하여 끊임없이 더 큰 집합체를 이룬다.

상품에서 도시까지. 교환의 장으로서의 시장은 끊임없이 변화하며 매 순간 도시의 시작점이 된다. 도시의 이미지가 아닌, 가장 원초적이고 날 것의 모습이 시장에는 있다. 오늘날의 서울은 어떠한 모습을 하고 있는가? 그 답은 오늘날의 시장에서 구할 수 있을 것이다.

Marketplaces serve as spaces where a wide array of components come together as one to give visitors an experience of Seoul that stimulates all five senses. Seoul's marketplaces are the very commodity of exchange within the city and also the place itself where that exchange occurs. Made up of many markets all over the city, Seoul is a place where exchange is turned into a commodity. Those who visit marketplaces can look forward to an experience that stimulates the senses in a variety of ways. Each market has its own unique set of smells, humidity, temperature, colors, brightness, and the bustling noise of ongoing conversations. All of these qualities come together from the stores and stalls, the items on display, and the interactions between buyers and sellers to create an array of microscopic, chemical reactions that give each marketplace its identity. Although individuals might have created markets, the unplanned, unrehearsed process of exchange is a naturally occurring phenomenon. The more spontaneous a traditional market is, the more natural are the exchanges that take place within them. In fact, the phenomenon of people from different backgrounds gathering and making a living within the marketplace is an example of society's fundamental tendency toward collectivism. This smaller collective system shares a basis in history, politics, economics, and technology, ultimately connecting to an even greater, overarching system.

From the products to the city, markets are ever-changing places of change and from where cities begin. These markets show the most fundamental, raw images of exchange outside of classic urban images. So what exactly is the image of Seoul today? The answer to this question can be found by visiting today's marketplaces.

1-01

서울 2045

오영욱

서울역사박물관

L02

가까운 미래의 서울을 상상한 입체지도
작업이다.
청계천의 지류들이 대부분 복원된 서울의
중심가의 모습과 인구가 줄고 새로운 기술이
도시에 유입되었을 때 서울은 어떻게
변화하였을 지를 예측하였다.
다른 지역과 마찬가지로 서울 역시
핫플레이스와 아닌 곳으로 나뉘어 발전의
양상이 달라질 것이다. 과거의 도시가 자연
위에 인간의 문명을 새겼다면 가까운 미래의
도시는 도시 위에 자연이 침투하는 모습일
것이다. 따라서 도태되는 동네는 자연의
품으로 돌아가게 될 것이다.
핫플레이스의 후보지는 조선시대의
고궁, 20세기의 정취를 지닌 단독주택가,
첨단문명이 집결된 업무지구 그리고
100년이상의 역사를 갖고 있는 재래시장이 될
것이다.
특히 21세기 초반 엄청난 침체를 겪었던
재래시장들이 중반에 이르러 다시금 주목 받은
모습을 그렸다.
본 작업에서는 추억에 잠겨 재래시장을
여행하는, 기술 문명에 지친 주민들과 전
세계에서 모인 호기심 많은 관광객들의
모습을 상상했다. 실제적인 물품 거래가
이루어지기보다는 기억과 인간 본성을
주고받는 장소로 살아남은 재래시장과 그
주변에 몰려 사는 사람들, 그리고 인구 감소와
기술 발전으로 많은 곳이 숲으로 돌아가게 된
도시를 그렸다.

SEOUL 2045

Young Wook Oh

Seoul Museum of History

L02

This is a three-dimensional map
imagining Seoul in the near future.
In particular, it presents the heart of
Seoul where most of the tributaries of
Cheonggyecheon are restored. It presents
a visualized forecast of how Seoul could
change with population decrease and new
technology engaged in the city.
Similar to just about anywhere else,
popular places in Seoul will undergo a
greater extent of change compared to
areas that are less popular. Eventually,
towns and neighborhoods that are not
often frequented will be forgotten and
slowly return to nature. While cities of the
past began from civilization occupying
nature, future cities could be nature
invading the built city.
Popular places include the palaces of the
Joseon Dynasty, the classic houses and
villas of the 20th century, the high rise
of business districts, and the traditional
markets that have witnessed 100 or
more years of history. In particular, the
local markets that experienced times of
hardship during the 21st century due to
the economic recession will be revitalized.
We imagine a people eventually tired
of their technology-driven civilization
and reminiscing about the past while
taking a stroll in a market mixed with
foreign tourists who dream of traveling to
traditional markets. In this future, people
will come to these markets to exchange
memories and personal experiences
rather than being absorbed in buying and
selling goods. We see people flocking to
these areas while other parts of the city
are reclaimed by forests owing to both
sparse population and technological
advancement.

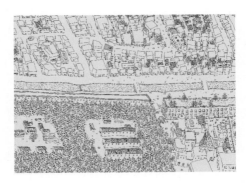

2019 DISPLAY_02
토마즈 히폴리토
서울역사박물관
L03

본 프로젝트는 공간과 그 안에서 발생하는
활동에 주목하며, 이로 인해 공간이 어떻게
변모해가는지 추적한다.
"행위의 지도화(Mapping Gesture)"에 관한
연구를 통해 새로운 영토를 발굴하고 구성원
사이의 간극을 정의하며, 그곳에서의 경험을
탐색하고자 하였다.
사진, 영상, 공연, 그림 등 다양한 매체를
활용해 작품에 대한 이해도를 높였으며 이
모든 과정이 프로젝트의 일부를 구성한다.
색다른 시도가 녹아든 토마스 히폴리토의
작품을 통해 그만의 신선하고도 독창적인
세계관을 엿볼 수 있다.

2019 DISPLAY_02
Tomaz Hipólito
Seoul Museum of History
L03

This project addresses the questions of
space, the activities occurring within that
space, and how the space transforms as
a result.
Through the research on "Mapping
Gesture," we sought to excavate a
new territory, define the gaps between
its constituent parts, and explore the
experience from that interval.
Multiple forms of media such as
photography, video, performance,
painting, and drawing are used to better
reveal the concept of the work. The entire
process becomes part of the work.
The unique approach of Tomaz Hipólito's
works offers a fresh and original view of
the world.

L03

동대문 시장과 배후기지
000간
서울역사박물관
L04

동대문 의류 시장은 1970년대 이후
내수시장의 확대와 의류수출 급증에 의해
의류산업이 한창 성장세를 띨 때 평화시장
일대를 중심으로 급속 팽창하였다. 동대문을
중심으로 제일평화시장, 동평화시장,
흥인시장, 광희시장 등이 잇달아 문을
열었고, 동대문시장은 남대문의 의류 원단,
부자재시장까지 흡수해 자재 조달에서
생산 및 판매까지 이루어지는 곳이 되었다.
청계천 일대의 생산공장들은 1980년대 이후
가내수공업의 소규모의 형태의 하청업체로
창신동, 신당동, 광희동 등지의 주택가로
분산되었다. 하지만 1990년대 중반 이후
대기업들의 의류 내수시장 진출 과 해외
공장 이전 등 동대문을 기반으로 한 의류
소상공인들의 의류 봉제산업이 침체되기
시작했다. 이에 따라 인접 지역의 생산자들과
밀접하게 관련 맺으며 생산기반으로 자리했던
배후생산지로써의 자산은 점차 사라져갈
위기에 처해 있다. 이번 전시는 동대문 시장과
인근 생산 기지들의 관계도와 함께 지역의
문제를 해결하는 '지속가능한 디자인'을 통해
지역의 자산이 새로운 가능성으로 변화해
나가는 실험을 보여주고자 한다.

DONGDAEMUN MARKET AND THE BACKSTREETS
000gan
Seoul Museum of History
L04

The Dongdaemun garment market
was first established as a result
of rapid increase in the domestic
market and export in the 1970s. Jeil
Pyeonghwa Clothing Market followed
by Dongpyeonghwa Market, Heungin
Market, and Gwanghee Market opened
around Dongdaemun(East Gate). Soon
after, Dongdaemun Market took over
Namdaemun's garment and subsidiary
markets, conducting everything from
production to sales. In the 1980s,
the manufacturing factories along
Cheonggyecheon transformed into
smaller-scale, subcontracted handicraft
factories scattered around residential
areas in Changshin-dong, Sindang-dong,
and Gwanghee-dong. However, starting in
the mid-1990s, small garment companies
like those based in Dongdaemun Market,
especially those making up the sewing
industry, faced economic hardships as
large conglomerates began to enter
the overseas market and outsource
manufacturing to overseas factories.
As a result, Dongdaemun Market and
manufacturers nearby were faced with
the risk of losing their business and the
hard work they had accomplished over
the years in establishing their network
of production operations. This exhibition
shows this intricate network of production
sites that make up Dongdaemun Market
in seeking solutions for local issues
through creating a 'sustainable design'
that transforms local assets in to future
possibilities.

시장의 초상
노경

서울역사박물관
L05

시장의 초상은 을지로를 중심으로 형성된 도심부 시장인 광장, 방산, 중부시장의 현 모습을 기록하는 작업이다. 동대문과 남대문 사이를 잇는 이 세 시장은 시작점과 주로 판매하는 물건이 다르지만, 도심부 시장으로써 많은 역사적 사건과 물리적 변화를 겪어왔다. 오랜 시간을 겪은 장소들을 다양한 시간대와 시점들의 영상으로 교차 시켜, 시장의 장소의 특징과 역사적 흔적을 엮으려 한다. 이 작업을 통해 관객들이 도심부 시장의 새로운 모습을 발견하고, 시장 공간과 을지로 지역에 대한 다양하게 해석할 수 있길 바란다.

PORTRAITS FROM THE MARKET
Kyung Roh

Seoul Museum of History
L05

Portraits from the Market is a project that aims to capture snapshots of the modern marketplace now, focusing mainly on Gwangjang, Bangsan, and Jungbu markets, the main marketplaces located in Seoul's Euljiro district that connects the space between Dongdaemun and Namdaemun. Although the items sold in these three markets and their histories are different, they have all faced similar historical events and physical changes throughout the years. This project translates the historical significance of these marketplaces and the stories they tell of transforming generations into video clips that express each market's unique characteristics and their respective remnants of the past. Through this project, we hope that audiences can rediscover the true identity of these marketplaces and interpret the meaning of these spaces and of the Euljiro district in various ways.

무엇이 가만히 스치는 소리
오재우

서울역사박물관

L06

무엇이 가만히 스치는 소리는 속삭임의 한국어 사전에 등재된 설명이다. 시장은 사람들이 모이는 곳에서 시작해 사람들이 빠져나가는 곳에서 끝난다. 사람들이 모이는 곳은 소리로 가득 찬다. 시장엔 물건을 사고파는 소리, 시장을 구성하는 사물의 소리가 시장의 아침과 낮 밤을 채운다. 새벽의 시장부터 저녁의 시장까지 머물다 보면 잊었던 기억들과 감각들이 살아난다. 버스와 자동차의 소음 혹은 조용한 공간에 돌아가는 환기시설의 차가운 소음과 다른 점이 있다. 시장에서는 사람들이 주고받는 소리가 그 안을 채우고 있는 것이다. 음식을 파는 이모는 말 한마디에 국수를 더 얹어준다. 새벽의 멸치 도매상들의 경매에선 말들이 넘쳐난다. 오가는 말들 안에서 판다 안 판다를 주거니 받거니 한다. 무덤덤한 아저씨는 어슬렁어슬렁하며 가격을 비교하고, 한마디 건네고, 종다리처럼 부지런한 아주머니는 이곳 저곳에서 물건값을 깎는다. 판매하는 사장님의 내치는 솜씨도 여간이 아니다. 물건의 가격이 모니터의 숫자가 아닌 종이의 숫자로 정해지고 인간의 목소리로 흥정 되는 시장의 소리. 개인적인 속삭임들이 아직 남아 있는 시장에서 발견한 무엇이 가만히 스치는 소리는 직접 깎은 나무들로 스피커를 제작해서 시장에서 일어나는 하루의 속삭임에 귀 기울여 볼 수 있는 작품이다.

THE SOUND OF GENTLY BRUSHING BY
Jaewoo Oh

Seoul Museum of History

L06

The Sound of Gently Brushing by is the description of the word "whisper" in the Korean dictionary. Markets open where the crowd is and close when and where people leave. And where there is a crowd there are sounds. What fills up the mornings, days and nights are the bustling noises of people selling and buying and the sounds of market goods. One can bring back forgotten memories and senses if he or she listens to a market from dawn to night. The sounds of a market are different from those of buses and cars, or the metallic noise of fans running inside a quiet space. Markets are filled with sounds of exchange. One word can make a lady selling food give you more noodles. Busy retailers flood an early morning anchovy auction with words as they discuss whether they will sell or not. A placid shopper quietly saunters around as he compares prices. The busy housewife is constantly bargaining here and there, but never always successful. The prices are handwritten on pieces of paper and bargained through spoken words. The Sound of Gently Brushing by allows the visitors to listen in on the daily whispers of a market through hand-crafted wooden speakers that are installed in a market.

데이터스케이프: 서울장의 형태

방정인, 둘셋

서울역사박물관

L 07

상품, 자본, 유통, 소비, 관광 등 현대 전통시장의 모든 것들은 '숫자'를 갖는 데이터로 존재하고, 이는 곧 시장의 또 다른 모습을 가시화한다. 이 '정보적 형상'에 주목하여 색다른 모습으로 서울의 전통시장을 시각화한다. 간단하게는 가게 하나가 생겨나고 소멸하는 과정의 연속이지만, 데이터스케이프는 전통시장의 과거의 모습과 오늘의 새로운 모습이 서로 중첩되는 시간 속, 균열하고 조화하는 여러 형상을 보여준다.

DATASCAPE: FORM OF SEOUL MARKET

Jeongin Bang, Studio Twothree

Seoul Museum of History

L 07

Everything about today's traditional markets — products, capital, distribution, consumption, tourism and more — exists as data in the form of numbers, and such data visualizes another form of the markets. The work focuses on "informational form" to create an extraordinary visualization of the traditional market. Simply put, it is a repeated process of the beginning and end of a shop. And Datascape takes the process to another level, exhibiting numerous forms where the past and present of the traditional markets crash and harmonize as they overlap in time.

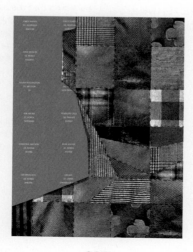

을地: 공존
윤현상재

협력 큐레이터.∵.최주연
다시세운광장, 세운상가 보행데크 – 청계대림
데크, 세운교
L08

윤현상재는 현장 프로젝트가 진행될 을지로
지역의 시간과 사람이 만들어낸 공존의 실체를
을地: 공존이라는 타이틀 아래 시장으로
재현하고자 한다. 원초적인 집합도시인
전통 시장의 개념을 다양한 관점을 통해서
들여다보고 시민들이 함께 즐길 수 있는
플랫폼을 만들고자 다양한 기획을 시도한다.
도시는 교환의 장소에서 시작되었음을 인지할
때 시대에 따른 시장의 변화가 현시대의 도시
문화를 투영시켜 주기도 한다. 서울의 시장은
거대 집합도시의 작은 축소판이라고 해도
과언이 아니다. 과거와 현재가 공존하고 있는
사이트 '을지로, 세운상가, 청계·대림상가'를
중심으로 디자이너와 상인이 모이고 시민들이
함께할 수 있는 프로그램을 기획한다. 이것은
공간과 사람을 엮는 스토리 기획을 통해
더불어 살아가는 집합도시의 한 풍경을
기록하는 과정이다. 윤현상재는 비엔날레
기간 동안 두 번의 시장을 기획한다. 을지로는
매력적이다. 사라져 가는 서울의 풍경이 그
속에 유지되고 있고 그 과거의 풍경과 현재의
모습이 조화롭게 공존한다. 그리하여 본
프로젝트는 시간, 사람, 그리고 사물... 서로
다른 누구와 무엇들이 뒤섞인 이 '공존'을
시장의 주제어로 선택한다. 예전에도
시장은 단순히 물건을 사고파는 곳뿐만
아니라 사람이 모여 이야기를 나누는
커뮤니케이션의 집합 장소였다. 윤현상재,
스페이스 비이(SPACE B-E)가 기획하는
을地: 공존이라는 시장은 '사람' 과 '사물'의
이야기를 담는다.

EULJI: COEXISTENCE
Younhyun Trading

Associate Curator.∵.Jooyeon Choi
Again Sewoon Square, Sewoon Plaza
Pedestrian Deck – Cheonggye Daelim
Deck, Sewoon Bridge
L08

As part of the Live Projects of the
2019 Seoul Biennale of Architecture
and Urbanism, Younhyun Trading is
developing a marketplace under the title
of Eulji: Coexistence, creating a space
of coexistence between the history and
people of the Eulji District in Seoul.
This project is an attempt at taking the
traditional market, a fundamental unit of
collective cities, and interpreting it from
various perspectives to offer a platform
that the public can enjoy. When it is
recognized that the city begins as a place
of exchange, changes to the markets
following from historical change can
also be seen to reflect the contemporary
culture of the city. Markets in Seoul
represent microcosms of city's identity
as a collective city. Sewoon Plaza in Eulji
District, and Daelim Plaza in Cheonggye
will be the platform where the designers
and vendors meet the citizens as they
are symbolic venues where the past and
present coexist. The goal is to tie together
the stories of both this space and
individuals to capture a snapshot of Seoul
as a living, breathing collective city. Over
the course of the 2019 Seoul Biennale,
Younhyun Trading has scheduled two
different marketplace projects. These
projects highlight the Eulji District's
unique qualities as a fascinating area
where remnants of Seoul's past coexist
with the ever-changing components
of the city's present. It is against this
backdrop that the overarching theme of
these markets highlights the coexistence
of time, people, objects, and stories within
these spaces. In the past, traditional
markets were not merely places where
items were bought and sold but,
rather, public hubs for communication
and exchange. It is with this in mind
that Younhyun Trading SPACE B-E is
conducting the Eulji: Coexistence market
project, aiming to bring the stories
between people and objects to life.

SEOUL CITY MARKET

Doc.01 사람의 취향

일정
2019년 9월 21일(토) – 22일(일)

기획의도
다양한 사람들의 공존을 시장의 키워드로 끌어온다. "디자이너들이 소장하고 있는 소장품과 브랜드의 제품 등 호기심의 스토리를 사물보다는 사람으로부터 먼저 시작해 보면 어떨까?"가 이번 기획의 시작이다. 함께 판매자이자 구매자가 되어 즐기는 도시장 (City Market) 을 만드는 것이 목적이다.

Doc.02 사물의 스펙트럼

일정
2019년 10월 26일(토) – 27일(일)

기획의도
이야기의 중점을 판매하는 아이템의 다양한 스토리에 두려고 한다. "예쁜 게 천지에 널린 이 시대에 우리가 바라보는 사물은 어떤 이야기가 담겨 있을까?" 사물에 대한 호기심을 가져보면 좋겠다. 책을 읽듯이 천천히 사물의 스토리를 들여다보는 것이 두 번째 시장의 키워드이다.

Doc.01 Personal Preference

SCHEDULE
Sep 21 (Sat) - 22 (Sun), 2019

PURPOSE
The core theme of the market emphasizes the coexistence of diverse people. The root of this project lies in shifting perspectives on the stories of the products and brands that designers create from objects themselves to the people behind them. The goal of this project is to create an enjoyable experience by establishing a city market that shares the identity of both buyer and seller.

Doc.02 The Spectrum of Objects

SCHEDULE
Oct 26 (Sat) - 27 (Sun), 2019

PURPOSE
The market takes an approach by focusing on the various stories behind the goods being sold. This project encourages audiences to consider how we as a society look at different things and the stories they tell, particularly in a generation that is filled with a countless number of beautiful objects. The goal of this market is to provide visitors with an experience that is much like reading a book, with new stories to be found among the pages of the various objects on display.

209

도시상회
어반플레이

협력 큐레이터∴홍주석
다시세운광장, 세운상가 보행데크 – 청계대림
데크, 세운교
L09

온라인 커머스가 시장을 장악하며 현대인들의
라이프스타일 변화를 가속화 시키는 현
상황 속에서 도심 시장의 의미를 재해석해
보고자 한다. 판매 행위가 오가는 시장의
기능을 넘어 사람과 사람이 만나 콘텐츠와
커뮤니티가 생성되고 새로운 임팩트가
창출되는 시장의 원리를 기반으로 다양한
전문 로컬 큐레이터들과 함께 미래 도시의
시장을 온·오프라인으로 재정의해 볼 수
있는 프로젝트를 진행하고자 한다. 이를 통해
무형의 콘텐츠가 지역을 기반으로 소통하여
발생시킬 수 있는 잠재적 가능성을 실험하고자
한다.

CITY MARKETS
Urban Play

Assistant Curator∴Hong Jooseok
Again Sewoon Square, Sewoon Plaza
Pedestrian Deck – Cheonggye Daelim
Deck, Sewoon Bridge
L09

This project encourages the
reinterpretation of city markets in the
context of e-commerce taking over
today's markets and how this has
catalyzed change in modern lifestyles.
In collaboration with various local expert
content creators, the ultimate goal of this
project is to redefine the online and offline
markets of future cities. This is done by
incorporating market principles to have
meaningful impacts and to facilitate the
creation of markets that are not simply
places for selling and buying goods but,
rather, spaces where people meet to
create new content and communities. All
of these efforts are an attempt to explore
the potential of using intangible content
in communication at the local level.

SEOUL CITY MARKET

일정

1차 2019년 10월 5일(토) – 6일(일)
2차 2019년 11월 2일(토) – 3일(일)

SCHEDULE

1st Oct 5 (Sat) – 6 (Sun), 2019
2nd Nov 2 (Sat) – 3 (Sun), 2019

아이템

도시건축 관련 책, 인테리어 소품,
조명, 빈티지 소품, 조경, 디자인 제품
등

ITEM

Books on urban architecture,
interior decoration, lights, vintage
items, landscaping, design items,
etc.

기획의도

도시 건축 기반의 다양한
크리에이터(건축가, 도시기획자,
메이커, 도시기획자, 디자이너,
아티스트 등)들의 유무형 콘텐츠를
중심으로 사람들의 다양한 취향을
공유할 수 있는 시장

PURPOSE

To create a market where urban
architecture content creators
(architects, urban planners,
designers, artists, etc.) share
items that attract people
with different interests and
preferences.

209

감각 場

UAUS: 국민대학교

다시세운광장
L10

본 프로젝트가 분석한 시장의 인상적인 이미지는 수많은 가판대의 집합이 자아내는 시각적 다양성, 그리고 그곳에 진열된 많은 상품과 음식 등 다양한 자극이다. 이 요소들을 재해석하여 표현하고자 한다. 전체를 구성하는 재료는 '죽부인'이다. 죽부인의 감촉과 시장의 역동적인 모습을 느끼게 하는 엮임, 그리고 닫힌 형태가 만들어 내는 가판대로서의 가능성에 집중했다. 절반가량으로 해체된 죽부인은 새로운 엮임 방식으로서 가판대의 역할을 넘어서 구조체로 연장된다. 새로운 형태로 변형된 죽부인에는 다이크로익 필름의 레이어가 추가된다. 다이크로익 필름은 특정 파장의 빛만 투과하거나 반사 또는 흡수하여 빛을 왜곡한다. 다채로운 색을 파빌리온에 입힘으로써 시장의 시각적 다양성을 표현하고 동시에 쇼윈도의 역할로서 내부의 상품을 들여다보는 창이 된다.

WEAVING DOME

UAUS: Kookmin University

Again Sewoon Square
L10

In trying to identify the most impressive and memorable aspect of marketplaces, we take a close look at the atmosphere created by the collection of multiple market stands and the diversity of food and other items being sold. We then reinterpret these factors that give marketplaces their unique identity and attempt to express them in a new way. We chose 'bamboo body pillows' as the main make-up of this space. We put a particular focus on the attempt to incorporate the feel of these traditional items along with the inclusiveness of marketplaces and the enclosed feeling of market stalls. This new design cuts the traditional bamboo pillow in half and uses a dichroic filter layer, which allows only certain wavelengths of light through, thus refracting other types of light that are filtered through splashing color to the structure. The new transformation of bamboo pillows brings a new visual atmosphere to the marketplace and also works as a display window to illuminate the items on display.

풍경재생
UAUS: 연세대학교
세운상가 보행데크 – 청계대림 데크
L 1 1

시장에서 다룰 상품으로 LP를 선택했다.
보통의 LP 앨범들은 수납을 위해 책장에
책처럼 꽂혀 있어 LP의 디자인된 케이스
가려진다. 따라서 LP자켓의 시각적 매력을
돋보이게 할 새로운 디스플레이 방식을
모색했다.
정방형의 LP 앨범들이 사선의 부재를 따라
내려가며 진열되고, 진열된 앨범을 벽돌에
꽂힌 투명한 아크릴 봉이 잡아 정면을
보이도록 한다. 결과적으로 뒤에 있는 앨범도
가리지 않고 그대로 보이며, 진열된 앨범이
공간의 일부가 되길 바란다. 진열을 고민하며
채택된 나무 프레임, 아크릴 봉, 벽돌이라는
재료는 공간으로 풀어져 관람 동선을 유도하게
된다.

PLAY-SCAPE
UAUS: Yonsei University
Sewoon Plaza Pedestrian Deck –
Cheonggye Daelim Deck
L 1 1

For this project, we selected vinyls (A.K.A
LP records) as the main object in the
context of marketplaces. Most vinyls are
stored like books on shelves, which often
means that the designs on their album
covers are hidden from plain sight. To
address this, we sought new ways to
display LP albums that emphasized the
unique visual features.
Here, the forward-facing vinyls are
displayed and held up by clear acrylic
bars that are placed in bricks along a
wooden frame. In this way the front cover
of the vinyls are fully displayed, avoiding
being stacked among one other. All the
components of the display, including the
wooden frame, the acrylic bars, and the
bricks, were carefully considered to utilize
the space in a way that leads the flow of
visitors throughout the exhibition.

L11

가설. 가설. 가설.
서승모

세운상가 보행데크 – 청계대림 데크
L12

'도로'와 '길'은 유사한 의미이지만 그 단어가 풍기는 어감은 사뭇 다르다. 기능을 위해 계획적이고 인공적으로 만들어진 것이 '도로'라면, '길'은 우발과 필요에 의해서 조금씩 덧붙여지고 응용되고 변형되는 공간이다. 신도시의 보행로를 걷는 것과 오래된 동네의 골목길을 거니는 것은 전혀 다른 체험이다. 가로수길이나 경리단길처럼, 단조로운 도로에 돌발적이고 흥미로운 요소들이 불쑥불쑥 등장할 때, 사람들은 "–길"이라는 이름을 붙인다.

가설. 가설. 가설은 길가에 불쑥불쑥 등장하는 평상이나 벤치처럼 다양하고 편안한 가구들을 배치하고, 친숙한 높이의 지붕과 처마를 조성하여 사람들이 앉거나 머물기도 하고, 무언가를 읽기도 하면서 쉴 수 있는 '길'을 만들고자 했다. 단순히 목적지와 목적지를 연결하는 동선 기능을 넘어서, 길 자체가 목적지이자 경유지가 되도록 하는 바람이다. 이를 구성하는 재료인 가설재(假設)는 규격화된 모듈 시스템에 의해 쉽게 설치와 해체가 가능하고, 다양하게 변주할 수 있다. 규격이 정해진 부재들의 결합과 반복에서 나오는 독특한 리듬감과 패턴이 길의 표정을 더 풍부하게 만들어 줄 수 있을 것이다. 또한 가설재가 갖고 있는 미완의 분위기가 그 곳을 찾는 사용자들이 품게 될 호기심과 상상력의 풍경이 되길 기대한다.

PERHAPS. PERHAPS. PERHAPS
Seo Seungmo

Sewoon Plaza Pedestrian Deck – Cheonggye Daelim Deck
L12

The words 'street' and 'road' have similar meaning with different nuances. While streets are man-made and planned to function, roads are often spaces unintended, in gradual tranformation. Think, for a moment, about how different it is to walk along a sidewalk in a new city as opposed to an alleyway in an old neighborhood. Just as with Garosu-gil or Gyeongridan-gil, areas characterized from unexpected surprises on the streets. These areas are given their names with the word 'road' or 'gil' (Korean word for road), at the end.

Perhaps. Perhaps. Perhaps aims to create a road by randomly placing benches and other types of comfortable street furniture such as awnings of familiar height inviting people to sit and stay to read a book or rest awhile. A road that is not only a connection from one destination to another but a destination in its own. The temporary structure unit that forms the project is made into a standardized module system that are easy to install, modify and de-stall. Which can also be used to create a sense of rhythm and various patterns that bring life to the roads and surrounding areas. Furthermore, the 'unfinished-look' of these units are expected to work as a backdrop for the curiosity and imagination of its users.

214

컵플라워
UAUS: 서울시립대학교
세운상가 보행데크 – 청계대림 데크
L13

사람들이 버리고 간 쓰레기가 상품이 될 수는 없을까? 우리는 그 중에서도 '일회용컵'에 주목하였다. 우리는 일회용 컵을 재활용하여 꽃을 담아서 파는 CUPLOWER라는 상품을 제안하며, 버려진 컵이 다시 채워지는 이야기를 전하고자 한다. 시민들은 우리의 파빌리온을 통해 버려지는 컵을 모으고 활용하여 상품을 만들어간다. 컵에 물구멍을 뚫고, 흙을 담고, 컵받침을 끼우고, 식물 모종을 심는 체험행위를 통해 CUPLOWER라는 새로운 상품을 만들어 갈 수 있고, 자연스레 버려지는 일회용 컵 역시 사라지게 될 것이다.

CUPLOWER
UAUS: University of Seoul
Sewoon Plaza Pedestrian Deck –
Cheonggye Daelim Deck
L13

Is it possible to give new life to what has been thrown out? We specifically looked at disposable cups to answer this question. Our proposal is to turn them into flowerpots, hence the name 'CUPLOWER'. The goal of this project is to collect used, disposable cups that would have otherwise been thrown away and repurpose them to create new products. Visitors can experience the whole process of making their own CUPLOWER from cutting holes into the bottom of the cups, filling them with soil, then placing cups under them to seeding. Naturally there will be no disposable cups that goes to waste.

L13

받히다, 바치다
UAUS: 가천대학교
세운상가 보행데크 – 청계대림 데크
L14

시장이라는 주제를 시장 속 소비자와 판매자라는 두 객체에 집중해서 풀어나갔다. 소비자의 경우 시장을 걸으며 디스플레이된 물건을 인지하고 시장에 접근해 물건을 사는 행위, 판매자는 디스플레이를 통해서 소비자들을 끌어들이고 물건을 판매하는 행위로 시장을 정의했다. 우리는 이러한 행위에서 물건을 '받히고' 소비자에게 물건을 '바치는' 역할을 파빌리온에 담으려 했다. 소주제는 음식이다. 음식하면 다른 소주제들과는 달리 '먹다' 라는 행위가 이루어진다는 것을 누구나 떠올릴 것이다. 이런 점에서 파빌리온에 먹는 행위가 일어날 수 있는 휴식공간을 제시함으로써 시장의 디스플레이와 함께 음식의 먹는 행위까지 나타나도록 했다.

CRATER: CREATE WITH CRATE
UAUS: Gachon University
Sewoon Plaza Pedestrian Deck – Cheonggye Daelim Deck
L14

In addressing the theme of marketplaces, we took a close look at the identities of both buyers and sellers within these areas. Buyers walk through markets observing items on display, approaching items and shops that they are interested in to ultimately make a purchase. On the other hand, sellers set up their displays to attract buyers into purchasing items. The relationship between these two roles is what creates and defines the atmosphere and system of the marketplace. This pavilion takes this relationship and emphasizes the important component of using displays to 'prop up' the items, and 'offer' them to consumers. The sub-theme is food. Different from other sub-themes, the theme of food would naturally evoke the act of 'eating'. Here, we have created a space where visitors can take time to relax and enjoy a meal while being surrounded by displays that are often seen within the marketplace.

L14

내가 만드는 키오스크
UAUS: 중앙대학교

세운상가 보행데크 – 청계대림 데크
L15

시장은 최근 그 형태에 있어 많은 변화가
일어났다. 플리마켓과 같이 누구나 판매자가
될 수 있고, 원하는 곳에 시장을 만들었다
없어지기도 한다. 이러한 변화에 대응하여
누구나 자신이 원하는 곳에 쉽고 빠르게
만들 수 있는 Kiosk KIT를 만들어 하나의
프로토타입을 제안하고자 한다. 변화하는
시장, IKIO KIT가 새로운 시장 유형의 시작이
되길 기대해 본다.

IKIO
UAUS: Chung-Ang University

Sewoon Plaza Pedestrian Deck –
Cheonggye Daelim Deck
L15

Modern markets have recently been in
significant transformation. Much like flea
markets, anyone has the opportunity to
sell anywhere. Taking this new model of
the modern market into consideration,
here we propose a prototype Kiosk KIT
that enables anyone to create their own
station for selling goods anywhere they
want in a quick and easy manner. The
IKIO KIT is the next step in opening up a
new chapter in the ever-changing model
of marketplaces.

L15

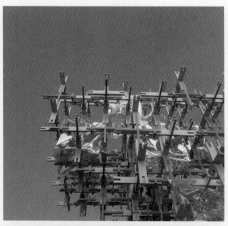

리: 커버
UAUS: 선문대학교

세운교
L16

시장의 목적인 거래에서 지금까지 이용되는
운송수단은 수레, 즉 리어카다. 리어카의
용도는 정말 다양하지만 그 중에서도 리어카에
간단한 조리기구를 담아 이동하며 판매하던
군것질 거리들에 대해서 주목했다. 리어카
위에서 시각과 후각적으로 사람들을 계속
유혹했던 간식거리인 주전부리를 메인
컨셉으로 잡고서 리어카의 재해석을 통해
우리의 시장을 표현했다. 우선 기성 리어카를
과감히 반으로 잘라내고 조인트로 연결성을
부여했다. 해체된 리어카는 세워져서 또
다른 재료들의 연결을 통해 각자 다른 역할
수행했고, 활동을 마친 리어카는 다시
정리되어서 원래 모습으로 되돌아간다.
우리는 리-어카의 리(Re)를 통해 재해석을,
담다의 의미인 COVER를 덧붙여서 "리어카의
재해석을 담다." RE:COVER이다.

RE: COVER
UAUS: Sun Moon University

Sewoon Bridge
L16

The most commonly used transportation
means in conducting trade in the
marketplace was the pushcart. Although
it has various uses, one of its most
notable functions is carrying simple
cooking devices and ingredients
for selling street food. RE:cover is a
reinterpretation of Korean traditional
markets by focusing on the luring sights
and scents of snacks and dessert piled on
these pushcarts. The pushcarts are cut in
half and linked back together with joints.
These split pushcarts are then set up and
equipped with different ingredients to
provide their own unique services, giving
these once-retired pushcarts a new role
in society that also reflects the past. The
name RE:COVER comes from the first
syllable of the Korean word for these
carts, 're', to create a play on words on the
word 'recover', referring to this project's
mission of bringing these pushcarts back
to life.

플로트폼
UAUS: 한양대학교

세운교
L17

우리는 사람들에게 잊혀진 선을 재조명하며
떠있는 선을 통해 파빌리온을 구축하고자
한다. 반복되는 모듈의 선들은 규칙을
가지거나 또는 익숙하지 않은 형태를 생성하여
정사면체 모듈과 정팔면체 모듈로서 표현된다.
두 개의 모듈이 만드는 형태들의 중첩과
병치를 통해 색다른 건축적 경험을 선사하며
그에 의한 공간은 선을 강조하게 되고, 이
선은 전시, 판매, 교류의 장이 된다. 간과된
요소를 의도적으로 드러내, 시장의 본질을
되새김한다. Floatform은 간과된 것들을
인지하는 파빌리온이다.

FLOATFORM
UAUS: Hanyang University

Sewoon Bridge
L17

Here, we attempt to create a pavilion that
uses floating lines to shine the spotlight
on line designs that are now forgotten.
Specifically, we use repetitive line
modules or unfamiliar shapes and express
them in the form of regular tetrahedrons
and octahedron modules. Through the
repetition and juxtaposition of the shapes
created using these two modules, we
present a new, unique architectural
experience and a space emphasizing the
use of lines. In turn, these lines become a
stage for exhibitions, selling goods, and
other forms of exchange. By deliberately
exposing factors that usually do not
receive much attention, we focus on the
fundamental characteristics of today's
marketplaces. In that sense, Floatform is
a pavilion that brings attention to such
things that often go unnoticed.

L17

서울시장산책
어반플레이

협력 큐레이터 ∴ 홍주석

IT와 도시문화적인 요인들로 인해 사람들의
도시 속 행태가 경험 위주로 재구성되어가고
있다. 다양한 콘텐츠를 기반으로 수시로
이벤트가 일어나는 골목이나 시장을 중심으로
사람들이 모이고, 이는 다시 온라인을 통해 더
많은 이들에게 새로운 콘텐츠로 전달된다.
특히 다양한 컨셉의 큐레이션 마켓에 대한
관심은 갈수록 높아지고 있는 상황에서
'서울시장산책' 프로젝트를 통해 한국만의
시간과 콘텐츠가 축적된 전통시장에 대한
의미를 재해석하여 현대 도시 속 전통시장의
가치와 지속 가능성에 대한 새로운 고민을
던지고자 한다. 전통시장만의 상점과 특화된
상품, 철학 있는 상인을 소개하고 이를 좀 더
가까이에서 체험할 수 있는 전통시장 도슨트
프로그램을 통해 전통시장만이 생산해 내는
한국만의 콘텐츠와 이를 둘러싼 도시 생태계에
대한 이해도를 높이고 그 잠재 가능성에 대해
논하고자 한다.

DISCOVERY SEOUL MARKET
Urban Play

Associate Curator ∴ Hong Jooseok

Thanks to advances in IT and
characteristics of urban culture,
tendencies in the mannerisms of those
in the city are mainly based on diverse
experiences. Improved access through
various content means that news of
events in markets or side streets reaches
the public quickly. By the same token,
those who visit these events often share
pictures and other content online for
others to see.
The Discovery Seoul Market project takes
into consideration the increasing interest
in the curating of different markets
with unique concepts. The goal of this
project is to reinterpret the significance
of Korea's traditional markets, which are
collections of the country's characteristics
and history, and rethink their value and
potential in the context of today's cities.
By introducing docent tours and other
hands-on programs, this project offers
opportunities for a closer look to learn
about the stores and specialized goods of
traditional markets and the philosophies
behind the traders who work there. In this
way, we hope to improve understanding
of the Korean contents produced within
and the urban ecosystem that surrounds
traditional markets, while also increasing
awareness and interest in their potential.

L-118

DISCOVERY SEOUL MARKET

1. **통인시장 투어**
2019년 9월 21일(토)
11:00 – 12:30
90분

통인시장은 1940년대 일제강점기에 생긴 공설시장으로, 6·25 전쟁 이후 서촌 지역의 급격한 인구 증가로 형성된 골목형 시장이다. 도시락 카페 '통' 등 다양한 먹거리를 체험한다.

2. **망원시장 투어**
2019년 9월 28일(토)
11:00 – 12:30
90분

망원시장은 1인 가구를 위한 소량포장으로 지역 주민에게 사랑 받고 있는 골목형 시장이다. 망원시장과 망원동 일대 투어를 통해 지역 소상공인과 교류한다.

3. **광장시장 투어**
2019년 10월 11일(금)
11:00 – 12:30
90분

광장시장은 1905년 주식회사로 시작된 우리나라 최초의 사설시장으로 포목과 미곡으로 시작하여 현재는 빈대떡, 마약김밥을 비롯한 먹거리로 전세계인의 사랑을 받고 있는 명실상부 대한민국을 대표하는 전통시장이다. 먹자골목을 비롯하여 100년 역사의 포목부, 국내 최대 수입구제상가, 국내 최고 디자이너들이 찾는 시장 책방 등을 방문한다.

4. **경동시장 투어**
2019년 10월 18일(금)
11:00 – 12:30
90분

경동시장은 서울에서 가장 큰 규모의 채소/청과물 시장으로 경동시장에서 구할 수 없는 식재료는 서울 어디에도 없다는 공식이 성립되는 그야말로 '서울의 부엌'이다. 전국 각지에서 올라온 신선한 식재료들과 약재를 만날 수 있는 기회를 제공한다.

1. **Tongin Market Tour**
11 AM – 12:30 PM,
Saturday Sep 21, 2019
90 minutes

Established during the period of Japanese colonial rule in the 1940s, Tongin Market is an alleyway market formed alongside the rapid population growth of the Seochon district after the Korean War. The tour offers various food experiences, including the lunch box café 'Tong'.

2. **Mangwon Market Tour**
11 AM – 12:30 PM,
Saturday Sep 28, 2019
90 minutes

An alleyway market, Mangwon Market is much loved by local residents as it offers smaller packages for single-person households. The tour takes participants around the market and Mangwon-dong, allowing them to interact with local small business owners.

3. **Kwangjang Market Tour**
11 AM – 12:30 PM,
Friday Oct 11, 2019
90 minutes

Kwangjang Market is Korea's first private market that took off as a corporation in 1905, providing textiles and rice. The market now offers bindaetteok (mung bean pancakes), mayak gimbap (addictive gimbap) and many more food items. Popular among visitors from around the world, the market serves as a representative traditional Korean market. Participants can visit the meokja golmok (food alleyway), a textiles alley that boasts 100 years of history, Korea's biggest vintage imports shops, and market bookstores that are popular among the best designers in Korea.

4. **Kyungdong Market Tour**
11 AM – 12:30 PM,
Friday Oct 18, 2019
90 minutes

Kyungdong Market is the largest vegetables and fruits market in Seoul. Also known as the Kitchen of Seoul, one can say if you can't find it in Kyungdong Market, you can't find it anywhere else in Seoul. The tour offers an opportunity to explore fresh ingredients and medicinal herbs from all parts of Korea.

L-18

서울마당
SEOUL
MADANG

서울마당은 서울건축도시전시관 지하3층에 위치하는 공간이다. 서울마당은 2019 서울도시건축비엔날레의 홍보관 역할을 하면서 "서울의 발견" 전시를 통해 서울의 다양한 모습을 담아내는 공간이다.

서울건축도시전시관의 입구를 들어서면 오른쪽 벽에 2019서울도시건축비엔날레를 홍보하는 영상을 접하게 되고 계단을 내려오면서 오른쪽 벽면에 3면의 영상을 통해서 비엔날레를 소개하게 된다. 서울 마당에 도착하면 터치스크린을 통해서 비엔날레에 관한 상세한 정보를 접할 수 있다.

"서울의 발견" 전시는 도시를 시민과 정부 및 지자체 그리고 전문가들이 집합적 노력을 통해서 만들고 그 도시를 시민들이 공평하게 누려야 한다는 2019 서울도시건축비엔날레의 주제인 "집합도시"에 대해 다양한 방식으로 시민들과 소통하려고 한다.

시민들이 직접 제작하여 응모하고 직접 당선작을 뽑은 "서울의 발견: 시민들이 좋아하는 공공 공간" 공모전의 결과를 전시한다. 시민들은 현장에서 집합도시에 대한 의견을 제안할 수 있고 직접 서울시를 디자인에 참여하는 체험을 할 수 있다. 또한 25개 구청이 자랑하는 도시의 공공공간으로 꾸민 "집합도시 서울풍경"을 통해 자신이 속한 구의 아름다운 공공공간은 물론 서울의 공공공간의 전체적인 풍경을 확인할 수 있다.

마지막으로 현재 진행 중인 도시 프로젝트들을 통해 미래의 집합도시 서울의 풍경을 점쳐 볼 수 있다.

총괄∵임재용
협력 큐레이터∵강민선
보조 큐레이터∵정진우

INTRO

Seoul Madang is located on the third lower level of the Seoul Hall of Urbanism and Architecture. It is not only a promotional space for the 2019 Seoul Biennale for Architecture and Urbanism but a home to the "Finding Seoul" exhibition.

Upon entering the space, there are videos introducing the 2019 Seoul Biennale projected to the right along the stairs down to the main hall. "Seoul Madang" begins with three touch-screens where visitors can access more information about the Seoul Biennale in greater detail. Then "Finding Seoul" communicates with the public on the topic of "collective cities," which is the overarching theme of the 2019 Seoul Biennale. It relays the message that cities should be collective spaces that offer equitable living opportunities. It also emphasizes that they are spaces created through collaboration among the public, the central government, local governments, and experts.

Visitors can participate on site to share their opinions and ideas on collective cities. Also, through the "magic wall," visitors can participate in designing the city. They can see all of the public buildings across twenty-five different districts throughout the city, including one's own neighborhood.

Finally, visitors can take a look into the future of Seoul by experiencing the ongoing projects of experts in different areas.

Director.∴Jaeyong Lim
Associate Curator.∴Minsun Kang
Assistant Curator.∴Jinwoo Jung

서울도시건축전시관
지하3층
Seoul Hall of
Urbanism &
Architecture B3F

M01 서울의 발견 (사진,영상)
M02 집합도시 서울
(25개구의 공공공간 전시)
M03 서울량반은 글 힘으로 살고..
M04 프로젝트 서울
(백사마을)
(강남권 광역복합환승센터)
(프로젝트 서울)
R01 비엔날레 인트로
R02 시민참여 프로젝트

M01 FINDING SEOUL
(PICTURE, VIDEO)
M02 COLLECTIVE CITYSCAPE OF
SEOUL
M03 TEXTUAL COLLECTIVE, SEOUL
M04 PROJECT SEOUL
(BAEKSA VILLAGE)
(GANGNAM COMBINED
TRANSFER CENTER)
(PROJECT SEOUL)
R01 BIENNALE INTRO
R02 CITIZEN PARTICIPATION
PROJECT

MAP

서울의 발견
(시민공모 사진 및 영상 전시)

시민들이 직접 서울비엔날레에 참여하는
방식의 하나로 온라인 사진 및 영상 공모전
서울의 발견: 함께 누리는 도시를 개최하였다.
총 사진 1,519개와 영상 100개의 작품이
접수 되었고 시민들의 투표를 통해서 최종
수상작이 결정되었다. 이 전시를 통해서
시민들이 생각하는 함께 누리는 좋은 공공
공간이란 무엇인지? 시민들은 어떠한 공공
공간을 꿈꾸고 있는지 확인할 수 있는 좋은
기회가 되었다. 도시의 주인은 시민이고
따라서 시민들은 그 도시를 공평하게 누릴 수
있어야 한다.

FINDING SEOUL
Photography and Videography
Exhibition Open to Public
Entries

Finding Seoul: Enjoying the City Together
was a pre-Biennale event encouraging
public participation before the opening.
A total of 1,519 pictures and 100 videos
were submitted, followed by a public vote
to select the winners. It revealed which
public spaces were popular by the masses
and what kind of public spaces citizens
dream of. After all, the city is of the people
and by the people, and everyone should
benefit equally from the city.

M01

집합도시 서울의 풍경

서울의 발견이 시민들의 관점에서 본 함께
누리는 도시의 풍경이라면 집합도시 서울의
풍경은 서울시 25개 구청이 선정한 시민들이
사랑하는 공공 공간에 대한 전시이다.
서울시의 공공 공간의 풍경은 25개 구에 속한
공공 공간의 집합으로 이루어진다. 서울시는
25개의 파편적 풍경을 하나의 집합적
풍경으로 묶어내는 새로운 전략이 필요하다.
전시는 각 구청 별로 32인치 모니터가
하나씩 주어지며 그 것을 이용하여 사진과
영상전시를 자유롭게 하게 된다. 전시장
바닥에는 서울시 지도가 설치 되어 모니터에
전시된 공공 공간들의 위치들을 확인할 수
있다. 관람객들은 먼저 개개인의 특별한
위치를 확인할 것이고 인근에 있는 시민들이
좋아하는 공공 공간들을 확인 할 수 있을
것이다. 모니터에서 펼쳐지는 서울의 풍경도
감상하면서 지도 위에서 펼쳐지는 공공 공간의
풍경도 같이 감상할 수 있다.

COLLECTIVE CITYSCAPE OF SEOUL

While Finding Seoul depicts Seoul from
the perspective of citizens and the ways
in which they can enjoy cities together,
the Collective Cityscape of Seoul displays
public spaces enjoyed by the public as
selected by twenty-five district offices
throughout the city. These districts come
together to comprise the landscape
of Seoul's public spaces. A strategy
is needed to connect these twenty-
five distinct, fragmented components
into a single, collective landscape.
For this exhibit, each district office is
represented by a 32-inch monitor that
displays selected pictures and videos.
Furthermore, there is a map of Seoul on
the floor of the exhibit so that audiences
can visualize the locations of the public
spaces shown on the monitors. Using this
map, not only can audiences see these
different spaces and their locations, but
also experience other popular spaces
that are in nearby areas. Through this
exhibition, visitors can enjoy the scenery
as shown on the monitors, along with the
images shown on the accompanying map.

M02

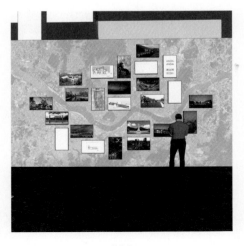

서울량반은 글 힘으로 살고..

강민선, 이마지나, 전수현 외
공동작업

> "도시는 장소를 넘어 시대물이다."
> — 패트릭 게데스

도시의 모습은 문자로도 그릴 수 있다.
속담, 슬로건, 영화나 드라마 제목, 시와
노래부터 현대의 신조어까지 다양한
매체에서 서울이라는 도시에 대한 개인 혹은
집합으로서의 시간, 기억, 경험, 욕망, 문화,
움직임과 느낌을 담고 있는 문구들은
무수하다. 길고 섬세한 묘사부터 짧지만
마치 주문처럼 우리의 기억을 바로 소환하는
문구까지, 만들어진 시기와 형식과 쓰임이
달랐지만 다함께 커다란 벽을 장식한다. 어떤
문구들은 여럿이 함께 공유하고 있는 것이
무엇인지 알려주고 또 어떤 문구들은 생소해서
궁금증을 불러 일으켰으면 한다. 관객의
참여(인터랙션)가 더해지면 벽에 뿌려져 있던
문구들이 재조합되어 도시의 특정한 모습을
보여 준다.

TEXTUAL COLLECTIVE, SEOUL

Collective Work by Minsun Kang, Imagina, Sooh and more.

> "But a city is more than a place in space, it is drama in time."
> — Patrick Geddes

The image of the city is also found in texts. Various texts in old and new media like proverbs, slogans, titles of films and TV dramas, poems, songs, and even modern-day slang hold its own time, memories, experiences, desires, and culture. They also reflect sentiments of the individual and the collective city. Some of these texts are long and detailed descriptions, some short and musical or even magical, conjuring up forgotten memories. These endless words, phrases and sentences from different times and places are collected to form a wide wall. And while some texts are shared by the masses, others are rare and foreign, likely to spark the viewer's curiosity. Finally with the visitors' sensual interaction the texts come alive to tell different collective stories.

프로젝트 서울

프로젝트 서울에서는 현재 서울시에서 진행중인 대표적인 도시 스케일의 공공프로젝트 두개를 소개한다. 하나는 백사마을 주거지 보존 프로젝트이고 다른 하나는 강남권 광역복합환승센터이다. 서울시에는 현재 수 많은 주택단지 프로젝트가 진행되고 있는데 백사마을 주거지 보존 프로젝트는 몇 가지 관점에서 차별화된다. 첫째, 대부분의 대규모 단지들이 기존 지형, 기존 마을의 공동체의 흔적을 물리적으로 밀어 버리고 시작하는 반면 백사마을은 기존 지형, 터, 마을의 풍경, 기존 공동체의 문화 및 흔적(삶의 풍경) 등을 최대한 지켜내려고 노력하고 있다. 둘째, 프로그램도 기존의 원주민은 물론 다양한 계층의 사람들을 위한 임대주택이다. 그저 물리적인 공간을 디자인하는 것이 아니고 그 들의 삶을 담을 수 있는 공동체를 디자인하고 있다. 그의 삶을 지원하게 되는 다양한 주민 공동 시설의 프로그램을 주민공동체와 같이 설계하고 있으며 그 운영방식도 같이 고민하고 있다. 마지막 차별점은 설계 방식이다. 도시 스케일의 공공프로젝트는 한명의 계획가가 단지 전체를 설계하는 경우가 많지만 백사마을은 10명의 건축가가 기존의 터를 20개로 나누어 설계하고 있다. 길과 터를 같이 설계하는 방식이다. 더불어 개개 프로젝트의 설계도 중요하지만 각각의 프로젝트가 만나는 경계가 더욱 더 중요하다. 많은 시간을 이런 경계지점에 할애하고 있다. 진정한 집합도시의 실천이다.

강남권 광역복합환승센터는 도심을 관통하는 다양한 교통 인프라들이 만나는 지하도시를 지상의 도시의 맥락과 연결시키는데 의미가 있는 작업이다. 도시의 공공 공간이 수평적으로만 확장되는 것이 아니고 수직과 수평 방향으로 동시에 확장되는 것이 이 프로젝트의 도시적 가능성이다.

마지막으로 '프로젝트 서울'에서는 2013년부터 진행되고 있는 서울시의 공공 프로젝트의 면면을 영상자료로 확인 할 수 있다.

PROJECT SEOUL

Project Seoul introduces two of the most definitive public projects that are currently being conducted in the urban level in Seoul: the Baeksa Village Residential Area Preservation Project and the Gangnam Intermodal Transit Center. The Baeksa Village Residential Area Preservation Project is different from other ongoing residential projects in Seoul. First of all, while most large-scale areas are built upon a foundation that strays away from local identity and traces of the previous community, Baeksa Village aims to preserve as much of the original characteristics, spaces, scenery, community, local culture, and its traces as possible. Second, this program is to create living spaces not only for local residents, but for incoming residents from all socio-economic backgrounds. And the purpose is not to simply design the physical spaces but to design a community that facilitates the lifestyles of the residents. Most importantly, those implementing this project are collaborating with the local community to develop various programs and facilities that meet local demands, as well as to find the best way to operate such infrastructure. Lastly, unlike most cases where a single individual is responsible for the entire project, ten architects divided Baeksa Village into twenty different sections. This method involves designing the roads and plots in tandem. Furthermore the border where plots meet were as important as each of the plots. On that note, a lot of time and attention are being paid to these boundaries as a real practice of a true collective city. The Gangnam Intermodal Transit Center aims to take infrastructure models for various modes of transportation that runs through the heart of the underground city and connect them with similar infrastructures above ground. The main function of this project is to help urban spaces break away from horizontal growth and explore ways to facilitate their vertical growth as well. Lastly, audiences can view video materials for a closer look at the various accomplishments of Seoul's public projects since 2013 as part of "Project Seoul."

시민참여
프로그램
PUBLIC
PROGRAM

2019 서울도시건축비엔날레(이하 서울비엔날레)는 각각의 도시가 시간적, 공간적, 사회적 요소들에 의해 끊임없이 변화되고 있는 과정에서 시민이 함께 참여하고 누릴 수 있는 새로운 집합유형을 모색하고자 '집합도시'란 주제를 담고 있다. 그리고 도시가 당면한 문제의 해법을 찾기 위해 세계 도시의 다양한 경험과 연구를 통해 공유하는 동시에 새로운 대안도 제시할 것이다. 이러한 실험과 논의는 시민참여프로그램을 통해 시민들에게 보다 쉬운 방식으로 구현되며, 이는 크게 교육, 투어, 영화영상 프로그램의 세 가지로 나뉜다.

교육프로그램은 기존의 학술적인 주제 강연을 비롯해 시민들이 쉽고 전방위적인 관점에서의 도시건축을 이해할 수 있도록 건축가뿐만 아니라 만화가, 미디어아티스트, 방송작가 등 여러 분야의 전문가들의 이야기를 들을 수 있는 특별강연으로 구성된다. 더불어 기존 서울비엔날레에서는 볼 수 없었던 게임형, 건축 키트 만들기형, 아이디어를 쏟아내는 토론형의 전시연계체험프로그램과 어린이를 대상으로 한 건축학교 등 시민이 직접 체험하고 즐길 수 있는 프로그램도 마련되었다.

또한 평소 개방이 되지 않는 대사관 및 관저를 방문할 수 있는 기회를 엿볼 수 있는 오픈하우스 서울 특별프로그램과 서울역사투어 및 서울테마투어, 도슨트투어 등 다양한 종류의 투어를 준비하여 시민이 참여하는 프로그램을 확대했다. 특히 건축계의 저명한 전문가로부터 직접 듣는 서울역사투어는 서울의 역사가 투영된 도시공간을 돌아다니며 과거와 미래를 살펴보는 시간이 될 것이며, 서울테마투어는 현재 서울의 도시건축 이슈 및 생활상을 들여다보는 동시에 숨겨진 공간과 전경을 감상할 수 있는 투어가 될 것이다.

올해 투어프로그램은 전문 도슨트를 초빙하여 콘텐츠 내용면에서도 더 알찬 내용을 접할 수 있다. 더불어 오디오가이드를 통해 관람객 스스로 전시를 즐길 수도 있고, 주말에는 도슨트를 통해

The 2019 Seoul Biennale of Architecture and Urbanism (hereinafter, Seoul Biennale), under the topic of "collective cities," explores a new, collective structure of cities that welcomes the participation of everyday citizens throughout the city's course of endless change as a result of different factors of time, space, and society. Also, in an effort to find a solution to the numerous problems that today's cities face, the Seoul Biennale provides a platform for representatives from cities all around the world to share their experiences as important information for further research. By including the public in this program, which is largely divided into the three smaller program areas of education, tours, and movie screenings, it is much easier to experiment with and discuss issues related to architecture and urbanism.

The education programs offer existing academic lectures to help participants better understand different perspectives on urbanism and architecture. These lectures are not only led by experts in the field architecture, but also by cartoonists, media artists, screenwriters, and many more. Furthermore, this year's Seoul Biennale will be the first of its kind to offer separate exhibitions for hands-on programs, in which participants can play games, build kits, and share their ideas, and for a children's architecture school.

This year's Seoul Biennale will also hold the Open House Seoul Special Program of embassies and official residences, which are normally not open to the public. There will also be numerous different tours of Seoul, including History Tours, Themed Tours, and Docent Tours. In particular, the Seoul History Tour will be a fascinating opportunity for participants to listen as a renowned architect takes them on a journey through urban spaces in Seoul that reflect the city's history, while they collectively consider its past and future. The Seoul Themed Tours take a different approach, as participants take a closer look at the living conditions of Seoul and the problems in urban architecture

직접 설명을 들을 수도 있어 관람객의 편의에
집중하였다. 영화영상프로그램은 제11회
서울국제건축영화제와의 다시 협업하여 영화제
기간 동안에는 비엔날레 입장권으로 영화
관람도 가능하게 연계하여 보다 많은 시민들이
비엔날레와 영화제에 참여할 수 있을 것이다. 올해
시민참여프로그램은 시민이 적극적으로 참여할
수 있는 프로그램으로 다양화하였고, 보다 더 쉽게
서울비엔날레를 이해할 수 있는 프로그램이 되게끔
노력하였다. 앞으로의 서울비엔날레가 더욱 더
다양한 시민참여프로그램으로 찾아올 수 있도록
많은 관심과 참여를 바란다.

총괄.'.'.'김나연
오픈하우스 서울 특별프로그램.'.'.'정혜린
출판, 영화영상프로그램.'.'.'천주현
시민참여프로그램(교육, 투어).'.'.'최예지

INTRO

that the city is facing today. At the same time, it is also an opportunity to see and appreciate the remarkable hidden spaces that are often missed in everyday life.

This year, for an upgraded experience, experts were invited to lead the docent tours and provide the highest quality content for the themed tours. Additionally, audio guided tours will be provided for the exhibition tours, so that visitors may enjoy the event on their own and at their own pace. Docent tours will be provided on the weekend as well to further accommodate our guests. Lastly, the movie screening programs were prepared in collaboration with the 11th Seoul International Architecture Film Festival (SIAFF) to grant Seoul Biennale ticket holders access to the SIAFF, encouraging the public to attend both events. For this year's citizen participation program, we made it our top priority to make it easier for the public to engage in the event and better understand the Seoul Biennale.

General Manager.∵.Nayeon Kim
Open House Seoul Special Program.∵.Hye Rin Jeong
Publication, Film & Video Program.∵.Ju Hyun Cheon
Public Program (Education, Tour).∵.Ye Ji Choe

INTRO

교육 프로그램 EDUCATION PROGRAM

	비엔날레 주제 강연 2019 Seoul Biennale of Architecture and Urbanism Lectures	특별강연 Special Lectures	전시연계 체험프로그램 Hands-On Exhibition Program Step #1	전시연계 체험프로그램 Hands-On Exhibition Program Step #2
	19:00 DDP 살림1관 나눔관 DDP Design Lab 1F Academy Hall P1 서울도시건축전시관 아카이브실 Seoul Hall of Urbanism and Architecture Seoul Archive P3	15:00 DDP 살림1관 나눔관 DDP Design Lab 1F Academy Hall P1 서울도시건축전시관 아카이브실 Seoul Hall of Urbanism and Architecture Seoul Archive P3	상시 운영 Throughout the Biennale period DDP 디자인 둘레길 DDP Design Pathway P1	10:30 – 12:00 13:00 – 14:30 DDP 살림1관 DDP Design Lab 1F P1
9.7 SAT				
9.8 SUN	프란시스코 사닌 / 집단성의 시급성과 당위성에 관하여 Franscisco Sanin / The Urgency and Agency of the Collective			나도건축가 I am an Architect, too!
9.11 WED	홍은주와 김형재 / 건축가와 함께 일하기 Eunjoo Hong and Hyungjae Kim / Working with Architects			
9.14 SAT				나도건축가 I am an Architect, too!
9.15 SUN				나도건축가 I am an Architect, too!
9.18 WED	홍주석 / 로컬 크리에이터가 바꾸는 도시의 미래 Jooseok Hong / The Local Creators Changing the Future of Cities		모두의 비엔날레 Biennale for Everyone (상시 운영, Throughout the Biennale period)	
9.21 SAT		이이남 / 건축 그리고 미디어 아트의 향기 Leenam Lee / The Scent of Architecture and Media Art		나도건축가 I am an Architect, too!
9.22 SUN				나도건축가 I am an Architect, too!
9.25 WED	최상기 / 집합도시를 이해하는 도구로서의 건축 Sanki choe / Architecture as a Tool to Understand Collective City			
9.28 SAT		김일현 / 공공과 일상 사이의 건축 Ilhyun Kim / Architecture Between Public Spaces and Everyday Life		나도건축가 I am an Architect, too!
9.29 SUN				나도건축가 I am an Architect, too!

CALENDAR

투어 프로그램 TOUR PROGRAM

전시연계 체험 프로그램 Hands-On Exhibition Program Step #3	서울시건축학교 Seoul Metropolitan City Architecture School	집합도시서울투어 Collecive City Seoul Tour	전시장도슨트투어 스탬프투어 Docent Tour Stamp Tour	연계투어 Related Tour
10:30 – 12:00 DDP 살림1관 DDP Design Lab 1F P1	13:00 – 17:30 서울도시건축센터 Seoul Center for Architecture & Urbanism P2	10:00 – 12:00 12:00 – 14:00 14:00 – 16:00		11:00 – 13:00 서울도시건축 전시관 Seoul Hall of Urbanism & Architecture
유레카! 서울 Eureka! Seoul		인스타시티성수 Insta-City Seongsu		정동역사탐방 Jeongdong History Tour
유레카! 서울 Eureka! Seoul				
유레카! 서울 Eureka! Seoul		을지로힙스터 지하도시탐험 Euljiro Hipster Exploring the City Underground		정동역사탐방 Jeongdong History Tour
			전시장도슨트투어 스탬프투어 Docent Tour Stamp Tour	
유레카! 서울 Eureka! Seoul	1 아이 1 도시 1 Child, 1 City	성문안첫동네 First Village within the City Gate		
유레카! 서울 Eureka! Seoul		서울생활백서 Seoul Living Guide		
유레카! 서울 Eureka! Seoul	1 아이 1 도시 1 Child, 1 City	한양-경성-서울 Hanyang-Gyeongseong-Seoul		
유레카! 서울 Eureka! Seoul		서울파노라마 Seoul Panorama		

CALENDAR

10.2 **WED**	조민석 / 밤섬 당인리 라이브 Minsuk Cho / Bamseom Danginri Live		
10.5 **SAT**		서현석 / 개인의 이상, 국가의 이념 Hyunseok Seo / Individual Ideals and National Ideology	나도건축가 I am an Architect, too!
10.6 **SUN**			나도건축가 I am an Architect, too!
10.9 **WED**	유아람 / 서울시장 이야기: 서울이라는 교환의 장소 Aram Yoo / The Story of Markets in Seoul: Seoul as a Trading Space		
10.12 **SAT**		김소현 / 사람이 살고 있습니다 Sohyun Kim / The People Who Live Here	나도건축가 I am an Architect, too!
10.13 **SUN**			나도건축가 I am an Architect, too!
10.16 **WED**	장영철 / 건축의 기획, 기획의 건축 Young Jang / The Plan- ning of Architecture, The Architecture of Planning		
10.19 **SAT**		박정현 / 발전 체제와 건축, 그리고 집합성 Junghyun Park / Developmental Regime, Architecture, and Collectivism	나도건축가 I am an Architect, too!
10.20 **SUN**			나도건축가 I am an Architect, too!
10.23 **WED**	임동우 / 집합적 결과물로서의 도시 Dongwoo Yim / City as Collective Consequences		
10.26 **SAT**		최호철, 박인하 / 만화, 손과 눈과 발로 그린 공간 Hochul Choi, Inha Park / Comics: Spaces Drawn by the Hands, Eyes, and Feet	
10.27 **SUN**			
11.9 **SAT**			
11.10 **SUN**			

모두의 비엔날레
Biennale for
Everyone
(상시 운영,
Throughout the
Biennale period)

유레카! 서울 Eureka! Seoul	1아이 1도시 1 Child, 1 City	조선-대한-민국 Joseon-Korea-Republic	
유레카! 서울 Eureka! Seoul		인스타시티성수 Insta-City Seongsu	
유레카! 서울 Eureka! Seoul	1 아이 1 도시 1 Child, 1 City	그림길겸재 Gyeomjae Art Street	
유레카! 서울 Eureka! Seoul		서울생활백서 Seoul Living Guide	
			전시장도슨트투어 스탬프투어 Docent Tour Stamp Tour
유레카! 서울 Eureka! Seoul	1 아이 1 도시 1 Child, 1 City	세운속골목 Alleyways of Sewoon	
유레카! 서울 Eureka! Seoul		을지로힙스터 Euljiro Hipster 지하도시탐험 Exploring the City Underground	
	1 아이 1 도시 1 Child, 1 City	타임슬립 Time Slip	
		서울파노라마 Seoul Panorama	

P1
동대문 디자인플라자
**Dongdaemun Design
Plaza (DDP)**

P2
돈의문박물관마을
**Donuimun
Museum Village**
서울도시건축센터
**Seoul Center for
Architecture &
Urbanism**

P3
서울도시건축전시관
**Seoul Hall of
Urbanism &
Architecture**

P4
이화여자대학교
아트하우스 모모
**Ewha Women's Univ
Art House Momo**

e

e   e   e
e
e

국립민속박물관
The National Folk Museum of Korea

a

국립현대미술관 서울관
National Museum of Modern and
Contemporary Art, Seoul

e

3
경복궁역
Gyeongbokgung Station

b

3
안국
Ang

←
P4

경희궁   j
Gyeonghuigung

a

5
광화문역
Gwanghwamun
Station

a

M A P

P2   j
c

l   l
l   P3
l   k   j   k
청계천
Cheonggyecheon

a

l

덕수궁
Deoksugung

a

2
을지로입구역
Euljiro 1-ga Station

5
서대문역
Seodaemun Station

k

서울시립미술관
f   Seoul Museum of Art

1/2
시청역
City Hall Station

j

a

4
명동역
Myeong

2/5
충정로역
Chungjeongno Station

f
f

4
회현역
Hoehyeon Stat   f

f

k

j

1/4
서울역
Seoul Station

j

f

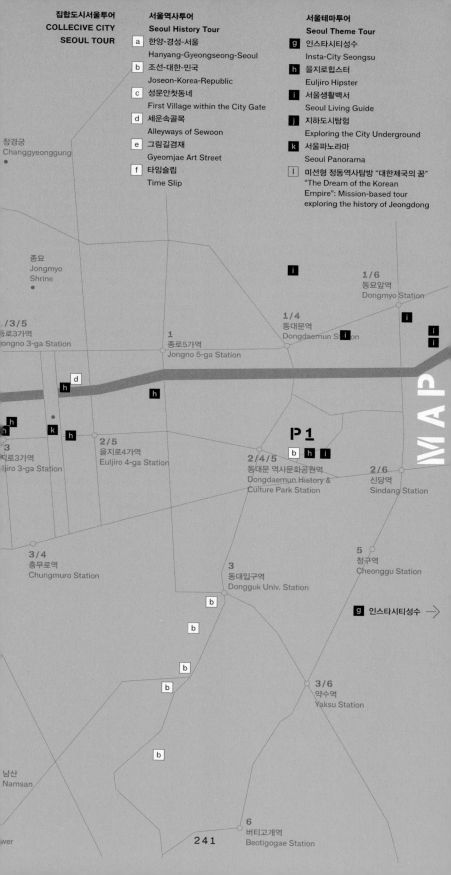

교육 프로그램
서울도시건축비엔날레 강연

DDP 살림1관 나눔관
서울도시건축전시관 아카이브실
P1 / P3

시민이 함께 고찰해보는 서울. 시장의
발달부터 도시의 생성과 성장에 이르기까지
강연자가 소개하는 다양한 집합형태를 통해
알아본다. 또한, 국가주도의 급진적 성장을
겪은 도시 서울이 직면하고 있는 '공공공간의
부재', '재개발', '공간의 상업화' 등 현 문제점과
방향성을 각 분야의 전문가들과 시민이
직접 탐구 및 제안한다. 강연자와 청중의
경계를 허물고 비엔날레 주제 '집합도시'를
다양한 방면에서 재해석한다. 더 나아가 일상
속에서 공감할 수 있는 여러 키워드를 가지고
도시라이프를 함께 이야기해본다.

참여 대상
전문가 및 일반시민 대상

참가비 / 예약
무료 / 자세한 정보와 참가신청은 홈페이지를
이용해 주시기 바랍니다.

Education Program
2019 SBAU LECTURES

DDP Design Lab 1F Academy Hall
Seoul Hall of Urbanism and Architecture
Seoul Archive
P1 / P3

In an effort to consider different aspects
of Seoul as a collective city with
everyday citizens, these lectures explore
different collective mechanisms from
the development of marketplaces to the
generation and growth of cities. Through
this platform, experts from different
fields and the general public will have the
opportunity to discuss and share personal
opinions on the absence of public spaces,
redevelopment, the commercialization
of space, and other problems faced by
Seoul as a city that has experienced rapid
growth due to national interventions.
These will also be accompanied by
discussions on the appropriate steps that
need to be taken in the future to address
these issues. During these lectures,
lecturers and the audience will have equal
opportunities to reinterpret the Seoul
Biennale's theme of "collective cities"
from various perspectives. Furthermore,
relatable concepts and keywords will be
used to lead discussions on city life.

Eligibility
Experts and the general public

Fee / Reservations
Free / Please check the official website
for more detailed information and to
make a reservation.

2019 비엔날레 주제강연

2019 서울비엔날레에 참여하는 전시 큐레이터, 작가, 디자이너가 직접 서울의 도시·건축, 비엔날레 주제 '집합도시'를 전시 기획의도 및 작품 등을 통해 강연한다. 시민들이 보다 쉽게 이해하고 관심도 증진 및 소통의 기회를 마련하는 자리이다. 총 8회 진행이다.

프란시스코 사닌
집단성의 시급성과 당위성에 관하여
2019. 9. 8. 일

본 강연에서는 2019 서울도시건축 비엔날레(SBAU)의 기본 전제이자 화두인 집단건축물로서의 도시에 대한 의견을 나누고자 하며, 이러한 논의의 시의성에 대해 알아보고자 한다. 본 강연에서는 출품인들의 활동을 소개할 뿐 아니라 인권으로써, 특히 공간적 권리로써의 도시와 이를 되찾기 위한 집단적 전략을 이해하고 실천하는 색다른 방안을 제시하고자 한다. 이를 통해 건축의 새로운 역할과 가능성을 알아볼 뿐 아니라, 당면한 과제를 짚어보고 오늘날 규율, 공언, 사회적 실천으로서의 건축의 책임에 대해 살펴보고자 한다.

홍은주와 김형재
건축가와 함께 일하기
2019. 9. 11. 수

"건축가와 함께 일 하는 것"이 다른 협업자들, 의뢰자들과 일하는 것에 비교해 특정 가능한 정도로 확연하게 다른 것은 아니다. 이 강연을 통해 "건축가"에 대해서가 아니라 "건축가와 함께" 주제에 접근해 간 과정과 방식에 대해 좀 더 구체적이고 세밀하게 살펴보고자 한다. 우리는 전시, 출판, 연구 등의 분야에서 건축 분야의 전문가들과 일한 경험을 갖고 있으며 이 경험들은 여러 의미에서 즐겁고 도전적인 것들이었다. 이 사례들을 공유함과 더불어 이번 비엔날레의 아이덴티티 디자인의 작업 과정과 결과에 대해 소개하고자 한다.

2019 SBAU Lectures

Several exhibition curators, authors, and designers will participate in the 2019 Seoul Biennale to give lectures on urbanism, architecture, and the topic of "collective cities" by explaining their exhibitions and their intent behind them. These lectures are meant to help the general public better understand and build interest in the topics of this year's Biennale, while also providing them with a space to communicate with experts and each other. There are a total of eight scheduled lectures.

Fransisco Sanin
The Urgency and Agency of the Collective
Sep 8, 2019, Sun

The lecture will introduce the basic premises and questions asked in the 2019 SBAU about the idea of the city as a collective artifact. And why it is urgent to talk about it today. It will not only introduce the work of the participants but attempt to put forward new possible ways of understanding and deploying the idea of the collective: strategies of reclaiming the city as a human right and particularly the notion of spatial rights. In that context it will question the new roles, challenges, possibilities and more importantly responsibilities for architecture as a discipline, profession and social practice today.

Eunjoo Hong and Hyungjae Kim
Working Together with Architects
Sep 11, 2019, Wed

This lecture compares collaborators and clients to show that they are not strikingly different from each other. Here, we approach the topic of working together with architects, rather than just dealing architects, in greater, more concrete detail. We have a diverse array of meaningful, enjoyable, and challenging experiences working with architecture experts to prepare exhibitions, publications, and research. During the 2019 Seoul Biennale, our aim is to share these experiences and introduce our work and results on an identity design project.

홍주석
로컬크리에이터가 바꾸는 도시의 미래
2019. 9. 18. 수

이제는 우리가 그 동안 가지고 왔던 소상공인에 대한 개념이 바뀌어야 하는 시기다. 동네 비즈니스가 곧 크리에이터 영역으로의 새로운 비즈니스의 잠재력을 내포하고 있기 때문이다. 급변하는 IT기술에 의한 라이프스타일의 변화로 인해 소상공인이 판매자가 아닌 크리에이터로서 작동해야 하는 시대가 오고 있는 것이다. 4차 산업 혁명시대에 우리가 명심해야할 것은 기술 개발을 넘어 '사람'과 '연결'이다. 사람과 사람과의 연결, 사람과 공간과의 연결, 사람과 지역과의 연결은 지역만의 콘텐츠를 기반으로 지속가능한 미래의 동네로 가는 시작점이 될 것이다.

최상기
집합도시를 이해하는 도구로서의 건축
2019. 9. 25. 수

함께 만들고 함께 누리는 '집합'의 도시를 형성해나가는 단서는 여러 사람들의 '집단'이 다져내는 관계에 따라 형성되는 건축의 패턴에서 찾을 수 있다. 즉, 거대 담론의 도시 계획이나 상업자본이 지배하는 도시형성의 논리에 어쩔 수 없이 종속되는 건축이 아니라, 생활 속에서 서로 반응하는 사람들의 행태가 집단적인 패턴으로 표출되는 공동주거, 시장, 혹은 공공 인프라스트럭처 등에서 발견되는 즉흥적이고 창의적인 형식을 연구의 대상으로 삼는다.

조민석
밤섬 당인리 라이브
2019. 10. 2. 수

최근 서울의 생태적인 시각과 산업공간의 전용에 대한 관심은 도시와 건축의 프레임워크에도 변화를 일으키고 있다. 서울도시건축 비엔날레 참여작인 '밤섬 당인리 라이브'는 도시의 자연적, 문화적 생태계에 관해 진행중인 매스스터디스의 두 프로젝트에 대한 수행적 현장 프레젠테이션이며 시민 참여를 수반해, 다양한 관점을 통해 '집합'의 개념을 풍부하게 하는 장이 될 것이다. 본 강연은 두 프로젝트를 소개하는 자리이자 10월 매주 토요일에 열릴 네 차례의 공개 미팅의 예고편이다.

P1/P3

Jooseok Hong
The Local Creator Changing the Future of Cities
Sep 18, 2019, Wed

Now is the time for us to change our past perceptions of small, local businesses, for they have the potential to succeed as new models for creative businesses. With rapidly developing IT technology that results in lifestyles that change just as rapidly, we are gradually entering an era where small businesses must be run not as salesmen but, rather, as creators. As we face the onset of the 4th Industrial Revolution, we must keep in mind the importance of connections and people when developing technologies. The only way for us to achieve sustainable cities of the future is to create local contents based on connections between all people, people and spaces, and people and regions.

Sanki Choe
Architecture as a tool to understand a collective city
Sep 25, 2019, Wed

Contrary to common belief, where the role of architecture is observed to be somewhat passive in our city, which has become more commercial and corporate-driven, how can a city become more relevant to the everyday life of its inhabitants? Instead of erroneously seeking the answer from top-down urban planning or from the real estate markets, the participating studios suggest new emerging forms that can be found in the collective patterns extracted from the spontaneous and responsive architecture that interacts with the inhabitants: communal housing, markets, urban infrastructure, and public buildings.

Minsuk Cho
Bamseom Danginri Live
Oct 2, 2019, Wed

Seoul's recent ecological perspectives and interest in post-industrial adaptive reuse have also brought upon changes in the urban and architectural frameworks. Minsuk Cho's contribution to the SBAU 2019 is titled Bamseom Danginri Live and is an on-site performative presentation of two in-progress projects, the Bamseom Ecological Observation Deck and the Danginri Cultural Center (Danginri Podium and Promenade), that each deal with the natural and cultural ecology of the city, respectively. It will involve public participation, investigation, and dialogue, enriching the notion of the 'collective' through multiple views.

유아람
서울 시장 이야기: 서울이라는 교환의 장소
2019. 10. 9. 수
　서울 안의 시장은 한 장소에서 독립적으로 형성된 것처럼 보인다. 하지만, 실은 동대문, 남대문, 을지로, 또 청량리를 포함한 수많은 상권들이 지속적인 영향관계를 이루며 존재해왔다. 사대문 안의 시장들은 시간과 공간 모두에서 사람들의 생활 속 깊이 자리 잡고 있다. 시장의 역사는 서울 시민의 삶의 역사이고, 서울사의 한 부분이다. 서로 주고받으며 매 순간 새로워지는 서울의 전통시장은 21세기에 또 한 번 다른 모습을 선보인다. 그 모습은 시장을 더 다채롭고 화려한, 새로운 시대의 삶까지 수용할 수 있는 공간으로 만든다.

장영철
건축의 기획, 기획의 건축
2019. 10. 16. 수
　건축의 감성은 건축가들의 전유물이 아니다. 건축을 전공하지 않는 다른 분야의 전공자들이, 기획에서의 건축의 가능성을 간파하고, 건축가 못지않은 공간 감수성을 발휘할 때가 있다. 물론 건축가들도 풍부한 상상력과 집요한 실행력으로 훌륭한 기획을 완성시키는 경우도 많다. 도시건축비엔날레도 건축의 기획이 어떻게 실행되는지 볼 수 있는 실례가 될 것이다. '이러한 기획에서 건축의 역할은 무엇이고, 건축으로 어떤 기획을 할 수 있을 것인가?'에 대한 큐레이터 스스로의 경험담과 생각을 정리하여 소통하는 것이 강연의 목적이다.

임동우
집합적 결과물로서의 도시
2019. 10. 23. 수
　집합적 결과물로서의 도시(City as a Collective Consequence)는 우리의 도시가 공간적, 시간적, 그리고 사회적 환경의 집합체임을 보여줌과 동시에, 경우에 따라서는 의도하지 않은 혹은 계획되지 않은 요소들의 개입으로 끊임없이 변화하는 생물임을 보여준다. 아무리 완벽하게 계획된 도시도 새로운 변수들이 개입이 되면서 새로운 결과물로 나타나기도 하며, 강력한 계획이 없는 도시에서도 도시 요소들간의 최적화 작용을 통해 새로운 질서를 만들어 나아가기도 한다.

Aram Yoo
The Story of Markets in Seoul: Seoul as A Trading Space
Oct 9, 2019, Wed
　Traditional markets in Seoul often appear as if they were somehow created on their own within a space. However, a number of well-known areas including Dongdaemun, Namdaemun, Euljiro, and Cheongnyangni share a close relationship with one other. The markets located near the four gates along the boundaries of the old city are deeply rooted in time and space of the daily lives. The history of traditional markets in Seoul is the history of the people and, in turn, a part of the overall history of the city. These markets are constantly engaging in acts of exchange, never the same at any given moment. Even in the 21st century, we see something different, one that is more vibrant and more diverse creating a space that welcomes a new era of life.

Young Jang
The Planning of Architecture, The Architecture of Planning
Oct 16, 2019, Wed
　There are cases where non architects understand the implications of architecture in curation and, knowingly or not, show exceptional architectural sensitivity. And of course architects often manage a great curatorial work with the power of execution and imagination. The 2019 Seoul Biennale of Architecture and Urbanism will be a good example to better understand the practice of curation of architecture. And the objective of this lecture is to communicate ideas and personal experiences of the curator on the questions of what the role of architecture is in curation and what architecture can help to curate.

Dongwoo Yim
City as collective consequences
Oct 23, 2019, Wed
　A city as a collective consequence indicates that our cities are a collective of spatial, temporal, and social environments and at the same time, organisms that constantly change due to the intervention of unintentional or unplanned factors. Even a city that has been perfectly planned can reveal a new consequence due to the intervention of new variables while a city devised without any solid plans creates a new order through the optimal interactions of the elements of the city.

特별강연

'건축 + α 분야'의 결합으로 비엔날레 주제 '집합도시'와 전시를 다양한 방면으로 폭넓게 이해한다.미디어아트, 건축영화, 만화, 다큐멘터리 등과 같은 다양한 분야와 도시건축을 접목하여 일상 속 공감과 흥미를 불러일으킨다. 총 6회 진행이다.

이이남
건축 그리고 미디어 아트의 향기

2019. 9. 21. 토

'파사드'란 프랑스어로 건물의 출입구로 이용되는 정면 외벽 부위 및 위치를 의미한다. 국어로는 '앞쪽'이라는 말로 부드럽게 풀 수 있다. 여기서 현대에 이르러서는 더 많고, 다양한 '앞쪽'이 생긴 건축에서는 '미디어 파사드'로서 소재가 된다. 건물자체를 관조하고 감상한다는 일은 건물의 모든 곳이 보이는 구도, 장소가 아닌 이상, 감상을 시도한다는 것은 쉽지 않다. 특히 도시의'건물'은 사람들에게 반복인식 되며 자연스레 하나의'물질'로서 인식되지 않는가. 실제로 사람들의 시선이 닿는 곳은 그 건물들을 같이 거니는 사람, 가게, 자신의 주변부에 빠져있다. 분명 건축가의 의도가 중첩되고 잘 쌓일수록 '건축'은 힘을 발휘한다.

김일현
공공과 일상사이의 건축

2019. 9. 28. 토

도시는 커다란 집이고 매번 다른 모습으로 각 사람에게 다가온다. 도시는 인간이 만들어낸 가장 큰 작품이다. 거기에 살고 살았던 사람들의 희망과 욕망으로 빚어낸 분신이기도 하다. 이렇게 대도시는 도시는 전위주의자들이 꿈꾸던 총체적 극장이 실현된 장소이다. 우리는 과연 일관된 한 사람이라고 자신 있게 말할 수 있을까?

Special Lectures

Taking into consideration that the topic of the 2019 Seoul Biennale is "collective cities," these lectures present audiences with a multi-disciplinary approach to better understand the different exhibitions. Experts in media art, architectural film, comics, and documentaries will present their work in relation to architecture and urbanism to promote interest and understanding of these concepts in everyday life. There are a total of six scheduled special lectures.

Leenam Lee
The Scent of Architecture and Media Art

Sep 21, 2019, Sat

In French, the term "façade" refers to the outer wall of a given structure that serves as its entrance. Modern architecture is characterized by a diverse array of façades, leading to a new category of "media façades." It is not always easy to appreciate a structure when its complete composition is not visible, especially when it comes to those in cities, which are often perceived by people as being repetitive and lacking any unique qualities. For the most part, people remember buildings and structures for the people, stores, and surroundings that characterize them. Although the architect's intended design and personal skills might contribute somewhat to a given structure's influence on beholders, the weather, viewing angles, and other variables inevitably impact the power of architectural design.

Ilhyun Kim
Architecture Between Public Spaces and Everyday Life

Sep 28, 2019, Sat

Cities are expansive houses that appear differently each time to everyone and are the greatest creation that man has ever made. They also serve as the alter ego of everyone who is living and anyone who has ever lived there. Large cities are complete manifestations of theaters long dreamed by avant-garde artists. Can we safely say that each of us is a person with a single, consistent identity?

P1 / P3

서현석

개인의 이상, 국가의 이념: 모더니즘 건축과 아시아 근대 국가의 형성

2019. 10. 2. 토

이 강연에서 서현석은 지난 5년간 본인이 진행해온 대한민국, 북한, 캄보디아, 스리랑카, 일본에서의 리서치를 기반으로 이 나라들이 근대 국가로 재탄생하면서 모더니즘 건축을 받아들인 궤적을 소개한다. 김수근, 김정희, 반 몰리반, 제프리 바와, 발렌타인 구나세카라, 단게 겐조 등 아시아 모더니즘을 이끈 건축가들의 개별적인 작품들을 영상으로 소개하고 그들의 업적을 역사적인 맥락에서 바라본다.

김소현

사람이 살고 있습니다: 동네한바퀴와 도시 이야기

2019. 10. 12. 토

오랫동안 자신의 자리를 지키며 성실한 땀을 흘리고 진심을 다해 살아온 평범한 소시민들이, 있어야 할 자리에 동네의 구성원으로 함께 무리지어 만들어내는 삶, 모여서 더불어 살아가는 모습은 삭막한 도시를 밝히는 '도시가 회복해야 할 가치'이며, 함께 만들고 누리는 도시, 그래서 인간이 중심이 되는 공동체로 가는 열쇠였다. 도시가 아름다운 건, 거기 사람이 살고 있기 때문이다. 더불어 사는 사람들이 있는 도시는 찬란하다. 정겹다. 따사롭다. 설렌다. 반갑다. 그립다. 풍요롭다. 굳건하다. 배우 김영철과 함께 천천히 걸으며 재발견한 도시와 동네의 가치, 행복한 집합도시로 가는 길에 관한 이야기를 동네 한바퀴의 여정을 통해 나누고자 한다.

박정현

발전 체제와 건축, 그리고 집합성

2019. 10. 19. 토

이 짧은 강연은 현재를 이해하기 위해 과거를 경유해야 한다는 것을 강조한다. 부족한 공공 공간, 집합성의 부재 등 현재 한국의 도시가 겪는 문제의 기원을 역사적으로 조망한다. 이를 위해 1960년대 이후 발전 체제의 정치와 건축의 관계를 다시 묻는다.

Hyunseok Seo

Individual Ideals and National Ideology: The Composition of Modernist Architecture and Modern Asian Nations

Oct 2, 2019, Sat

In this lecture, Dr. Seo Hyunseok incorporates his research over the past five years on the Republic of Korea, the Democratic People's Republic of Korea, Cambodia, Sri Lanka, and Japan to retrace their journeys toward rebirth as modern, sovereign nations, highlighting their adoption of modernist architecture along the way. Through his video productions, Dr. Seo introduces Kim Swoo Geun, Kim Jeonghee, Van Molyvann, Geoffrey Bawa, Valentine Gunasekara, and Kenzo Tange, leading architects in modernist architecture, along with their most notable works to discuss their importance in this historical context.

Sohyun Kim

The People Who Live Here: Around the Neighborhood with Kim Youngcheol and Local Stories

Oct 12, 2019, Sat

Indeed, the key to communities is the mindset of individuals to band together in creating a living space through hard work and dedication and bringing values that shine happiness on these otherwise bleak urban spaces. In other words, it is the locals living in a given city that are responsible for its beauty, creating within this space a sense of brightness, warmth, excitement, welcome, abundance, fondness, and stability. Here, we explore the journeys shown in Around the Neighborhood with Kim Youngcheol, strolling through these spaces to rediscover the local values of cities and neighborhoods against the backdrop of happy, collective cities.

Junghyun Park

Developmental Regime, Architecture, and Collectivism

Oct 19, 2019, Sat

Importance of looking to the past in order to understand the present. In particular, we look at history to find the source of the urban issues affecting Korea today, including the lack of public spaces and the absence of collectivism. To begin, we must first question the relationship between politics and architecture in the developmental regime following the 1960s.

최호철, 박인하

만화, 손과 눈과 발로 그린 공간

2019. 10. 26. 토

월 아이스너는 만화를 '시퀀셜 아트'로 정의했다. '연속예술'로 번역되지만 시퀀스와 시퀀스가 더해가는 순차적 예술로도 번역할 수 있다. 연속예술에서는 '연속'에 집중하지만, 만화는 하나의 시퀀스에서 출발한다. 만화에서 시퀀스는 하나의 공간에서 벌어지는 연속적인 사건이다. 때문에 만화는 항상 공간을 고민할 수밖에 없다.

Hochul Choi, Inha Park

Comics: Spaces Drawn by the Hands, Eyes, and Feet

Oct 26, 2019, Sat

Will Eisner defined comics as a "sequential art." While the Korean translation of this term focuses on the "sequential" aspect, it is important to note that comics are a form of art that begin from a given sequence. When it comes to comics, a sequence is a sequential event that begins from a single space. This is why spaces must carefully be considered when creating any comic.

교육 프로그램

전시연계체험프로그램

기획 / 운영.˙.˙. 권현정(아키에듀)
DDP 살림1관
P 1

서울 비엔날레는 전통적인 도시건축 전시의
모습을 넘어서는 종합적인 문화예술의 장을
지향하며, 다양한 분야의 전문가들이 다각적인
매체를 통해 도시건축의 메시지를 전달한다.
전시연계체험 프로그램은 다양한 연령대,
직업, 관심사들을 담을 수 있도록 폭 넓은
프로그램을 구성하여 각계각층 시민들이
각자의 취향에 맞는 프로그램에 참여 할 수
있으며, 이를 통해 보다 쉽게 전시를 이해하고
공감할 수 있도록 하였다.

1.   **모두의 비엔날레**
     전 연령이 참여할 수 있는 '놀이형
프로그램'으로, 현대 도시를 구성하는
네트워크, 커뮤니티, 도시환경 등의 다양한
요소를 보드게임을 통해 알아가고, 경험할 수
있다.

일정	2019. 9. 7. – 11. 10., 비엔날레 기간 동안 수시 운영
소요 시간	약 30분
참여 대상	전 연령 가능
참가비	무료

2.   **나도 건축가**
     수업 참여자들과 함께 도시를 이루는
최소한의 단위인 집을 만들고, 가상의 마을을
구상해봄으로써 공적 공간과 사적 공간의
영역성, 사회적 요소, 일조, 조망, 이웃과의
관계 등 마을 속에서 이루어 지는 다양한 삶과
마을만들기에 대해 진지하게 고찰해 보는
'만들기형 프로그램'이다.

일정	2019년 9. 8. – 10. 20., 매주 토요일, 일요일
수업 시간	오전 10:30 – 12:00 (매주 토요일, 일요일) 오후 13:00 – 14:30 (* 9. 8., 9. 14., 9. 22., 10. 6., 10. 20. 운영)
참여 대상	전 연령 가능 (14세 미만의 어린이는 부모님과 함께 수업에 참여할 수 있습니다.)
참가비	5천원 (재료비)
예약	자세한 정보와 참가신청은

Education Program
HANDS-ON EXHIBITION PROGRAM

Coordination / Operations.˙.˙.
Kwon Hyun-jeong (ArchiEdu)
DDP Design lab 1F
P 1

The 2019 Seoul Biennale of Architecture
and Urbanism aims to provide a platform
that offers more than just an exhibition
on architecture and urbanism, serving
as a comprehensive space for arts and
culture. In collaboration with experts
from numerous different fields, this
event presents the underlying message
of architecture and urbanism through a
wide spectrum of media. In particular,
the Hands-On Exhibition Program is
composed of various activities and
exercises to appeal to people of all ages,
professional areas, and interests, in the
hope that all members of society can
find something that suits their personal
taste. In this way, more people can better
understand and appreciate exhibitions
on architecture and urbanism.

1.   **The Biennale for Everyone**
     Participants in this activity
will have the opportunity to learn
more about the different aspects of
modern urbanism, including networks,
communities, and environments, through
board games. This program allows people
from all age groups to learn more about
this topic in a fun, playful setting.

Schedule	Throughout the Biennale period, Sep 7 – Nov 10, 2019
Program Length	Approx. 30 min.
Eligibility	All age groups
Fee	Free

2.   **I Am an Architect, Too!**
     Individuals in this class will
work together to build houses, the
smallest unit of cities, and create their
own virtual city while considering the
different characteristics of public and
private spaces, including social factors,
landscapes, and the relationships
between people who share such spaces.
As part of this creative class, participants
will make close observations on different
lifestyles in making their imaginary cities.

Schedule	Every Saturday and Sunday, Sep 8 – Oct 20, 2019
Class Times	10:30 AM – 12:00 PM (Every Saturday and

홈페이지를 이용해 주시기
바랍니다.

3.   유레카! 서울: 더 나은 세상 만들기

서울의 더 나은 미래에 대해 다각적인
관점에서 바라보는 '아이디어 제안형'
수업이다. 전시의 주요 키워드를 바탕에 두고,
창의적 사고를 유도하는 다양한 방법으로 토론
수업을 진행, 참여자들의 적극적인 아이디어
제안으로 수업이 진행된다. 또한, 전 세계의
다양한 학생들이 제안한 ' 더 나은 세상을
만드는 방법'에 대한 실천 사례들을 살펴본다.

일정	2019. 9. 8. – 10. 20., 매주 토요일, 일요일
수업 시간	오전 10:30 – 12:00 (매주 토요일, 일요일)
참여 대상	전 연령 가능 (14세 미만의 어린이는 부모님과 함께 수업에 참여할 수 있습니다.)
참가비	5,000원 (재료비)
예약	자세한 정보와 참가신청은 홈페이지를 이용해 주시기 바랍니다.

Sunday)
1:00 – 2:30 PM
(On Sep 8, 14, 22 and
Oct 6, 20)

Eligibility	All age groups (children under 14 must attend with a guardian)
Fee	5,000 KRW (material costs)
Reservation	Please check the official website for more detailed information on reserving a spot for this class.

3.   Eureka! Seoul: Making a Better World

This class consists of using the
exhibition's keywords and encouraging
participants to share their ideas through
various methods of discussion and
creative thinking. As part of this class,
participants will review various cases
proposed by children from around the
world on the topic of "ways to make the
world better" and share their ideas using
a wide range of different perspectives.

Schedule	Every Saturday and Sunday from Sep 8 – Oct 20, 2019
Class Times	10:30 AM – 12:00 PM (Every Saturday and Sunday)
Eligibility	All age groups (children under 14 must attend with a guardian)
Fee	5,000 KRW (material costs)
Reservation	Please check the official website for more detailed information on reserving a spot for this class.

교육 프로그램

서울시 건축학교 프로그램

기획 / 운영.∵.∵. 권현정(아키에듀)
서울도시건축센터
P 2

역사적으로 볼 때, 도시는 오랜 기간 다양한
사회 가치와 지배구조를 통해 진화했다.
비엔날레 기간에 운영되는 서울시 건축 학교는
학생들이 이번 도시전에 참여하는 도시를
주제별로 살펴보고, 탐구하는 과정 속에서
도시의 물리적 성장과 변화를 유도했던 사회적
상황과 의사 결정자의 의도, 유토피아적
비전과 그의 현실화 과정 등 다양한 진화
요소들을 통해 깊이 있게 도시를 탐구하고자
한다. 학생들은 개인의 관심도에 기반하여
주제와 도시를 선정하는 서울시 건축 학교의
자기주도형 수업을 통해 도시에 대한 흥미를
더욱 키울 수 있을 것이다.

1 아이, 1 도시

학생들은 자신이 선택한 도시와
다른 여러 도시가 어떻게 성장하고, 공유되고,
연결되는지 시간적 요소, 공간적 요소, 사회적
요소를 전시물을 통해 자신의 필터로 찾아
볼 것이다. 이 활동을 통해 각각의 도시에서
중요하게 생각하는 이슈를 살펴보고, 결과를
서로 공유하면서 전세계 도시를 탐구할 수
있다. 또한 세계 여러 도시가 각자 나름의
가치와 이슈 등 다양한 면으로 서로 연결되어
있다는 점을 수업을 통해 알아가게 된다.

일정	2019. 9. 21. – 10. 26. 매주 토요일 (각 수업은 3주 연속 수업임)
수업 시간	13:00 – 17:30
참여 대상	초등 저 학년반 , 초등 고학년 반, 중학생 반
참가비	3만원
예약	서울시 공공예약 사이트 http://yeyak.seoul.go.kr 수업 사전등록 필수

Education Program
SEOUL METROPOLITAN CITY ARCHITECTURE SCHOOL PROGRAM

Coordination / Operations.∵.∵.
Kwon Hyun-Jung (ArchiEdu)
Seoul Center for Architecture & Urbanism
P 2

From a historical perspective, all cities
have evolved over a long period of time
through various sets of social values
and governance structures. There are
various factors that can impact this
process of evolution including the
physical development of cities, the social
circumstances that lead to changes, the
intentions of the final decision makers of
such change, utopian visions of change,
and the process of achieving that change
overall.
During the 2019 Seoul Biennale of
Architecture and Urbanism, the Seoul
Metropolitan Architecture School will
review and select the participating cities
by theme and then proceed to explore and
learn more about them in greater detail.
This program offers classes that are, for
the most part, self-led, with students
selecting topics and cities that they are
interested in and then learning about
them through different exhibits.

1 Child, 1 City

Students choose their own city
and then explore how that city and other
cities develop, exchange, and connect
with each other. In this process, they will
search for relevant factors of time, space,
and society through exhibition materials.
Students will then review the important
issues that each city currently faces
and share their results, thus creating an
opportunity for all participants to learn
more about cities around the world.
Through this course, students will also
have the opportunity to learn about how
the various issues and value systems of
different cities are linked to one another.

Schedule	Every Saturday from Sept 21 – Nov 9, 2019 (each class is conducted over the course of three weeks)
Class Times	2:00 PM – 5:00 PM
Eligibility	All elementary and middle school students
Fee	30,000 KRW
Reservation	Seoul Metropolitan City public reservation site: http://yeyak.seoul.go.kr

P 2

영화영상프로그램
제11회 서울국제건축영화제

아트하우스 모모
DDP 살림1관 (특별상영)
P4 / P1

Film and Video Programs
THE 11TH SEOUL INTERNATIONAL ARCHITECTURE FILM FESTIVAL

Arthouse Momo
DDP Design Lab 1F (Special Filming)
P4 / P1

서울도시건축비엔날레와 연계하여 진행되는 제 11회 서울국제건축영화제는 도시라는 거대한 사회적 유기체 속에서 건축의 의미를 살펴보고, 이를 대중에게 친숙한 '영화'라는 매개체를 통해 소개하여 도시와 건축에 대한 시민들의 관심을 제고하고자 한다.
이번 영화제는 6개의 섹션으로 나누어 진행되는 영화상영은 물론, 영화 관람 전 후로 이어지는 관객과의 대화, 호스트 아키텍트 포럼 및 특별 대담으로 이루어진다. 특히, 건축 전문가뿐만 아니라 도시와 건축에 관한 작품을 만들었거나 행사를 기획한 다양한 분야의 예술가들을 초대하여 진행하는 '관객과의 대화' 프로그램은 시민들의 영화에 대한 이해를 돕고 좀 더 풍부한 관람경험을 제공한다.
건축이라는 소재의 강점을 살려 영화를 보는 '공간'에 대한 새로운 모색을 꾀하고자 하는 이번 영화제는 서울도시건축비엔날레와의 연계를 통해 더욱 풍성한 도시와 건축에 대한 해석을 제안하며, 새로운 차원의 건축 담론을 제공하는 그릇이 될 것이다

The 11th Seoul International Architecture Film Festival, co-presented with the Seoul Biennale of Architecture and Urbanism, will explore the meaning of architecture in the context of the city as a huge social organism and introduce it through the medium of 'film' which is familiar to the public, raising their interest toward 'urbanism' and 'architecture'.
The Film Festival consists of 6 sections of 'screening', 'Guest Talk' before and after the film, 'Host Architect Forum' and 'Special Talk'. In particular, 'Guest Talk' which invites not only architects but also artists from various fields who have created their artworks related to city and architecture will help citizens to understand the films and provide a wide range of watching experience.
By taking advantage of the strength of the architecture, the Festival aims to explore new perspective on 'space' where the movies are screened to suggest a rich interpretation on architecture and urbanism to be an arena providing new dimension of discourse.

상영시간은 변동될 수 있으니 홈페이지를 통해 최신 정보를 참고하시기 바랍니다.

Please check the following websites as the film schedule may change.

http://seoulbiennale.org
http://www.siaff.or.kr/2019

상영일	시간	게스트 토크	상영작	감독	러닝 타임	연도	장르	국가
9월 25일 (수)	20:00		바르셀로나 파빌리온: 미스의 숨결을 따라서	사비 캄프레시오, 펩 마르틴	57	2018	다큐멘터리	스페인
9월 26일 (목)	11:00		렌조 피아노 건축 워크숍	프란체스카 몰테니	34	2018	다큐멘터리	이탈리아
	12:30		더 나은 미래를 향한 건축: 라틴 아메리카의 실험	카테리나 클라이와덴코, 마리오 노바스	84	2017	다큐멘터리	에콰도르, 독일
	14:30		이창	알프레드 히치콕	112	1954	픽션	미국
	17:00	●	로저 다스투스: 캐나다의 모더니스트	에티엔느 데로지에	103	2016	다큐멘터리	캐나다, 미국
	20:00	●	도시를 꿈꾸다	조셉 힐렐	81	2019	다큐멘터리	캐나다
9월 27일 (금)	11:00		할머니와 르 코르뷔지에	마욜렌 노르미에	58	2018	다큐멘터리	프랑스
	13:00		디자인 캐나다	그렉 듀렐	76	2017	다큐멘터리	캐나다
	15:00	●	모두를 위한 궁전	보리스 미시르코프	90	2018	다큐멘터리	불가리아, 독일, 로마니아
	17:30		실험적 도시	채드 프라이드릭스	96	2017	다큐멘터리	미국
	19:00	●	바르셀로나 파빌리온: 미스의 숨결을 따라서	사비 캄프레시오, 펩 마르틴	57	2018	다큐멘터리	스페인
9월 28일 (토)	11:00		도시를 꿈꾸다	조셉 힐렐	81	2019	다큐멘터리	캐나다
	13:00		도시, 인도를 짓다	프렘짓 라마찬드란	74	2009	다큐멘터리	인도
	15:30		초고층 빌딩: 하늘을 향한 경쟁	크리스 뱀포드	60	2019	다큐멘터리	호주
	17:00		이창	알프레드 히치콕	112	1954	픽션	미국
	20:30		디자인 캐나다	그렉 듀렐	76	2017	다큐멘터리	캐나다
9월 29일 (일)	11:00		모두를 위한 궁전	보리스 미시르코프	90	2018	다큐멘터리	불가리아, 독일, 로마니아
	13:00	●	월드 트레이드 센터, 그 후	바시아 마이진스키, 레오나르드 미신스키	57	2018	다큐멘터리	미국
	15:00	●	더 나은 미래를 향한 건축: 라틴 아메리카의 실험	카테리나 클라이와덴코, 마리오 노바스	84	2017	다큐멘터리	에콰도르, 독일
	18:00		착륙, 아모레퍼시픽 빌딩	시린 사바히	21	2018	다큐멘터리	독일
	20:00		할머니와 르 코르뷔지에	마욜렌 노르미에	58	2018	다큐멘터리	프랑스

P4 / P1

Arthouse Momo Theater 1

Date	Time	Guest Talk	Title	Director	Running Time	Year	Genre	Country
Sep 25 Wed	20:00		Mies On Scene. Barcelona In Two Acts	Xavi Campreciós, Pep Martín	57	2018	Documentary	Spain
Sep 26 Thu	11:00		The Power of the Archive: Renzo Piano Building Workshop	Francesca Molteni	34	2018	Documentary	Italy
	12:30		Do More with Less	Katerina Kliwadenko and Mario Novas	84	2017	Documentary	Ecuador, Germany
	14:30		Rear window	Alfred Hitchcock	112	1954	Fiction	USA
	17:00	●	Roger D'astous	Etienne Desrosiers	103	2016	Documentary	Canada, USA
	20:00	●	City Dreamers	Joseph Hillel	81	2019	Documentary	Canada

Sep 27 Fri	11:00		Bonne Maman et Le Corbusier	Marjolaine Normier	58	2018	Documentary	France
	13:00		Design Canada	Greg Durrell	76	2017	Documentary	Canada
	15:00	●	Palace for the people	Georgi Bogdanov, Boris Missirkov	90	2018	Documentary	Bulgaria, Germany, Romania
	17:30		The Experimental City	Chad Freidrichs	96	2017	Documentary	USA
	19:00	●	Mies On Scene. Barcelona In Two Acts	Xavi Campreciós, Pep Martín	57	2018	Documentary	Spain
Sep 28 Sat	11:00		City Dreamers	Joseph Hillel	81	2019	Documentary	Canada
	13:00		Doshi	Premjit Ramachandran	74	2009	Documentary	India
	15:30		Building To The Sky Episode Six: The Future Is Now 2012 – 2020	Chris Bamford	60	2019	Documentary	Australia
	17:00	●	Rear window	Alfred Hitchcock	112	1954	Fiction	USA
	20:30		Design Canada	Greg Durrell	76	2017	Documentary	Canada
Sep 29 Sun	11:00		Palace for the people	Georgi Bogdanov, Boris Missirkov	90	2018	Documentary	Bulgaria, Germany, Romania
	13:00	●	Leaning Out	Basia Myszynski, Leonard Myszynski	57	2018	Documentary	USA
	15:00	●	Do More with Less	Katerina Kliwadenko and Mario Novas	84	2017	Documentary	Ecuador, Germany
	18:00	●	Landing	Shirin Sabahı	21	2018	Documentary	Germany
	20:00		Bonne Maman et Le Corbusier	Marjolaine Normier	58	2018	Documentary	France

아트하우스 모모 2관

상영일	시간	게스트 토크	상영작	감독	러닝 타임	연도	장르	국가
9월 25일 (수)	20:00		바르셀로나 파빌리온: 미스의 숨결을 따라서	사비 캄프레시오, 펩 마르틴	57	2018	다큐멘터리	스페인
9월 26일 (목)	13:00		도시, 인도를 짓다	프렘짓 라마찬드란	74	2009	다큐멘터리	인도
	17:00		나의 아저씨	자크 타티	116	1958	픽션	프랑스, 이탈리아
	19:30	●	공사의 희로애락	장윤미	89	2018	다큐멘터리	한국
9월 27일 (금)	13:00		월드 트레이드 센터, 그 후	바시아 마이진스키, 레오나르드 미신스키	57	2018	다큐멘터리	미국
	17:30	●	렌조 피아노 건축 워크숍	프란체스카 몰테니	34	2018	다큐멘터리	이탈리아
	19:30	●	플러스 사이즈, 새로운 아름다움	조반나 모랄레스 바르가스	105	2018	다큐멘터리	캐나다
9월 28일 (토)	13:00		실험적 도시	채드 프라이드릭스	96	2017	다큐멘터리	미국
	17:30		팬티 이야기	스테판 프리조	60	2019	다큐멘터리	벨기에
9월 29일 (일)	13:00	●	나의 아저씨	자크 타티	116	1958	픽션	프랑스, 이탈리아
	17:30	●	초고층 빌딩: 하늘을 향한 경쟁	크리스 뱀포드	60	2019	다큐멘터리	호주
	19:37		사일로 468	안티 세페넨	37	2013	다큐멘터리	핀란드

Arthouse Momo Theater 2

Date	Time	Guest Talk	Title	Director	Running Time	Year	Genre	Country
Sep 25 Wed	20:00		Mies On Scene. Barcelona In Two Acts	Xavi Campreciós, Pep Martín	57	2018	Documentary	Spain
Sep 26 Thu	13:00		Doshi	Premjit Ramachandran	74	2009	Documentary	India
	17:00		Mon Oncle	Jacques Tati	116	1958	Fiction	France, Italy
	19:30	●	Under construction	JANG Yunmi	89	2018	Documentary	South Korea
	13:00		Leaning Out	Basia Myszynski, Leonard Myszynski	57	2018	Documentary	USA
	17:30	●	The Power of the Archive: Renzo Piano Building Workshop	Francesca Molteni	34	2018	Documentary	Italy
Sep 27 Fri	19:30	●	A Perfect 14	Giovanna Morales Vargas	105	2018	Documentary	Canada
	13:00		The Experimental City	Chad Freidrichs	96	2017	Documentary	USA
	17:30		The Story of a Panty, and of Those Who Make It	Stéfanne Prijot	60	2019	Documentary	Belgium
	13:00	●	Mon Oncle	Jacques Tati	116	1958	Fiction	France, Italy
Sep 28 Sat	17:30	●	Building To The Sky Episode Six: The Future Is Now 2012 – 2020	Chris Bamford	60	2019	Documentary	Australia
Sep 29 Sun	19:37		Silo 468	Antti Seppänen	37	2013	Documentary	Finland

P4 / P1

투어 프로그램
집합도시 서울투어
서울비엔날레 현장 및 서울일대

서울이라는 도시를 크게 역사투어와
테마투어로 나누어 걸어본다. 반복되는 일상
속에서 단편적으로 지나쳤던 서울의 곳곳을
다양한 키워드와 방식을 통해 비일상화 할
수 있는 기회로, 시민들은 일상과 비일상의
경계에서 마치 여행을 떠나는 설렘으로 서울을
마주하고 비엔날레 주제 '집합도시'를 이해할
수 있을 것이다. 더불어, 현재 도시 서울의
이슈인 재생건축, 재개발과 관련된 현장을
방문하고 전문 도슨트와 시민들의 소통을 통해
나아가야 할 미래 서울의 모습을 그려본다.

참가 대상
일반시민 대상

참가비
무료

예약
자세한 정보와 참가신청은 홈페이지를 이용해
주시기 바랍니다.

Tour Program
COLLECTIVE CITY SEOUL TOUR
Seoul Biennale of Architecture and
Urbanism, Seoul Metropolitan Area

This program is a walking tour of Seoul,
focusing on different historical themes
throughout the session. The goal of this
tour is to take the mundane, everyday
places in Seoul that we tend to pass by
without much thought and presenting
them in a different way. By blurring the line
between everyday life and new adventures,
participants will join this exciting trip
through Seoul and learn more about the
Biennale topic of collective cities along
the way. Furthermore, this tour will give
participants a first-hand look at urban
regeneration construction and urban
redevelopment, which are both important
issues currently in Seoul. Through
communication between the tour guide
and the participants, this program is an
opportunity for the public to share their
thoughts and ideas on steps forward in
Seoul's urban development.

Eligibility
All members of the public

Fee
Free

Reservation
Please check the official website for more
detailed information on reserving a spot.

서울역사투어

근, 현대 서울의 도시·건축은 물론, 한국적 건축 모더니즘에 대해 역사의 흐름을 따라 진행되는 투어이다. 작은 골목길부터 대도시에 이르기까지 다양한 규모와 형태의 집합양상을 역사의 현장과 시간이 중첩된 곳에서 직접 눈으로 보고 느껴본다. 그리고 겸재의 그림길을 따라 서울의 과거와 현재를 걸어본다. 총 6회 진행.

a 한양-경성-서울

일정	2019. 9. 28. (토) 14:00 – 16:00
도슨트	안창모
루트	경복궁 → 육조거리, 광화문네거리 → 서울광장 → 숭례문 → 서울도시건축전시관

일제강점기로 인해 산업혁명 없이 근대사회에 진입한 한국, 특히 서울은 근대에 발전한 다른 도시와 변화의 양상이 크게 달랐고 도시의 구조와 건축, 그 중에서도 집합주택의 모습에서 두드러졌다. 근대 국가의 틀을 갖춰가는 과정에서 역사적 사건이 도시공간과 건축에 어떻게 압축적으로 집합되어 투영되었는가를 살펴보고자 한다.

b 조선-대한-민국

일정	2019. 10. 5. (토) 10:00 – 12:00
도슨트	안창모
루트	광화문 → 장충공원, 박문사터 → 남산2호터널 → 유관순동상 → 자유센터,국립극장 → DDP

c 성문안첫동네

일정	2019. 9. 21. (토) 14:00 – 16:00
도슨트	조정구
루트	돈의문박물관마을

아름답고 오래된 동네들이 사라지고 있다. 교남동 또한 재개발구역에 속해 근린공원이 될 운명이었지만 '돈의문박물관마을'이라는 낯선 이름을 얻었다. 투어를 통해 시민들은, 지금은 서울시 재생사업의 시험장이 된 '돈의문박물관마을'에서 기록과 연구 작업을 통해 마을의 조성과정, 의미, 재생방향을 함께 고민할 수 있을 것이다.

Seoul History Tour

This tour explores the modern and traditional architecture and urbanism of Seoul, from the small side streets to the heart of the city. Participants will have the opportunity to see the differing scales and styles that have developed over time to create Seoul's urban identity. This tour follows Gyeomjae Art Street, passing through the contrasting landscapes of Seoul's past and present. The tour will be a held for a total of six sessions.

a Hanyang-Gyeongseong-Seoul

Schedule	Saturday Sep 28, 2019, 2:00 AM – 4:00 PM
Docent	Changmo Ahn
Route	Gyeongbokgung Palace → Yukjogeori, Gwanghwamun Intersection → Seoul Plaza → Namdaemun → Seoul Hall of Urbanism & Architecture

Different greatly from other cities, Seoul entered its era of modernism without the influence of Japanese colonial rule or an industrial revolution. Taking that into consideration, the city's structure and architecture, particular its multifamily houses, are remarkable, to say the least. This tour explores the process through which Seoul developed the modernized structure it possess today, as well as how urban spaces and architecture combined to create Seoul's identity as a collective city.

b Joseon-Korea-Republic

Schedule	Saturday Oct 5, 2019 10:00 AM – 12:00 PM
Docent	Changmo Ahn
Route	Gwanghwamun Plaza → Jangchungdan Park, Former site of Bangmoonsa → Namsan Second Tunnel → Ryu Gwansun Statue → Freedom Center, National Theater of Korea → DDP

c First Village within the City Gate

Schedule	Saturday Sep 21, 2019 2:00 PM – 4:00 PM
Docent	Junggoo Cho
Route	Donuimun Museum Village

The beautiful and long-standing villages of the past are gradually disappearing. Included on the list of areas set for redevelopment, Gyonam-dong will most likely be turned into a neighborhood park, and is now referred to as Donuimun Museum Village. Nowadays, with Seoul becoming the site for several urban regeneration projects, it is an important time for utilizing records and research to carefully consider the composition, meaning, and direction of regeneration projects for villages down the road.

d 세운속골목

일정   2019. 10. 19. (토)
      14:00 – 16:00
도슨트 조정구
루트   세운상가 및 일대

세운상가와 일대는 수백 년 된 골목 속에
작은 공장과 가게들로 마련된 삶의 터전이다.
서울시장은 이 터전의 재생을 생각하며
2019년 철거를 멈추게 했고 '노포의 가치'를
말한다. 국가적 개입으로 계획된 세운상가의
과거부터 현재에 이르기까지 골목과 속 골목,
필로티와 데크, 옥상을 누비며 서울의 시간 위
중심에 서있어 본다.

e 그림길겸재

일정   2019. 10. 12. (토)
      14:00 – 16:40
도슨트 임형남
루트   경복고 → 청운초 → 청풍계
      → 현대家 → 옥인동 군인아파트,
      윤비친가 → 수성동계곡 → 배화여고
      필운대

경복궁 서쪽지역은 인왕산과 백악산이 만나는
곳이자, 조선시대 권세가들의 거주지였고 많은
인물과 예술가들이 태어난 곳이다. 그러나
현재, 이 곳에는 겸재가 남기고 간 그림들과
전해 내려오는 이야기, 서민들의 생활만이
들어차 있다. 묘한 상상력과 함께 겸재의
그림길을 탐험해본다.

f 타임슬립

일정   2019. 10. 26. (토)
      14:00 – 16:30
도슨트 임형남
루트   드라마센터 →
      중앙정보부(구) → 서울예술대학
      → 와룡묘 → 남산신궁 →
      회현시민아파트 → 후암동적산가옥

남산은 서울의 안산이며 다양한 층위를 가지고
있다. 한강을 바라보는 산의 남쪽 한남동,
이태원과 서울의 정궁을 바라보는 산의 북쪽
지역은 다른 역사적 배경을 가지고 있다.
남산자락을 걸으며, 조선시대와 근대, 현대의
중첩된 시간을 체험하고자 한다.

d Alleyways of Sewoon

Schedule Saturday Oct 19, 2019
          2:00 – 4:00 PM
Docent   Junggoo Cho
Route   Sewoon Plaza and the
          surrounding area

Sewoon Plaza and its surroundings serve
as a space for factories and shops, created
as a result of several centuries of history
within the alleyways of this area. In 2019,
the mayor of Seoul halted the tearing down
of this area and stressed the importance
of old stores against the backdrop of urban
regeneration. Participants in this tour will be
able to see how government interventions
helped Sewoon Plaza develop from the
past until now to develop into a unique
area sitting at the center of Seoul's history,
characterized by its alleyways within the
alleyways, rooftops, decks, and pilotis.

e Gyeomjae Art Street

Schedule Saturday Oct 12, 2019
          2:00 PM – 4:40 PM
Docent   Hyoungnam Lim
Route   Kyungbock High School
          → Chungwoon Elementary School
          → Chungpoonggye → Hyundae
          Family Residents → Okin-dong
          Military Apartment, Yoonbichinga
          → Suseong Valley → Paiwha Girls'
          High School Pirundae

The area west of Gyeongbok Palace is where
Inwang and Baegak Mountains meet and,
also, where those with authoritative power
during the Joseon Dynasty lived. It is also the
birthplace of many famous individuals and
artists. However, the images that Gyeomjae
left behind and the history that they share
are filled only with images of the daily lives
of the common people. Visitors on this tour
will explore the artworks of Gyeomjae as they
stimulate their imaginations.

f Time Slip

Schedule Saturday Oct 26, 2019
          2:00 PM – 4:30 PM
Docent   Hyoungnam Lim
Route   Drama Center → Former
          KCIA → Seoul Institute of the Arts
          → Waryongmyo → Joseon Shrine
          → Hoehyeon Citizen Apartments
          → Former Residents of Japanese
          Colonists in Huam-dong

Namsan is one of Seoul's most prominent
mountains and boasts a number of different
views. The south side overlooks the Han
River, while the north side overlooks
Hannam-dong, Itaewon, and the ancient
palaces. On this tour, participants will
walk along the different historical views of
Namsan, while also taking a journey through
time, exploring the city's past from the
Joseon Dynasty, transition into modernism,
and its identity as a modern city.

서울테마투어

도시를 살아가는 데 우리가 주목하는 것 중 하나가 도시 라이프스타일이다. 시대별 트렌드에 따라 도시 라이프스타일이 바뀐다. 이번 비엔날레에서는 2030세대의 놀이문화를 투어의 주 테마로 잡고 서울의 일상 속 숨어있던 여러 장소를 방문하고자 한다. 도시의 시작점인 시장에서부터 힙스터들이 누비는 재생건축의 성지이자 소위 '핫플레이스'로 불리는 곳에 이르기까지 다양한 체험활동을 통해 서울을 즐기고 서울의 미래를 그려본다. 총 11회 진행.

g 인스타시티성수

일정	2019. 9. 8. (일)
	2019. 10. 6. (일)
	14:00 – 16:30
도슨트	심영규
루트	대림창고, 성수연방 → 카페어니언 → 드림인쇄소 → 오르에르, WxDxH → 우란문화재단 → 서울숲, 붉은벽돌재생지역 → 블루보틀 → DDP

서울의 대표적인 준공업지역에서 힙스터 타운이 된 성수. 매캐한 금속냄새로 가득했던 공업지역은 이제 서울숲의 영향을 시작으로 낡고 방치된 공간에 갤러리와 카페가 들어서면서 개성과 정체성을 갖추게 되었다. 도시재생의 중심지를 핫플레이스, 인더스트리얼, 도시인프라 3가지 층위로 나눠 핵심적인 공간을 둘러본다.

h 을지로힙스터

일정	2019.09.15.일요일 / 10.20.일요일 14:00- 16:30
도슨트	이희준
루트	DDP → 4FCAFÉ, 방산시장 → 금속공장, N/A갤러리 → 세운상가 → 을지로OF → 만선호프 → (DDP)

과거의 것, 올드한 것으로 치부되었던 시장이 '뉴트로', '힙스터'등의 단어를 통해 호기심과 새로움으로 다가왔다. 과거와 현재가 공존하는 인쇄 및 포장 전문의 방산시장과 을지로 힙스터 성지를 방문함으로써 과거 도시 형성의 기초와 현재를 시대적 정서에 기반하여 의의를 되새겨 본다.

Seoul Themed Tour

One of the most important things to consider when living in cities is urban life, which changes with every passing trend across every generation. With the keyword of the 2019 Seoul Biennale of Architecture and Urbanism being leisure and fun in 2030, the themed tours aim to introduce participants to various types of daily lives in the city. On these tours, participants will have the opportunity to enjoy Seoul and envision the city's possible future, visiting everything from marketplaces to reconstructed areas popular with hipsters and other modern hot spots. This tour will be held for a total of 11 sessions.

g Insta-City Seongsu

Schedule	Sunday Sep 8, 2019 Sunday Oct 6, 2019, 2:00 PM – 4:30 PM
Docent	Youngkyu Shim
Route	Daelim Changgo, Seongsu Yeonbang → Cafe Onion → Dream Printing → Or.Er, WxDxH → Wooran Foundation → Seoul Forest, Red Brick Urban Regeneration Zone → Blue Bottle Coffee → DDP

What was once Seoul's representative heavy industry town, Seongsu-dong is now famous for being a hipster hot spot. Seoul Forest played an important role in transforming this area from a place filled with the smell of burning metal and old, neglected spaces to one filled with galleries and cafes, creating a newfound identity and unique style. The tour splits this center of urban regeneration into three categories: popular places, industrial areas, and urban infrastructure.

h Euljiro Hipster

Schedule	Sunday September 15th / Sunday October 20th, 2019 2:00-4:30PM
Docent	Heejun Lee
Route	DDP → 4FCAFÉ, Bangsan Market → Metal Factory, N/A Gallery → Sewoon Plaza → EuljiroOF → Mansun Pub → DDP

Once considered old and things of the past, markets have made suddenly become the centers of social interest and attention under the new catchphrases of "hipster" and "newtro." Through this tour, participants will better understand the meaning behind these markets of the past, such as Bangsan Market, renowned for its printing and packaging services, as they find new identities in today's society.

ⓘ 서울생활백서

일정	1 2019.09.22.일요일 10:00-12:00
	2 2019.10.13.일요일 12:00-14:00
도슨트	이희준
루트	1 마장축산시장, 마장키친 → 서울풍물시장 → 동묘벼룩/창신완구시장 → DDP
	2 마장축산시장 마장키친 → 창신동채석장전망대 → 이음피음봉제역사관 → DDP

테마가 있는 각 시장들을 돌아보며 비엔날레 주제 '집합도시'를 이해한다. 의,식,주 테마에 맞춰 쿠킹클래스, 봉제체험, 벼룩시장 등 다양한 프로그램들을 직접 체험해봄으로써 시장을 통한 도시의 구성 및 발전, 집합성과 의의를 되새겨본다.

ⓙ 지하도시탐험

일정	2019.09.15.일요일 / 10.20.일요일 10:00-12:30
도슨트	김선재
루트	돈의문박물관마을 → 경희궁방공호 → 서소문성지역사박물관 → 뮤지스땅스(10. 20.에만) → 여의도 SeMA벙커 → 서울도시건축전시관

지상의 도시·건축뿐만 아니라 지하도시 인프라에도 시선을 두어 탐험해 보는 시간이다. 서울의 과거 흔적을 따라, 그 중에서도 서울의 숨겨진 지하공간을 개방하여 보존 및 변경된 공간의 의의와 현재를 느껴본다.

ⓘ Seoul Living Guide

Schedule	1 September 22nd, 2019, 10:00AM-12:00PM
	2 October 13th, 2019, 12:00-14:00PM
Docent	Heejun Lee
Schedule	
Route	1 Majang meat market, Majang Kitchen → Seoul Folk Flea Market → Dongmyo Flea Market, Dongdaemoon Stationery and Toy Market → (DDP)
	2 Majang meat market Majang Kitchen → Changshin-dong Quarry Viewing observatory → Iumpium Sewing History Center → DDP

This tour guides visitors through the markets of Seoul, which serve as the building block of any city. By introducing the structures and histories of these markets, participants will gain a stronger understanding of Seoul's identity as a collective city. This tour offers participants with several options for those interested in fashion, food, and shopping, including a cooking class, sewing lesson, and flea market tour. The goal of this tour is to explore Seoul's markets and demonstrate the meaning behind cities and their collectiveness, development, and composition.

ⓙ Exploring the City Underground

Schedule	Sunday September 15th / Sunday October 20th, 10:00AM-12:30 PM
Docent	Sun Jae Kim
Route	Donuimun Museum Village → Gyeonghui Palace Shelter → Seosomun Shrine History Museum → Musistance Theater (10.20only) → Yeoido SeMA Bunker → Seoul Hall of Urbanism & Architecture

This tour explores the aspects of urbanism and architecture above ground, but also the infrastructure of the city underground. Tracing the steps of Seoul's past, participants will have the opportunity to visit the hidden underground spaces of the city and better understand the meaning and current state of preserved and transformed spaces in Seoul.

k 서울파노라마

<u>일정</u>	2019.09.29.일요일 10:00-12:00, 14:00-16:00 / 10.27.일요일 14:00-16:00
<u>도슨트</u>	김나연
<u>루트</u>	서울로7017,윤슬 → 서소문청사정동전망대 → 서울도시건축전시관 서울마루 → 서울도서관 하늘뜰 → 세운, 청계상가 옥상

다양한 높이와 위치, 범위에서 펼쳐지는 서울의 풍경을 바라보고자, 평소 접하기 어려운 곳에서 서울 곳곳의 도시 마스터플랜과 인프라를 찾아 경험한다. 비엔날레 전시장을 비롯하여 투어를 통해 둘러보게 될 각 스팟에서 내려다보거나 올려다보는 서울의 모습은 매우 색다를 것이다.

k Seoul Panorama

<u>Schedule</u>	Sunday September 29th, 10:00AM-12:00PM, 2:00-4:00PM / Sunday October 27th, 2019, 2:00-4:00PM
<u>Docent</u>	Nayeon Kim
<u>Route</u>	Seoullo7017, Yunsl: Manridong reflecting Seoul → Jeongdong Observatory → Seoul Hall of Urbanism & Architecture SEOUL MARU → Seoul Metropolitan Library Sky Garden → Sewoon, Cheonggye Plaza Rooftop

This tour takes participants to different locations with differing heights and spaces, showing Seoul's landscape from areas that are not easy to visit on a daily basis. On this tour, participants will be able to see and experience the master plan of the city and its detailed infrastructure, enjoying never before seen views of Seoul from above and below, starting from the Biennale exhibition hall.

투어 프로그램
전시장 도슨트투어
DDP, 돈의문박물관마을
프로그램 내용

관람객 입장에서 전시를 쉽게 이해할 수
있도록 전시장 도슨트 설명을 제공한다.
토요일, 일요일에 운영된다. 최대 20명
사전예약. 1일 5회(한국어: 11시 / 14시 /
16시, 영어: 13시 / 15시).

오디오가이드 어플리케이션을 활용하여
전시를 감상할 수 있다. 관련 사항 안내소 및
홈페이지 참고

일정
2019. 9. 7. – 11. 10.
비엔날레 행사기간 내

장소
DDP, 돈의문박물관마을

참여 대상
일반시민

Tour Program
EXHIBITION HALL DOCENT TOURS
DDP, Donuimun Museum Village

Docent tours will be provided to visitors in
order to help them better understand the
exhibitions. These guided tours will begin
from point of entry and will be available
only on the weekends. There will be five
tours per day (Korean: 11 AM / 2 PM /
4 PM, English: 1 PM / 3 PM).

Replacing our old format of human guided
tours, we have begun implementing audio
guide application tours to provide more
accurate, higher quality explanations of
the exhibits. Please refer to Information
Desk or our website for more information.

Schedule
Throughout the Biennale period
(Sep 7 – Nov 10, 2019)

Venue
DDP, Donuimun Museum Village

Eligibility
General public

P1/P2

투어 프로그램
개인자유투어(스탬프투어)
비엔날레 각 전시장

전시장 및 현장 별로 독특한 디자인의
스탬프를 비치하여 방문기록을 남길 수 있다.
안내소에 비치된 비엔날레 지도를 각 스팟에서
스탬프로 완성시킨 후 기념품과 교환할 수
있다.

일정
2019. 9. 7. – 11. 10.
비엔날레 행사기간 내

참여 대상
일반시민

Tour Program
INDEPENDENT TOURS (STAMP TOURS)
Each Biennale Exhibition Hall

We have designed unique stamps for each
site and hall throughout the exhibitions
of the Biennale so that visitors keep a
record of where they have been. Visitors
can receive a map from the Information
Desk that they can use to collect these
stamps. Upon collecting them all, the map
can be exchanged for a special prize as a
souvenir.

Schedule
Throughout the Biennale period
(Sep 7 – Nov 10, 2019)

Eligibility
General public

투어 프로그램
연계투어 정동역사탐방

미션형 정동역사탐방 "대한제국의 꿈"
(탐방 + 거점연극 + 체험)
정동 일대

2019년은 3.1운동 100주년, 임시정부수립
100주년입니다. 의미가 깊은 해인만큼 정동의
근대역사와 문화에 대한 이해를 높이고자
기획된 투어 프로그램입니다.
투어는 고종의 밀사 역할을 맡은 배우들이
직접 등장해 스토리텔링을 하고, 시민들과
함께 소통하며 진행됩니다. 탐방 중에
진행되는 거점연극 아관파천, 을사늑약을
통해 자연스럽게 정동의 공간을 이해할 수
있습니다.
본 투어는 2019 정동역사재생활성화사업에
선정되어 2019 서울도시건축비엔날레
연계투어 프로그램으로 진행됩니다.

일정 및 소요시간
2019. 9. 8., 15. (일)
11:00 – 13:00
(약 2시간 소요)

투어 루트
서울도시건축전시관 → 고종의길1 →
정동마루 → 구러시아공사관 → 고종의길2 →
중명전 → 정동마루 → 서울도시건축전시관
(*투어루트는 당일 현장 상황에 따라 변동될
수 있습니다.)

참여 대상
8세 이상

참가비
무료

참가신청 방법
예약 문의 전화 02-741-3581

Tour Program
JEONGDONG HISTORY TOUR

"The Dream of the Korean Empire":
Mission-based tour exploring the history
of Jeongdong (Exploration + Theater
Performance + Hands-on Experience)
Jeongdong Area

2019 is an important year in Korean
history, marking the 100-year anniversary
of the March 1st Movement and the
establishment of the provisional
government of the Republic of Korea.
On that note, this program aims to raise
awareness and understanding of the
modern history and culture of Jeongdong.
As part of the tour, actors playing the role
of envoys to the Great Emperor Gojong
interact and communicate with visitors
through storytelling. Also, two plays,
Emperor Gojong Seeks Refuge at the
Russian Legation and The Korea-Japan
Treaty of 1905, are given in the middle of
the tour to help visitors better understand
Jeongdong and its historical significance.
This tour was selected as part of the 2019
Jeongdong History Revitalization Project
and will be introduced as a program for
the 2019 Seoul Biennale of Architecture
and Urbanism.

Dates and Time
Sunday, Sep 8th and 15, 11:00 AM –
1:00 PM (Approx. 2H)

Tour Route
Seoul Hall of Urbanism & Architecture
SEOUL MARU → First Road of Emperor
Gojong → Jeongdong Maru → Former
Russian Legation → Second Road of
Emperor Gojong → Jungmyeongjeon
→ Jeongdong Maru → Seoul Hall of
Urbanism & Architecture SEOUL MARU
(* Route is subject to change on the day
of the tour)

Eligibility
ages 8+

Fee
Free

Reservations
Call 02-741-3581 for reservations and
other inquiries.

연계전시 및 행사
COLLATERAL EXHIBITIONS AND EVENTS

평양다반사

큐레이터
윤혜정, 심상훈

기간
2019. 9. 7. – 11. 10.

장소
서울도시건축전시관

주최
서울특별시 도시재생실

일상의 평양의 모습을 담고 전시함으로써, 서울시민들에게 이질적인 평양의 모습이 아니라 공감대를 형성할 수 있는 평양의 모습을 보여준다. 이를 통해 남북간의 동질성을 회복하고 교류의 방안을 모색하는데 토대가 될 수 있는 분위기를 형성한다. "평양의 방"에서는 우리에게 친숙한 공간을 세팅해 놓고, 프로젝션 기법을 활용하여 서울의 풍경을 보여준다. 이는 관람객이 참여하는 순간 서울의 풍경이 평양의 풍경으로 바뀌며, 방 안의 사물들 역시 북한의 물건들이 프로젝션 되는 식으로 변화하게 된다. 한편 "평양 속의 나"에서는 크로마키 스크린 등 다양한 장치를 통해 관람객의 모습이 평양의 일상의 공간 안에 투영되어 마치 관람객이 평양에서 평양 주민들과 함께 걸어 다니고 있는 듯한 착각이 들게 한다. 마지막으로 "내가 그리는 서울과 평양"에서는 관람객이 직접 종이에 건물을 그리고, 그것이 서울 혹은 평양의 공간에 디지털화되어 나타나도록 한다. 이를 통하여 시민들은 현재의 서울 혹은 평양이 아니라 그들이 그려 나가는 각 도시의 모습을 볼 수 있다.

DAILY LIFE IN PYONGYANG

Curator
Hyejung Yoon, Sanghoon Shim

Period
7th September – 10th November, 2019

Venue
Seoul Hall of Urbanism & Architecture

Hosted by
Seoul Metropolitan Government Office of Urban Regeneration

This exhibition displays everyday life in Pyongyang to show an image of the North Korean city that Seoulites can relate to, rather than emphasizing its striking differences. This lays the groundwork for the exhibit's goal of restoring a sense of connection and homogeneity between North and South Korea and exploring approaches for mutual exchange. In the "Pyongyang Room", a familiar image of Seoul is shown on a projector screen, which then changes to an image of Pyongyang once visitors enter the room. Even the objects in the room are replaced with projected images of everyday objects found in Pyongyang. Meanwhile, the "Me in Pyongyang" exhibit uses a chromakey screen and other equipment to project the image of visitors onto various backdrops of everyday life in Pyongyang. This provides participants with the virtual experience of walking the streets of Pyongyang alongside North Korean citizens. Lastly, at the "Drawing Seoul and Pyongyang" exhibit, visitors have the opportunity to draw buildings on paper and watch their drawings come to life through digitalization technology. In this way, visitors can use their imaginations to create unique images of Seoul and Pyongyang that may differ from how the two cities actually look today.

건축자산의 새로운 시선

큐레이터
전진흥, 최윤희

기간
2019. 9. 7. – 11. 10.

장소
서울도시건축전시관

주최
서울특별시 도시재생실

"건축자산의 새로운 시선"은 한옥 위주에서 근현대 건축물 등 건축자산 전반으로 확장된 서울시 도시재생 정책의 새로운 변화를 소개한다.

건축자산이란 비문화재이지만, 서울의 특성과 시대적 층위를 보여주는 역사문화적 가치가 있는 건축물, 공간환경, 기반시설을 말한다. 최근 체부동 생활문화지원센터, 대선제분, 그리고 사직동 선교사주택이 우수건축자산으로 등록을 마쳤다. 이를 시작으로, 서울시는 서울의 기억과 일상이 어우러진 건축자산이 창의적으로 활용되고 재생될 수 있는 건축자산 진흥 정책을 추진한다.

전시는 도시를 활력 있게 만들고, 미래에도 유효한 도시기능을 담당하는 건축자산의 다양한 단면들을 수집하고 재구성하여 입체영상으로 표현된다. 그 환경 속에서 관람객들은 한옥 등 건축자산에 대한 새로운 시선을 마주하게 될 것이다. 또한 참여형 시민포럼으로 확장시켜 서울시민의 삶, 일상에서 건축자산의 역사적, 사회적, 경관적 가치를 재발견하는 시간도 마련한다.

"건축자산의 새로운 시선"은 역사도시 서울과 공존하고 있는 건축자산이 과거와 현재를 잇는 미래의 탄생이라는 메시지를 전달하며, 서울의 지속가능한 미래를 시민들과 함께 꿈꾸는 기회를 제공할 것이다.

NEW PERSPECTIVES ON ARCHITECTURAL ASSETS

Curator
Jinhong Jeon,
Yunhee Choi

Period
7th September –10th November, 2019

Venue
Seoul Hall of Urbanism & Architecture

Hosted by
Seoul Metropolitan Government Office of Urban Regeneration

"New Perspectives on Architectural Assets" highlights the expanded scope of preservation and promotion by the Seoul Metropolitan Government from just Hanok to include all architectural assets such as contemporary architecture. Although not cultural assets, architectural assets include buildings, spatial environment and infrastructure that display Seoul's progression over time. Recently, the Chebudong Community Arts Support Center, the Daesun Flour Mill Factory and the Sajikdong Missionary House were designated as exemplary architectural assets, setting in motion Seoul's efforts to transform and re-use various forms of older buildings and spaces in creative ways.

Various facets of architectural assets which contribute to the diversity and dynamic nature of Seoul are collected and reconstructed to create a visual experience. In that environment, visitors will gain a new perspective on the potential values of the architectural assets. Also, there will be a public forum inviting the residents of Seoul to discover the historical, social, scenic values of the architectural assets in their daily life. The overarching goal of this exhibition is to put forward sustainable future of Seoul by sharing with visitors the message of connecting past and future, delivered by the architectural assets that coexist as part of Seoul's historical urban identity.

P3

267

2019 서울건축문화제

총감독
천의영

기간
2019.9.6. - 9.22.

장소
문화비축기지 T6

주최
서울특별시

서울이 열린 공간에 주목하다
지난 4-5세기의 도시 공간 진화의 역사를
살펴보면, 특정 공간의 번성은 보다 많은
이들에게 참여와 혜택의 기회를 제공하는
열린 공간의 성장과 직접적으로 관련이
있었다. 하지만 우리 주변에 새로 생기는
많은 주거들이 게이티드 커뮤니티의 섬들로
만들어 지면서 사회와의 연결을 줄이고
있다. 이는 바람직한 모습은 아니며 더욱이
우리가 추구하는 미래도시의 모습도
아니다. 눈을 돌려보면, 우리가 익히 들어온
일본의 소도시인 다케오의 시립도서관은
여러 논란에도 불구하고 인구가 줄어드는
작은 도시에서 예산을 절감하고 365일
운영하면서도, 서비스의 질을 개선시킨다는
점에서 민간이 운영하는 열린 공공 공간의
혁신 모델을 보여주고 있다. 미국 뉴욕
맨해튼이나 미국 동부의 도시들은 역사적
제약을 갖고 있지만 서부의 실리콘밸리와
시애틀이 기존 도시의 제약과 한계를 벗어난
디지털 도시로서 새로운 가능성들을 열어주는
또 다른 '열린 공간'의 모델이 되고 있다. 미국
서부의 이런 도시들은 상대적으로 기존 사회와
다른 개념을 시도하기가 자유롭기 때문일
것이다. 이러한 점에서 집합도시(Collective
City)라는 서울 건축비엔날레의 주제와 큰
틀에서 함께 호흡하며 서울이라는 대도시 속에
만들어지거나 만들어질 가능성이 있는 작은
'열린 공간'에 주목하며 서울건축문화제의
전체행사를 준비하고자 한다.

2019 SEOUL ARCHITECTURE FESTIVAL (SAF)

Co-directors
Chun, Eui Young

Period
6th September –
22th September, 2019

Venue
Oil Tank Culture Park T6

Hosted by
Seoul Metropolitan Government

New Open Space
According to the history of urban spatial
evolution in the last 4th and 5th centuries,
the prosperity of a particular space was
directly related to the growth of new open
spaces, which provided more people
with opportunities for participation and
benefits. However, many of the new
dwellings around us are being made
into the gated community and it caused
reducing our connection to society. It
is not a desirable and, moreover, not a
future city we seek. The municipal library
of Takeo, a small Japanese city, shows an
open, privately run model of public space
in that it will cut budgets and improve
the quality of service while operating 365
days in smaller, shrinking cities. New York
and other US cities on the East Coast
have historical constraints, but Silicon
Valley and Seattle on the West Coast
are becoming models of another "open
space" that opens up new possibilities
as digital cities beyond the constraints
and limitations of existing cities. It may
be because cities on the West Coast are
relatively free to try new things different
from existing societies. In this regard,
Seoul Architecture and Culture Festival is
in line with the theme of Seoul Biennale
as a Collective city, and aims to prepare
for the event by paying attention to the
open space in the city of Seoul.

하버드대학교	Hanyang University
한양대학교	Hanyang University ERICA
한양대학교 에리카	Hongik University
홍익대학교	University of Hong Kong
홍콩대학교	Chinese University of Hong Kong
홍콩중문대학교	RISD Architecture & Canon Foundation
로드 아일랜드 스쿨 오브 디자인	Generalitat de Catalunya,
카탈루냐 자치정부, 바르셀로나 광역행정청	Area Metropolitana de Barcelona
태국디지털경제홍보처,	"depa X SHDT" (Digital Economy
쇼우헝디자인기술주식회사	Promotion Agency of Thailand and
아랍사회과학협의회	Shouheng Design and Technology Inc)
비판적 방송을 위한 MIT 연구소	ACSS (Arab Council for Social Sciences)
하버드 건축대학원, 앤드류 멜론 재단	MIT Critical Broadcasting Lab (Ana
퀸즐랜드 대학교	Miljacki, Faculty lead)
뉴욕 주립 대학교 버펄로 건축대학	Harvard University Graduate School of
취리히 연방 공과 대학교	Design, Andrew W. Mellon Foundation
뉴욕 공과 대학교, 시드니 공과 대학교	The University of Queensland
평등주의와 도시를 위한 미시간 멜론	University at Buffalo School of
프로젝트, 미시간 대학교 터브먼 건축대학	Architecture and Planning
덴마크 예술 재단, 주한 덴마크 대사관	ETH Zürich (Eidgenössische Technische
SAC, 하인츠와 지젤라 프리드리히 재단 /	Hochschule Zürich)
인스부르크 대학교	New York Institute of Technology (NYIT),
개성공단재단	University of Technology Sydney (UTS)
서매틱 콜라보레이티브, 예일대학교 건축대학	Michigan Mellon Project for
선전 인택트 스튜디오(영상 후반 작업: 선전	Egalitarianism and the Metropolis,
위롱 테크놀로지)	Taubman College University of
자카르타특별지구, 인도네시아 창조경제부,	Michigan
카르손 & 사갈라 건축디자인, noMaden &	Danish Arts Foundation, Embassy of
Pupla 프로젝트	Denmark in Korea
런던대학교 바틀릿 건축대학, 포스터+	SAC & Heinz und Gisela Friedrichs
파트너스, 웨스턴 윌리엄슨+파트너스	Stiftung / University of Innsbruck
르노-닛산 자동차, 포드 자동차	Gaesong Industrial District Foundation
메데인 시	Somatic Collaborative/ Yale School of
밀라노 시	Architecture
그레이엄 예술고등교육재단, 컬럼비아 대학교	Intact Studio, Shenzhen (video post
건축대학원	production: Yulong Technology,
주한 코스타리카 대사관	Shenzhen
하다드 재단	Capital Special Region of Jakarta, Badan
로열 멜버른 공과대학교 건축&도시설계대학	Ekonomi Kreatif (BEKRAF), Kalson
오토데스크(Autodesk), 펜실베이니아 대학교	Sagala Architecture & Design (KSAD),
울산발전연구원	noMADen & Pupla Project
비엔나 비지니스 에이전시, 빈 마케팅,	The Bartlett School of Architecture,
BKA(오스트리아 문화예술부), 오타크링거	Foster + Partners, Weston Williamson
양조장 GmbH, 요제프 마너 &Comp. AG	+ Partner
뉴욕 공과 대학교, 웰링턴 빅토리아 대학교	Renault Nissan; Ford
영주시	Alcaldía de Medellin
암스테르담 시	City of Milano
프로 헬베티아	Graham Foundation for Advanced Studies
서던캘리포니아 대학교 프라이스 행정대학	in the Fine Arts, and a Columbia
공간분석 연구소	University Graduate school of
이음피움 봉제역사관	Architecture, Planning and Preservation
뮤지스땅스	Embajada de Costa Rica en Corea
배화여자고등학교	Haddad Foundation
서소문성지역사박물관	RMIT School of Architecture & Urban
	Design
	Autodesk, University of Pennsylvania
	Ulsan Development Institute
	Vienna Business Agency, Wien Marketing,
	BKA (The Arts and Culture Division of
	the Federal Chancellery of Austria);
	Ottakringer Brauerei GmbH, Josef
	Manner & Comp. AG (inquired)
	New York Institute of Technology (NYIT) &
	Victoria University of Wellington (VUW)

서울시립미술관
서울도시건축센터
서울시 경제정책실
서울시 도시공간개선단
서울시 도시재생실
서울시 역사도심재생과
서울시 푸른도시국
서울시 행정국

City of Yeongju
City of Amsterdam
Pro Helvetia
SLAB: Spatial Analysis Lab of USC Price
    School of Public Policy
Iumpium Sawing History Center
Musistance
Paiwha Girls' High School
Seosomun Shrine History Museum
Seoul Museum of Art
Seoul Center for Architecture & Urbanism
Seoul Economic Policy Office
Seoul Urban Improvement Bureau
Seoul Urban Regeneration Office
Seoul Historic City Center Regeneration
    Division
Seoul Green Seoul Bureau
Seoul Administrative Services

2019 서울도시건축비엔날레 가이드북

초반 1쇄 발행∴2019년 8월 31일

펴낸 곳∴서울특별시 도시공간개선단
편집∴임여진
영문 교정∴앨리스 김, 리비아 칭 왕(주제전)
국문 교열∴임여진, 천주현
디자인∴홍은주, 김형재 (전수민 도움)
인쇄·제책∴(주)으뜸프로세스

ISBN∴979-11-6161-718-3 (03540)

2019 Seoul Biennale of Architecture and
Urbanism

First Printed on∴August 31, 2019

Published by∴Seoul Urban Space
Improvement Bureau
Edited by∴Jean Im
English Copy Editing∴Alice S. Kim, Livia
Qing Wang (Thematic Exhibition)
Korean Proofreading∴Jean Im, Ju Hyun
Cheon
Design∴Eunjoo Hong & Hyungjae Kim
(Assistant Sumin Jeon)
Printed and Bound by∴Top Process, Ltd.

ISBN∴979-11-6161-718-3 (03540)